Wetlands Conservation

Wetlands Conservation

Current Challenges and Future Strategies

Edited by

Sanjeev Sharma
Jawaharlal Nehru University, New Delhi, India

Pardeep Singh
PGDAV College, University of Delhi, New Delhi, India

WILEY Blackwell

Registered Offices
John Wiley & Sons, Inc., 111 River Street, Hoboken, NJ 07030, USA
John Wiley & Sons Ltd, The Atrium, Southern Gate, Chichester, West Sussex, PO19 8SQ, UK

Editorial Office
9600 Garsington Road, Oxford, OX4 2DQ, UK

For details of our global editorial offices, customer services, and more information about Wiley products visit us at www.wiley.com.

Wiley also publishes its books in a variety of electronic formats and by print-on-demand. Some content that appears in standard print versions of this book may not be available in other formats.

Library of Congress Cataloging-in-Publication Data

Names: Sharma, Sanjeev (Assistant professor in environmental sciences), editor. | Singh, Pardeep, editor.
Title: Wetlands conservation : Current challenges and future strategies / edited by Sanjeev Sharma, Pardeep Singh.
Description: Hoboken, NJ : Wiley-Blackwell, 2022.
Identifiers: LCCN 2021008931 (print) | LCCN 2021008932 (ebook) | ISBN 9781119692683 (hardback) | ISBN 9781119692669 (adobe pdf) | ISBN 9781119696322 (epub)
Subjects: LCSH: Wetlands conservation.
Classification: LCC QH75 .W4656 2022 (print) | LCC QH75 (ebook) | DDC 333.91/8–dc23
LC record available at https://lccn.loc.gov/2021008931
LC ebook record available at https://lccn.loc.gov/2021008932

Cover Design: Wiley
Photograph: Chandertal Lake, Ramsar Wetland located in Lahaul-Spiti District, Himachal Pradesh-India
Photo Credit: Sanjeev Sharma

Set in 9.5/12.5pt STIXTwoText by Straive, Pondicherry, India
Printed and bound by CPI Group (UK) Ltd, Croydon, CR0 4YY

C9781119692683_090921

Contents

Preface *xiii*
List of Contributors *xvii*

1 Global Wetlands: Categorization, Distribution and Global Scenario *1*
1.1 Wetlands Definition, Categorization and Classification Criteria *1*
1.1.1 Wetlands- Categorization and Classification *3*
1.1.2 Human- Made Wetlands *5*
1.2 Importance of Wetland Ecosystem *5*
1.3 Spatial Distribution and Potential of Global Wetlands *7*
1.4 Status and Impacts on the Wetlands Ecosystem *8*
1.4.1 Conservation Measures and Future Strategies *10*
1.4.2 Conclusion and Recommendation *11*
Acknowledgements *13*
References *13*

2 Ramsar Convention: History, Structure, Operations, and Relevance *17*
2.1 Background *17*
2.2 The Ramsar Convention *18*
2.3 The Convention Text *19*
2.4 Wetland Definition and Classification *19*
2.5 Mission of the Convention *22*
2.6 Structural Framework of the Convention *22*
2.7 Operational Framework of the Convention *25*
2.7.1 Convention Membership *25*
2.7.2 Ramsar Regions *26*
2.7.3 National Ramsar Committees *30*
2.7.4 The Montreux Record *31*
2.7.5 Ramsar Strategic Plan *31*
2.7.6 Three Pillars of Ramsar Convention *31*
2.7.7 The Convention Budget *32*
2.8 External Partnerships and Synergies *33*
2.9 Education and Outreach *35*
2.9.1 Communication, Education, Participation, and Awareness (CEPA) *35*
2.9.2 World Wetlands Day *36*

2.10 Legal Status *36*
2.11 Effectiveness of the Convention *37*
References *38*

3 Ecological Importance of Wetland Systems *40*
3.1 Introduction *40*
3.2 Importance of Wetlands in Flood Control *40*
3.3 Role of Wetlands in Groundwater Replenishment *41*
3.4 Role of Wetlands in Stabilization and Storm Protection of Shorelines *42*
3.5 Role of Wetlands in Sediment and Nutrient Retention *43*
3.6 Role of Wetlands in Water Purification *44*
3.7 Biodiversity of Wetlands *45*
3.8 Wetland Products *46*
3.9 Sociocultural Values of Wetlands *46*
3.10 Wetlands in Relation to Recreation and Tourism *47*
3.11 Wetland and Climate Change *48*
3.12 Summary *49*
Acknowledgments *50*
References *50*

4 Ecological and Societal Importance of Wetlands: A Case Study of North Bihar (India) *55*
4.1 Introduction *55*
4.2 Geographical and District-Wise Distribution of Wetlands in North Bihar *58*
4.2.1 Kabartal *60*
4.2.2 Baraila Jheel *60*
4.2.3 Kusheshwar Asthan *62*
4.2.4 Jagatpur Wetland *62*
4.2.5 Moti Jheel *63*
4.2.6 Gogabeel Pakshi Vihar *64*
4.3 Wetlands: Promoters of Sustainable Livelihood and Services *64*
4.4 North Bihar Wetland Biodiversity: Status and Role *65*
4.5 Urbanization, Pollution, and Climate Change Impacts *71*
4.6 Legal Framework, Policies, and Challenges *77*
4.7 Conclusion *79*
Acknowledgments *80*
References *80*

5 Recognizing Economic Values of Wetland Ecosystem Services: A Study of Emerging Role of Monetary Evaluation of Chandubi Ecosystem and Biodiversity *87*
5.1 Introduction *87*
5.2 Methodology of Ecosystem Valuation *90*
5.2.1 Market Prices – Revealed Willingness to Pay *90*
5.2.1.1 Market Price Method *91*
5.2.1.2 Productivity Method *91*

5.2.1.3 Hedonic Pricing Method *92*
5.2.1.4 Travel Cost Method *93*
5.2.2 Circumstantial Evidence – Imputed Willingness to Pay *94*
5.2.2.1 Damage Cost Avoided, Replacement Cost, and Substitute Cost Methods *94*
5.2.3 Surveys – Expressed Willingness to Pay *95*
5.2.3.1 Contingent Valuation Method *95*
5.2.3.2 Contingent Choice Method *96*
5.3 Ecosystem Services of Wetland *97*
5.4 Chandubi Wetland: Introduction, Impact, and Introspection *97*
5.5 Scaling up Wetland Conservation, Wise Use, and Restoration for Achieving Sustainable Development Goals *103*
5.6 Wetlands' Role in Achieving SDGs *104*
5.7 Conclusion *108*
Acknowledgments *109*
References *109*

6 Ecosystem Services of Lagoon Wetlands System in India *111*
6.1 Introduction *111*
6.2 Chilika Lagoon *112*
6.3 Ecosystem Services Provided by Chilika Lagoon *112*
6.3.1 Provisioning Services *114*
6.3.1.1 Commercial Fisheries *114*
6.3.1.2 Other Flora and Fauna of Chilika Lagoon *114*
6.3.1.3 Navigation *115*
6.3.2 Regulating Services *115*
6.3.3 Cultural Services *116*
6.3.4 Supporting Services *116*
6.4 Threats and Management of Chilika Lagoon *117*
6.5 Pulicat Lagoon *118*
6.6 Ecosystem Services Provided by Pulicat Lagoon *119*
6.6.1 Provisioning Services *119*
6.6.1.1 Fisheries in Pulicat *119*
6.6.2 Aquatic Flora and Fauna of Pulicat *120*
6.6.3 Regulatory Services Provided by Pulicat Lagoon *120*
6.6.4 Historical and Cultural Importance of Pulicat Lagoon *120*
6.6.5 Supporting Services Provided by Pulicat Lagoon *121*
6.6.6 Threats and Management of Pulicat Lagoon *121*
6.7 Conclusion *123*
Acknowledgments *124*
References *124*

7 Sustainable Practices for Conservation of Wetland Ecosystem *129*
7.1 Introduction *129*
7.2 Role of Wetlands in the Ecosystem *130*
7.3 Challenges to Conserve Wetlands *133*
7.4 Wetland Management and Sustainable Development *134*

7.5 Future Strategies for Wetland Conservation *135*
7.6 Development of the Legal Framework *135*
7.7 Technology Intervention with Baseline Data for Wetland Conservation *136*
7.8 Development of National Action Plans *136*
7.9 Promotion of Research for Conservation Setup *136*
7.10 Conclusion *136*
References *137*

8 Assessing the Benefits, Threats and Conservation of Reservoir-Based Wetlands in the Eastern Himalayan River Basin *140*
8.1 Introduction *140*
8.1.1 RBWs' Significance and Ignorance *141*
8.1.2 RBWs in India *142*
8.1.3 The RBWs in the Eastern Himalayas *143*
8.2 The RBWs in the Tista Basin *144*
8.3 Benefits of Reservoirs as Wetland *145*
8.3.1 Ecosystem Services Provided by the RBWs *145*
8.4 Assessment of Ecosystem Services in the Tista Basin Provided by the RBWs *147*
8.5 Adverse Impact of RBWs *149*
8.5.1 Construction and Function of RBWs Across the World *149*
8.5.2 Adverse Impact of RBWs in the Eastern Himalayas *149*
8.6 Assessment of Impact on the Tista basin *150*
8.7 Potential Challenges and Threats to RBW *152*
8.7.1 Anthropogenic Activities *152*
8.7.2 Variations in Water Level *153*
8.8 Climate Change *153*
8.9 Management and Conservation of RBWs *154*
8.10 Conclusion *155*
References *156*

9 Spatiotemporal Evaluation of Causes and Consequences of Wetland Degradation *162*
9.1 Introduction *162*
9.2 Classification of Wetlands *162*
9.3 Causes of and Consequence of Wetland Degradation *164*
9.3.1 Natural Causes *164*
9.3.1.1 Storms Surge *165*
9.3.1.2 Disintegration of Barrier Islands *165*
9.3.1.3 Flooding and Salinization *165*
9.3.1.4 Herbivory *166*
9.3.1.5 Climate Change *166*
9.3.1.6 Major Shifts in a River's Course *166*
9.3.2 Anthropogenic Causes of Wetland Loss *166*
9.3.2.1 Infrastructure Development *167*
9.3.2.2 Land Conversion *167*

9.3.2.3 Water Withdrawal *168*
9.3.2.4 Eutrophication and Pollution *168*
9.3.2.5 Overharvesting and Overexploitation *168*
9.3.2.6 Introduction of Invasive Species *168*
9.3.2.7 Others *169*
9.4 Consequences of Wetland Loss *170*
9.4.1 Loss of Biodiversity *170*
9.4.2 Decrease in Water Level *171*
9.4.3 Loss of Habitat *171*
9.4.4 Climate Change *171*
9.4.5 Emission of Greenhouse Gases *171*
9.4.6 Erosion of River Delta *172*
References *172*

10 The Status of Current Knowledge, Distribution, and Conservation Challenges of Wetland Ecosystems in Kashmir Himalaya, India *175*
10.1 Introduction *175*
10.2 Wetlands Over North-Western Kashmir Himalaya *176*
10.2.1 Current Status *176*
10.2.2 Wetland Classification *178*
10.2.2.1 High Altitude Wetlands (HAWs) *182*
10.2.2.2 Mid-Altitude Wetlands (MAWs) *182*
10.2.3 Wetland Distribution and Extent in Kashmir Himalaya *182*
10.3 Wetland Functions and Values *184*
10.3.1 Regulatory functions *184*
10.3.1.1 Regulation of Global Climate *184*
10.3.1.2 Groundwater Recharge and Discharge *184*
10.3.1.3 Water Purification *185*
10.3.1.4 Natural Hazard and Flood Control *185*
10.3.1.5 Sediment Retention *185*
10.3.2 Provisioning Functions *185*
10.3.2.1 Food Resources *185*
10.3.2.2 Raw Materials *186*
10.3.2.3 Medicinal Resources *186*
10.3.3 Cultural Functions *186*
10.3.3.1 Tourism, Aesthetics, and Recreation *186*
10.3.3.2 Scientific and Educational Information *186*
10.3.4 Supporting Functions *187*
10.3.4.1 Biodiversity Habitats *187*
10.3.4.2 Nutrient Cycling *187*
10.3.5 Economic Values *187*
10.4 Drivers of Wetland Degradation *187*
10.4.1 Land System Changes *188*

10.4.2 Pollution *189*
10.4.3 Floating Agriculture *190*
10.4.4 Siltation *190*
10.4.5 Roads and Railways *190*
10.4.6 Plantations *190*
10.4.7 Overexploitation *191*
10.4.8 Weed Infestation *191*
10.4.9 Hunting and Poaching *191*
10.4.10 Land Reclamation *191*
10.5 Wetland Conservation in Kashmir Himalaya *191*
10.5.1 Legal Framework *192*
10.5.2 Conservation Challenges *193*
10.5.3 Conservation Strategies *193*
10.5.4 Knowledge Gaps *193*
10.6 Conclusion *195*
Acknowledgments *195*
References *195*

11 Heavy Metal Pollution in Coastal Environment and Its Remediation Using Mangroves: An Eco-sustainable Approach *201*
11.1 Introduction *201*
11.2 Pollution in Mangrove Habitats: A Global Concern *202*
11.3 Heavy Metal Cycling in the Mangrove Ecosystem *203*
11.4 Heavy Metal Transport, Uptake, and Release *204*
11.5 Bioavailability and Concentration of Heavy Metals in the Sediments *204*
11.6 Factors Affecting Heavy Metals in the Sediment *205*
11.7 Heavy Metal Accumulation in Mangrove Plants *210*
11.8 Heavy Metal Remediation Potential of Mangroves *210*
11.9 Distribution of Heavy Metals in Different Plant Tissues of Mangrove Species *214*
11.10 Application of Phytoremediation to Coastal Pollution Remediation *214*
11.10.1 Phytoremediation Using Constructed Wetlands (CWs) Technology *214*
11.10.2 Phytoremediation Using Constructed Floating Bed *216*
11.11 Eco-remediation Technologies as Sustainable Natural Treatment Systems for Waste Water Treatment *217*
11.12 Conclusion and Future Prospects *217*
References *218*

12 Mangrove Forests: Distribution, Species Diversity, Roles, Threats and Conservation Strategies *229*
12.1 Introduction 229
12.2 Mangrove Species Diversity *230*
12.3 Geographical Distribution of Mangroves Across the Globe and India *237*
12.4 Important Roles of Mangroves *237*

12.4.1 Mangrove Forests are the Richest and Most Biodiverse Ecosystems on Earth *241*
12.4.2 Aquaculture: Shrimp and Fish Cultivation *242*
12.4.3 Protection from Natural Disasters: Mangroves Act as Natural Bioshields Against Natural Disasters *242*
12.4.4 Medicinal Value of Mangroves *243*
12.5 Threats to Mangroves *243*
12.5.1 Human Settlements and Other Developmental Activities *244*
12.5.2 Excessive Extraction of Wood *245*
12.5.3 Conversion of Mangrove Forests for Farming and Related Activities *245*
12.5.4 Conversion of Mangrove Forests for Aquaculture *245*
12.5.5 Global Warming, Climate Change, and Sea Level Rise *246*
12.5.6 Limits to Landward Movement *246*
12.5.7 El Niño and La Niña Events *247*
12.6 Strategies for the Conservation of Mangroves *247*
12.6.1 Increased and Focused Research on Understanding Mangroves *247*
12.6.2 Implementation of Mangrove Conservation-Related Laws, Guidelines, and Other Initiatives *247*
12.6.3 Strengthening Conservation Mechanisms *251*
12.6.4 Targeting Land Ownership-Related Issues *251*
12.6.5 Involvement of Local Communities *251*
12.7 Conclusion *252*
Acknowledgments *253*
References *253*

13 Wetland Conservation and Restoration *272*
13.1 Introduction *272*
13.2 Wetlands: Role and Importance *274*
13.3 Wetland Loss Leading to Ecological Imbalance *275*
13.4 Wetland Management Strategies: Current Status *277*
13.5 Wetland Restoration and Sustainability *280*
13.5.1 4th Ramsar Strategic Plan 2016–2024 (Source: Ramsar Secretariat 2016) *280*
13.6 Conclusion *281*
Acknowledgments *281*
References *281*

Index *285*

Preface

Wetlands are critical natural resources and most productive ecosystems on earth. They are rich in terrestrial as well as aquatic flora and fauna. Wetlands hold significance for biodiversity, water conservation and ecosystem services. The critical ecosystems of wetlands support livelihoods and form a lifeline for billions of people living within wetlands and in their periphery. Majority of the global population is directly and indirectly dependent on these wetlands for drinking water, irrigation and other essential natural ecosystem services. Wetlands are known to provide habitats to as much as 20% of the planet's various life forms. Wetlands form habitats for endangered and threatened species of wildlife, aquatic flora and fauna. The wetlands are also feeding, breeding and resting areas for a host of migratory birds which travel vast distances over many continents and across different geographical zones and biomes. Wetlands are referred to as biological supermarkets and *'kidneys and lungs of the landscapes where they exist'* due to their ecological interface with the hydrological, chemical and biological cycles.

Many wetlands are a direct source of drinking water and many rivers also originate from wetlands. Wetlands have supported human existence since antiquity. The nomadic populations across the globe have settled nearby these wetlands since ancient times as wetlands provided ample pastures for livestock rearing. Wetlands have social, economic and cultural significance for human beings. For example, in the Himalayan region of India wetlands have socio-religious and cultural importance for the inhabitants. People of these regions worship the wetlands and consider these water bodies as sacred. Since 1900, around half of the world's wetlands have disappeared due to exploitation of natural resources by human beings in the wetlands. Intense human activities in the wetlands have led to degradation and loss of these areas. In addition to anthropogenic pressures, climate change, land use patterns, land cover change, urbanisation, global warming modern agricultural practices etc. have affected the wetland system adversely, across the globe. Preservation and restoration of wetlands will help to conserve these bodies and will also lead to upholding the social, cultural, economic and religious significance of the wetlands. Wetlands hold immense significance in controlling and mitigating floods. For example mangrove wetlands reduce wave energy and thus protect coastal communities from floods etc. Through the process of nitrogen recycling, these wetlands help in improving the downstream water quality. They also help in climate change mitigation and adaptation.

All through in the twentieth century wetlands have faced neglect and disregard. The significance of these wetlands ecosystems were realised by people after second half of the twentieth century onwards and across the globe consciousness rose about the consequences of losing these valuable natural resources. 1971, a Convention on Wetlands called the Ramsar Convention was held in Ramsar, Iran for the conservation and sustainable use of wetlands. This Convention upholds the distinction of being the first modern treaty between countries aimed at conserving natural resources. This Convention necessitated signatories to draft policies for wetland conservation and their wise use. Following the Ramsar Convention on Wetlands, a number of policy instruments were formulated for the protection and conservation of these resources at regional, local and global levels. Though a legal and policy framework is existing on the wetlands, still these resources are not adequately conserved and managed and a large number of wetlands continue to be in deplorable state.

Many governments and different NGO's are working for wetlands conservation at the local, global and regional level. Some of the International organisations working for conversation and protection of the wetlands include *Wetlands International, Ramsar Convention Bureau, WWF, Wildfoul & Wetlands Trust, US Fish and Wildlife Services, The Wetlands Institute, The Wetlands Initiative, Society of Wetlands Scientist, IUCN, Birdlife International* and many state and national wetland authorities.

This book aims at elucidating both policy and management practices for wetland managers and indigenous technical knowledge for the management of wetlands. This book contains 13 chapters representing different geographical regions ranging from the coastal regions to high altitude Eastern and Western Indian Himalayas. Different types of studies on wetlands such as those of coastal, mangroves, lagoon, reservoir-based wetlands and high-altitude wetlands studies from are highlighted in this book. This book provides baseline information on wetlands, along with definition, categorization, classification and global status of wetlands as well conservation policies. The Ramsar Convention, legal frameworks and effectiveness of the instruments of the Convention are also highlighted in the book. It also describes the different tools and methods used for recognition of economic evaluation of wetlands. Case studies from Eastern Himalayan, Western Himalayan, plains of India and coastal regions in this edited book provide an in-depth understanding of these diverse ecosystems, along with their current status, challenges and future strategies for conservation. People's participation and action learning through multidisciplinary and holistic approaches for wetland conservation and restoration are also suggested in the book. In addition the book also attempts to highlight sustainable practices for conservation of wetlands ecosystems along with the development of national plans and research for their conservation. Various initiatives have been carried out by different bodies for conservation of wetlands, but a significant reason for the failure of conservation initiatives is lack of accurate scientific information on wetlands. Thus, science based participatory conservation actions have been strongly advocated in the book. It is recommended in the book that genuine research be conducted, which can lead to desirable policy interventions. This book cover insightful scientific and socio-cultural aspects related to the wetlands. It is aimed that this book will be a useful resource for researchers, ecologists, geographers, policymakers and other stakeholders working on the natural resources, wetland and water management.

We are indebted to all the individual authors for their expertise and contributing valuable information in form of chapters to complete this book. We would like to express our gratitude to our Publisher Wiley Blackwell, John Wiley & Sons for accepting the book proposal and for their continuous support, encouragement. We acknowledge and express our appreciation for all reviewers, Content Refinement Specialist Vimali Joseph and Senior Commissioning Editor Andrew Harrison, Senior Managing Editor Rosie Hayden from Wiley for their valuable constructive suggestions, continuous support and encouragement.

Editors

Date: July 1 2021

Sanjeev Sharma & Pardeep Singh

List of Contributors

Thattantavide Anju
Department of Plant Science
School of Biological Sciences
Central University of Kerala
Kasaragod
Kerala
India

Vijay Archa
Department of Plant Science
School of Biological Sciences
Central University of Kerala
Kasaragod
Kerala
India

Alvia Aslam
Department of Environmental Science
Central University of South Bihar
Gaya
Bihar
India

Sami Ullah Bhat
Department of Environmental Science
School of Earth and Environmental Sciences
University of Kashmir
Hazratbal
Srinagar
Jammu and Kashmir
India

Moharana Choudhury
Voice of Environment (VoE)
Guwahati
Assam
India

Poulomi Chakraborty
Indian Institute of Technology Kharagpur
Kharagpur
West Bengal
India

Meenakshi Chaurasia
Department of Botany
University of Delhi
New Delhi
India

Shahid Ahmad Dar
Department of Environmental Science
School of Earth and Environmental Sciences
University of Kashmir
Hazratbal
Srinagar
Jammu and Kashmir
India

Kausik Ghosh
Department of Geography
Vidyasagar University
Midnapore
West Bengal
India

Ginkuntla Saikiran Goud
Department of Plant Science
School of Biological Sciences
Central University of Kerala
Kasaragod
Kerala
India

Shalini Gupta
School of Environment and Sustainable Development
Central University of Gujarat
Gandhinagar
Gujarat
India

Zoha Jafar
Department of Civil Engineering
Faculty of Engineering
Jamia Millia Islamia
New Delhi
India

K.C. Jisha
Department of Botany
MES Asmabi College, P. Vemballur
Kerala
India

Arun Kumar Kashyap
Department of Biotechnology
Government E. Raghavendra Rao PG Science College
Bilaspur
Chhattisgarh
India

Komal
Department of Higher Education
Government Degree College
Bani
Jammu and Kashmir
India

Ajay Kumar
Department of Plant Science
School of Biological Sciences
Central University of Kerala
Kasaragod
Kerala
India

Deepak Kumar
United Nations Development Programme (UNDP)
New Delhi
India

Raj Kumar
Centre for Study of Regional Development
School of Social Sciences
Jawaharlal Nehru University
New Delhi
India

Rakesh Kumar
Department of Environmental Sciences
University of Jammu
Jammu
Jammu and Kashmir
India

Sushil Kumar
Department of Botany
Government Degree College
Ramnagar
Jammu and Kashmir
India

Vijay Kumar
Department of Applied Sciences & Humanities
Faculty of Engineering & Technology
Rama University
Kanpur
Uttar Pradesh
India

Sughosh Madhav
School of Environmental Sciences
Jawaharlal Nehru University
New Delhi
India

Gauhar Mahmood
Department of Civil Engineering
Faculty of Engineering
Jamia Millia Islamia
New Delhi
India

Shilpi Nagar
Department of Environmental Studies
University of Delhi
New Delhi
India
and
Centre for Fire, Explosive and Environment Safety (CFEES)
Defence Research Development Organization (DRDO)
New Delhi
India

Sadaf Nazneen
Department of Civil Engineering
Faculty of Engineering
Jamia Millia Islamia
New Delhi
India

Purushothaman Parthasarathy
Department of Civil Engineering
College of Engineering and Technology
Faculty of Engineering and Technology
SRM Institute of Science and Technology
Chengalpattu
Tamil Nadu
India

Kajal Patel
Department of Botany
University of New Delhi
New Delhi
India

Mahika Phartiyal
Centre for the Study of Regional Development
School of Social Science
Jawaharlal Nehru University
New Delhi
India

Jos T. Puthur
Plant Physiology and Biochemistry Division
Department of Botany
University of Calicut
Malappuram
Kerala
India

Arun M. Radhakrishnan
Department of Plant Science
School of Biological Sciences
Central University of Kerala
Kasaragod
Kerala
India

Nirala Ramchiary
School of Life Sciences
Laboratory of Translational and Evolutionary Genomics
Jawaharlal Nehru University
New Delhi
India

Deeksha Ranjan
Department of Applied Sciences & Humanities
Faculty of Engineering & Technology
Rama University
Kanpur
Uttar Pradesh
India

Rajesh Kumar Ranjan
Department of Environmental Science
Central University of South Bihar
Gaya
Bihar
India

Irfan Rashid
Department of Geoinformatics
School of Earth and Environmental Sciences
University of Kashmir
Hazratbal
Srinagar
Jammu and Kashmir
India

Rohit Rattan
Western Himalayas Conservation Programme
World Wide Fund for Nature India
New Delhi
India

Krishna Rawat
School of Environment and Sustainable Development
Central University of Gujarat
Gandhinagar
Gujarat
India

Vijay Saigal
Department of Law
University of Jammu
Jammu
Jammu and Kashmir
India

Bharti Sharma
School of Biosciences and Biotechnology
BGSB University
Rajouri
Jammu and Kashmir
India

Sanjeev Sharma
Centre for the Study of Regional Development
School of Social Sciences
Jawaharlal Nehru University
New Delhi
India

Sudeep Shukla
Environment Pollution Analysis Lab
Bhiwadi
Alwar
Rajasthan
India

Pardeep Singh
Department of Environmental Studies
PGDAV College
University of Delhi
New Delhi
India

Jitendra Kumar Singh
School of Environment and Sustainable Development
Central University of Gujarat
Gandhinagar
Gujarat
India

Shuchita Singh
Lachoo Memorial College of Science and Technology
Jodhpur
Rajasthan
India

Indu Tripathi
Department of Environmental Studies
University of Delhi
New Delhi
India

Khushaboo Verma
Faculty of Education
Banaras Hindu University
Varanasi
Uttar Pradesh
India

Reeta Verma
School of Environment and Sustainable Development
Central University of Gujarat
Gandhinagar
Gujarat
India

Vidya P. Warrier
Department of Plant Science
School of Biological Sciences
Central University of Kerala
Kasaragod
Kerala
India
and
Department of Biotechnology
Indian Institute of Technology
Madras
Chennai
Tamil Nadu
India

Amit Kumar Yadav
School of Environment and Sustainable Development
Central University of Gujarat
Gandhinagar
Gujarat
India

1

Global Wetlands

Categorization, Distribution and Global Scenario

Sanjeev Sharma[1,*], *Mahika Phartiyal*[1], *Sughosh Madhav*[2], *and Pardeep Singh*[3]

[1] *Centre for the Study of Regional Development, School of Social Science, Jawaharlal Nehru University, New Delhi, India*
[2] *School of Environmental Sciences, Jawaharlal Nehru University, New Delhi, India*
[3] *Department of Environmental Studies, PGDAV College, University of Delhi, New Delhi, India*
[*] *Sanjeev Sharma, Email: sanjeevsharma@mail.jnu.ac.in; Cell: +91 9418613054*

1.1 Wetlands Definition, Categorization and Classification Criteria

Wetlands form a crucial part of the socio-ecological system as they are a storehouse of numerous ecosystems services. Wetland systems manage hydrological processes and preserve the natural ecological system to regulate the ecological balance and well-being of humanity. Economic interests of the environment of the wetlands have not been recognised yet by policy planners and decision-makers, neither at the global level, nor at regional levels. Wetlands are distributed in all geographical regions and climate zones of the planet earth. Globally and nationally, many attempts have been made to assess and identify the wetlands systems (Hu et al., 2017; Cowardin et al. 1979; Briggs 1981; Paijmans et al. 1985; Scott 1989; Gopal 1977; Gopal and Sah 1995). These wetlands systems are the most resourceful and important ecosystem on terrestrial as well as aquatic systems. Various elements of wetland, including geomorphology, hydrology, vegetation, water chemistry or substratum characteristics, have been emphasised by different classification schemes.

Globally, wetland systems are classified according to their origin, use, hydrology, composition, water level, physical and chemical characteristics (Gopal et al. 1990). Stanton (1975) suggested that Queensland wetlands be categorised into inland wetlands and coastal wetlands, which can be further divided on the basis of vegetation (i.e. mangroves, salt water meadows, salt marshes and salt mudflats), flood length and frequency. Cowardin et al. (1979) introduced a hierarchy consisting of wetland structures, subsystems, and groups, which is the most detailed classification scheme developed till date. The techniques and subsystems are based on geomorphological principles, while the groups usually stress the existence of substratum and physiognomy of the vegetation. Hydrology, water chemistry, and soil features are used as modifiers at the class and subclass level. A significant

Wetlands Conservation: Current Challenges and Future Strategies, First Edition.
Edited by Sanjeev Sharma and Pardeep Singh.

drawback of the classification developed by Cowardin (1997) is its uncertainty for practical purposes. Briggs (1981) classified the wetlands on the basis of vegetation. Paijmans et al. (1985) arranged them based on vegetation and hydrology. In an attempt to categorise the wide range of wetlands covered by the Ramsar definition, Scott (1989) identified 30 groups of natural and man-made wetlands. Gopal and Sah (1995) proposed wetlands classification in India based on vegetation types that define specific hydrological regimes.

Wetland is a general term used for all kinds of ecosystems that stay wet for a period that is necessary for them to act as habitats. The term 'WETLAND' was first used officially in 1956 in the U.S. newspaper. The operation for Fish and Wildlife (Martin et al. 1953; Shaw and Fredine 1956; Tiner 2005) defined "*wetlands as lowlands covered with shallow and sometimes temporary or intermittent waters referred to by such names as marshes, swamps, bogs, wet meadows, potholes, sloughs, and river-overflow lands*". Many scientists across the world have defined and written widely about wetlands after 1953. The most popular wetland definitions are: '*The wetlands are lands where saturation with water is the dominant factor determining the nature of soil development and the types of plant and animal communities living in the soil and on its surface* (Cowardin *et al.* 1979)'. This definition was later adopted by the U.S. Fish and Wildlife Service. Wetlands are defined as '*lands transitional between terrestrial and aquatic ecosystem systems where the water table is usually at or near the surface, or the land is covered by shallow water*' (Mitsch and Gosselink 1986*).* '*A wetland is an ecosystem that arises when inundation by water produces soils which are dominated by anaerobic processes, which, in turn, forces the biota, particularly rooted plants, to adapt with flooding*' (Keddy 2010).

The Australian Convention (Hart et al. 1990; Semeniuk and Semeniuk 1995) defines wetlands as:

> "*Areas of seasonally, intermittently, or permanently waterlogged soils or inundated land, whether natural or otherwise, fresh or saline.*"

The Canadian wetland classification system (Zoltai and Vitt 1995; Warner and Rubec 1997) defines wetlands as:

> "*Land that is saturated with water long enough to promote wetland or aquatic processes as indicated by poorly drained soils, hydrophytic vegetation, and various kinds of biological activity which are adapted to a wet environment.*"

The Department of Conservation, New Zealand, (Johnson and Gerbeaux 2004) defines wetlands as:

> "*Permanently or intermittently wet areas, shallow water, or land water margins that support a net ecosystem system of plants and animals that are adapted to wet conditions.*"

Classification of wetlands and deep water habitats of the United States (Cowardin et al. 1979) defines wetlands as:

> "*Lands transitional between terrestrial and aquatic systems where the water table is usually at or near the surface or the land is covered by shallow water.*"

The Ramsar Convention on Wetlands gave the most widely recognized and accepted definition of wetlands. It defines wetlands as: '*The areas of marsh, fen, peatland or water, whether natural or artificial, permanent or temporary, with water that is static or flowing, fresh, brackish or salt, including areas of marine water the depth of which at low tide does not exceed six meters*' (Finlayson and Moser 1991). Wetlands being transitional areas tend to possess characteristics of both terrestrial and aquatic systems, which are unique ecosystems in themselves.

1.1.1 Wetlands- Categorization and Classification

Different types of wetlands have some common characteristics like:

a) Hydrology that causes wet or flooded soils
b) Soils influenced by anaerobic processes and
c) Life forms, especially rooted vascular plants, adapted to life in flooded ecosystems

There are many wetlands habitats, including human constructed and natural reservoirs and ponds, paddy fields, marshy and swampy lands.

The most important primary productive ecosystem in the world are marshes and swamps. They are recharged naturally by the availability of water sources and different forms of precipitation.

Marshes are dominated by herbaceous vegetation and non-wooded wetlands while swamps are dominated by woody plants and trees.

The subtypes of marshes and swamps are:

i) **Tidal Marshes:** Tidal marshes are commonly found near the shorelines and have high salt content. They sequester millions of tonnes of carbon every year. Major tidal marshes are existing in Eastern North America and Coastal Europe.
ii) **Freshwater Marshes:** The freshwater marshes account for more than 90% of the wetland area in the U.S. Florida Everglades is one of the world's most extensive marshes.
iii) **Swamps** are found in still water areas and around floodplains. They include forested systems, the herbaceous systems and the Papyrus systems.
iv) **Mangrove**: The World's Mangrove Swamp forest-covered around 14 million hectares by the end of the 20th century. Majority of mangroves are existing in the economically impoverished regions of the world. These regions are mainly around the Indian Ocean, West Pacific region, Niger and the Sunderban Deltas.
v) **Peatlands:** Peatlands can be marshes or swamps, depending upon the vegetation. Bogs are nutrient-poor peatlands since they are isolated from the groundwater. Whereas fens are peatlands rich in nutrients as they receive inputs from the groundwater. Peatlands covered at least 500 million hectares including bog, moor, fen, muskeg are some landscapes formed by peats.
vi) **Peat soil:** Peat soil is concentrated in the Yellow sea region. They occupy 3% of the land surface and store large carbon sink (Ramsar Convention on Wetlands 2018).

The most generally recognised classification system for wetlands has been devised by the Convention on Wetlands of International Significance (Ramsar Convention). This classification method splits wetlands into three main groups (i.e. marine, inlands and human-made), divided into 42 types of wetlands and summarized in Table 1.1. The codes approved

Table 1.1 Classification of wetlands as defined by *Ramsar Convention Bureau, 2010*

A - Permanent shallow marine waters in most cases less than six m deep at low tide; includes sea bays and straits
B - Marine sub-tidal aquatic beds; includes kelp beds, sea-grass beds, and tropical marine meadows
C - Coral reefs
D - Rocky marine shores; includes rocky offshore islands, sea cliffs
E - Sand, shingle or pebble shores; includes sand bars, spits and sandy islets; includes dune systems and humid dune slacks
F - Estuarine waters; permanent water of estuaries and estuarine systems of deltas
G - Intertidal mud, sand or salt flats
H - Intertidal marshes; includes salt marshes, salt meadows, saltings, raised salt marshes; includes tidal brackish and freshwater marshes
I - Intertidal forested wetlands; includes mangrove swamps, nipah swamps and tidal freshwater swamp forests
J - Coastal brackish/saline lagoons; brackish to saline lagoons with at least one relatively narrow connection to the sea
K - Coastal freshwater lagoons; includes freshwater delta lagoons. Zk (a) – Karst and other subterranean hydrological systems, marine/coastal
L - Permanent inland deltas.
M - Permanent rivers/streams/creeks; includes waterfalls.
N - Seasonal/intermittent/irregular rivers/streams/creeks.
O - Permanent freshwater lakes (over 8 ha); includes large oxbow lakes.
P - Seasonal/intermittent freshwater lakes (over 8 ha); includes floodplain lakes.
Q - Permanent saline/brackish/alkaline lakes.
R - Seasonal/intermittent saline/brackish/alkaline lakes and flats.
Sp - Permanent saline/brackish/alkaline marshes/pools.
Ss - Seasonal/intermittent saline/brackish/alkaline marshes/pools.
Tp - Permanent freshwater marshes/pools; ponds (below 8 ha), marshes and swamps on inorganic soils; emergent vegetation water-logged for at least most of the growing season.
Ts - Seasonal/intermittent freshwater marshes/pools on inorganic soils; includes sloughs, potholes, seasonally flooded meadows, sedge marshes.
U - Non-forested peatlands; includes shrub or open bogs, swamps, fens.
Va - Alpine wetlands; includes alpine meadows, temporary waters from snowmelt.
Vt - Tundra wetlands; includes tundra pools, temporary waters from snowmelt.
W - Shrub-dominated wetlands; shrub swamps, shrub-dominated freshwater marshes, shrub carr, alder thicket on inorganic soils.
Xf - Freshwater, tree-dominated wetlands; includes freshwater swamp forests, seasonally flooded forests, wooded swamps on inorganic soils.
Xp - Forested peatlands; peat swamp forests.
Y - Freshwater springs; oases.
Zg -Geothermal wetlands
Zk (b) – Karst and other subterranean hydrological systems, inland

Source: Ramsar wetland-type classification retrieved from: https://www.environment.gov.au/water/wetlands/ramsar/wetland-type-classification © Department of Agriculture, Water and the Environment(Southern Australia). Licensed under CC BY 4.0.

by Recommendation 4.7 and modified by Resolutions VI.5 and VII.11 of the Contracting Parties Conference would represent different types of wetlands. This classification aims to provide a detailed framework to encourage the rapid identification of the key wetland ecosystems at each *location (Ramsar Convention Bureau, 2010; website-Department of Agriculture, Water and the Environment, Southern Australia)*

1.1.2 Human- Made Wetlands

1) Aquaculture or aqua farming in both freshwater and salt water (e.g., fish farming, aquatic plants, algae and other organisms)
2) Ponds less than 8 ha of lands; agricultural ponds small tanks.
3) Artificial irrigation, including canals, small sprinkler systems and paddy fields.
4) Seasonally flooded agricultural land, including flood river bed, meadows and pasture lands.
5) Salt exploitation sites; salt pans, salines, etc.
6) Water storage system over 8 ha; reservoirs/barrages/dams/impoundments.
7) Excavations; gravel/brick/clay pits; borrow pits, mining pools.
8) Wastewater treatment area including sewage farms, settling ponds, oxidation basins, etc.
9) Canals and drainage channels, ditches.

1.2 Importance of Wetland Ecosystem

Wetlands are vital for human prosperity and regulating the environmental system. Since the beginning of human civilization on this planet earth, wetlands have been a lifeline for human civilisations. Civilizations have flourished mainly along the periphery of the wetland ecosystems. The wetlands are ecosystem stems; they serve a range of purposes for the well-being of humans and support the natural flow of ecosystem system, such as management and monitoring of the natural ecological and environmental processes (Figure 1.1). The regulatory system is essential for the protection of the human and planet health (Schuyt and Brander 2004). Wetlands are considered to be natural surface water and groundwater purifiers, water rechargers, recyclers of nutrients and human waste. They are essential to protect watersheds, to monitor climate and disasters such as floods, droughts, ecosystems for rich biodiversity, to stabilise shorelines, to desynchronize flood flows to constitute a major source of carbon sinks etc. (Kusler et al. 1994; Kraus 1995). Environment roles arise from a range of biological and physical interactions and interdependence of the socio-ecological system. The wetland systems are a class of land cover that is ecologically fragile and highly susceptible to anthropogenic pressure and climate change. Through potential or close potential evapotranspiration, wetlands directly influence the global and local/regional ecosystems by exchanging water, heat and energy with surrounding as well as regional atmosphere composition and climate (Fan et al. 2010). Russi et al. (2013) observed that wetlands get engaged in the global biogeochemical cycle through greenhouse gas emissions and carbon dioxide sequestration and indirectly influence the environment from local to global scale. The collaborative environment of biological and geochemical processes in the wetlands contribute to emission of greenhouse gases. The wetlands systems are the largest source of methane (CH_4) in the world, and the only one dominated by the atmosphere

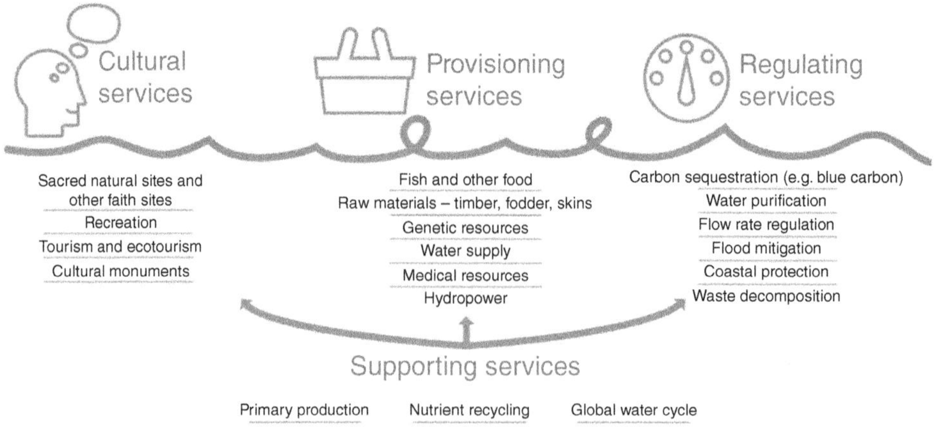

Figure 1.1 Ecosystem services from wetlands system *Source:* Ramsar Convention (2018) and Gardner and Finlayson (2018) © Ramsar Convention Secretariat.

(Prigent et al. 2001; Bousquet et al. 2011). Many of these tasks are often essential for individuals, directly or indirectly, and are therefore continually influenced by human activities (Reis et al., 2017 & 2019).

Wetlands are resourceful for humankind as they perform many significant functions. Wetlands are essential for the provision of ecosystem system services for human beings' and they are survival and safeguard of the natural ecosystems. Wetlands system are world's most biological productive ecosystems and account for 47% of the global ecosystem's value. They provide essential tangible and non-ecosystem services for biotic and abiotic components of the environment (MEA 2005; Russi *et al.* 2013). Wetlands, including natural and artificial wetlands, ponds, rivers, swamps, marshes, peatlands, mangroves, and coral reefs, are major sources of ecosystem services and contribute in regulating the ecological system and people's livelihoods. Wetlands are also known as "*the kidney of the earth system*" and believed to be the cradle of animals and plants. The wetlands are the most biologically diverse ecosystems and act as source and purifier of water on earth. Wetlands can filter pollution from the water and remove toxins and pollutants from the aquatic system due to their high and long-term capacity to filter pollution from the water. These wetlands are helping to protect humanity by conserving natural resources from natural hazards like floods, droughts and many other disasters. Wetlands are the major storehouse of carbon than any other system and the production of food, fibre and ecotourism services (Mitsch and Gosselink 2000; Keddy 2010; Junk *et al.* 2013).

Since 1970, more than 35% of the wetlands have been lost which is more than three times greater than forest loss (Global wetlands Outlook-2018). In controlling the hydrological process, wetlands play a vital role. The wetland systems regulate the movement of water supply, replenish groundwater, purify surface water, underground water, and manage and monitor the hydrological cycle phase. The ecology of wetlands plays a crucial role in well-being of human beings and all other species of flora, fauna, climate change adaptation, biodiversity conservation, hydrology and soil conservation, and protection of the health of the planet (Ramsar Convention Bureau 2001). Centred on the principle of conventional medical practices and providing help for human health, 80% of the world's population relies primarily on health care services in these wetlands.

Wetlands are universally considered to be one of the primary natural pools of greenhouse gas methane (CH_4) emission as they contributing 20–40% of the total annual emissions to the atmosphere, which adds a robust radiative forcing from CH_4 (Bousqet al. *al.* 2011; Qin et al. 2014). Wetland ecosystem plays an extremely significant part in influencing the global climate system by biogeochemical feedback mechanisms (Seneviratne *et al.* 2010; Fisher *et al.* 2011).

1.3 Spatial Distribution and Potential of Global Wetlands

Globally, wetlands spread from the tropics to the tundra region across all climatic zones and are the most productive and vital habitats (Mitsch and Gosselink 1993). The spatial distribution of global wetlands might have occurred before the establishment of human civilization. Understanding the global distribution of wetlands will help to promote our understanding of the sustainable growth of wetlands, help in improving the wetlands, and the decision-making processes to preserve and protect this important part of the ecological system. Globally seasonal inland wetlands account for about 6% of the world's terrestrial area, and about 89% of them are unprotected (as specified by IUCN I–VI and Ramsar protected areas) and provide <40 % of the global Species (Ramsar 2001; Mitra *et al.* 2003; Reis 2017).

Global wetland area range from 0.54 to 21.26 million km^2, and the class-specific spatial consistency of wetlands is less than 1% (Shenget *et al.*, 2016). Ecologically wetlands, natural or artificial, represent one of the planet's most valuable and vital parts of ecosystems and are also one of the most vulnerable and threatened natural resources. As per the Global Wetland Outlook: state of the world's wetlands and their services to people Ramsar Convention 2018, "*The most recent estimate of global inland and coastal wetland area is more than 12.1 million km^2, an area almost as large as Greenland. Of this, 54% is permanently inundated, and 46% seasonally flooded. An estimated further 5.2 million km^2 are intermittently or occasionally inundated, but this is believed to include areas of formerly converted wetlands affected by extreme storm events.* Approximately 93% of wetlands are inland systems, with 7% marine and coastal – although this coastal estimate does not include a number of wetland groups, such as near shore sub tidal wetlands, which also fall under the Ramsar definition. Global wetland areas are small in comparison: wetlands occupy an estimated 0.3 million km^2 and rice paddy 1.3 million km^2 (Davidson et al. 2018; Davidson and Finlayson 2018). The size of the global wetlands has increased considerably since the 1980s, primarily due to advancement in remote sensing, geographical information system tools and mapping methods but there has not been an increase in the actual size of the wetlands (Davidson *et al.* 2018). The largest wetlands region at the continental level are distributed in (Figure 1.2) Asia (32%), North America (27%) and Latin America & Caribbean (16 %), Europe (13%), Africa (10%) and Oceania (3%) (Davidson et al. 2018).

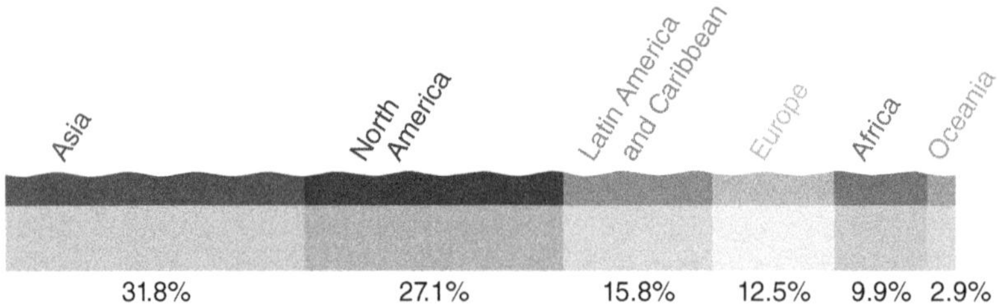

Figure 1.2 Regional distribution (%) of the wetland area *Source:* Davidson et al. (2018) and Ramsar Convention on Wetlands; WORLD OUTLOOK, 2018 (2018). © Ramsar Convention Secretariat.

1.4 Status and Impacts on the Wetlands Ecosystem

The wetland ecosystem is a very sensitive aspect of the ecological landscape. The health of flora and fauna species is associated with the conditions and distribution of the wetlands. Any change on a small and large scale in the wetland ecological system impacts the biodiversity and habitat of the species. The changes in wetlands leads to changes in number of species, distribution of the species, alteration in migratory bird routes, and changes in community structures and species interconnection (Erwin 2009). Rapid changes in population dynamics or any other environmental or climate change often result in wetland encroachment due to increasing demand for ecosystem services in wetlands (Mitchell 2013; Davidson 2014; Dixon *et al.* 2016). These wetlands are increasingly degraded, and the provision of wetlands services is at serious risk. Based on an analysis of 189 reports on change in wetlands system, Davidson outlined in 2014, *that 'over a long period wetlands have been converted to other land-use practices and have lost more than 50% since 1900 AD and more than 87% since the beginning of the 18th century'.* Trustworthy information on the spatial distribution of wetlands would help academics, policymakers, scientists, and environmental conservationists understand the location, area, distribution and current physio-biological status of global wetlands and develop efficient conservation policies measures and management plans. However, globally (Davidson 2014; Dixon *et al.* 2016) and at regional levels some work has been done on recent changes in global wetlands, distribution, loss and status.

Wetlands are currently at higher risk of climate change for example alternations have been noted in coral reefs, mangroves and tropical forests, sub-arctic forests, prairies and arctic/alpine areas. Inland freshwater wetlands are likely to be affected mainly by precipitation patterns and more frequent or intense droughts, storms and floods. Changes in rainfall patterns and volume entering river systems would change water availability to coastal wetlands including deltas and estuaries, affecting salinity and sediment and nutrient supply. Melting will cause mountain snowfields and glaciers that feed many of the world's largest river and wetland systems will disappear. Water quality will be harmed as a result of rising water temperatures, floods, and droughts, which will exacerbate various forms of pollution. Many semi-arid regions are uniquely vulnerable to climate change impacts and are expected to undergo severe anthropogenic and natural pressures for water resources and wetlands. There may be many other minor hazards specific to different wetlands, such as deterioration and loss of habitat, conversion for agriculture and infrastructure development, changing land use and land cover, urbanisation, pollution from agricultural runoff, sewage and solid waste disposal, encroachment and remediation of wetlands, degradation of wetlands, the infestation of weeds and alien species invasions, poaching of wildlife and man-wildlife conflict.

Xu et al. (2019) has identified 2414 sites of international importance wetlands covering 254543972 ha and have been recognized globally significant. According to Xu et al. (2019), 2303 wetlands of global significance are distributed unevenly across the globe's continents. Many wetlands have been damaged and degraded due to rapid farming and urbanisation in North America and Europe, leading to loss of most floodplain regions (Tockner and Stanford, 2002). Increasing pressure on water resources by population is also very dominant in these regions, which increases biodiversity loss. Such adverse impacts on wetlands are high in developing countries in the continents of Asia and Africa (Vorosmarty *et al.,* 2010). Some parts of South America, Australia and Northern Europe have remained less affected by the controlled and regulating anthropogenic activities as compared to other parts of the globe. As a result of greater water regulation for agricultural growth and urban and industrial expansion, ecological qualities have deteriorated in recent decades in Australia (Finlaysonet *et al.,* 2013). Based on 153 reports published in 63 different journals, Kandus et al. 2018 found that most wetlands in South America have been conserved and are in relatively excellent condition. China's wetlands cover 53.42x106 acres, accounting for 10% of worldwide wetlands and 5.58 % of the country's total land area. Multiple contributing variables are constantly putting China's wetlands at jeopardy. Since the 1950s, a considerable number of wetlands have been lost or damaged. 33% of China's total wetlands have been lost in the time span of 1978 -2008. Cropland expansion has majorly led to the loss of vegetated wetlands in China during 1990 to 2010 (Mao et al. 2021).

Loss of vegetation due to invasive species has resulted in the spread and displacement of native species. Adversely vegetation loss is also affected by coastal wetlands due to anthropogenic pressure and climate change impacts. Global warming and climate change are of serious concern to coastal and high-altitude wetlands. Flood plains are no longer considered wetlands when flood management control fails. Other factors, such as agricultural development, land use, land cover changes, urbanisation, infrastructure development and desertification, are also equally responsible for the degradation of wetlands. According to a study by Zelder and Kercher (2005), reported over 5000 km^2 of wetland is lost annually in Asia due to agriculture, dam construction, and other factors, and about 50% of the global wetland area has been lost as a result of human activity. Over the last 100 years, most continents have experienced significant wetland losses. A study conducted by Xu et al. (2019) on the basis 2303 wetlands of international importance found that largest numbers of wetlands are located in Europe, and Africa has the largest area of land; out of these study locations, more than 50% of the locations are influenced by contamination (54%), biological resource use (53%), natural system alteration (53%) and farming and aquaculture (42%). Major wetlands depend on fish, water birds and tortoises in severe decline, with 1/4th of them threatened extinction, especially in the tropics and wetlands site in United Kingdom and Irelands are least affecting among all over the continents (Xu et al. 2019). According to the Global Wetlands Outlook (2018) since, 1970, 81% of inland wetland populations and 36% of coastal and marine species have declined and more than >10 percent of inland species are subject to global risk assessment from the IUCN Red List. The highest degree of the global threat of extinction are marine turtles (100% globally threatened), wetland-depend mega fauna (62%), freshwater reptiles (40 %), marine molluscs (37%), amphibians (35%), corals (33%) and crabs and crayfish (32%): coral reef-dependent parrotfish and surgeonfish (2%), and dragonflies (8%) have low globally threatened status (Global Wetlands Outlook, 2018). Loss of vegetation due to invasive species has resulted in the spread and displacement of native species. Vegetation loss is also affected by coastal wetlands.

Wetlands are linked to malaria and have been drained for this reason as well. Water quality of the wetlands has also deteriorated due to increasing pollution, drainage, weed infestation and siltation. It's expected to deteriorate further. The main sources of pollution are untreated industrial waste, waste and wastewater from power plants, chemical, fertiliser and pesticide runoff from agricultural land, etc. Eutrophication causes algal blooms, which reduces the water's oxygen content, leading to the extinction of other species. Salinity has also been increased due to pollution. At least 5.25 trillion persistent plastic particles are afloat in the World's oceans and have enormous impacts on coastal waters (Ramsar Convention on Wetlands, 2018). Successive droughts have led to the early departure of many migratory birds.

Finlayson and Valk (1997) claim that the decrease of wetlands coincides with growing scientific evidence of their value to humans, biodiversity, and the environment's quality. At present, the poor consciousness of the significance of the wetland ecosystem by decision-makers and underestimating the values and benefits of wetlands are also a major concern. The protection of wetlands is now becoming a problem and challenging task. It complicates attempts to establish strategies to reduce risks and protect these important resources. Promoting a wetland wise use initiative is crucial to human livelihoods and survival. Finlayson and Valk (1997) argue that the decline in wetlands coincides with increasing scientific evidence of their importance to people, wildlife, and the environment's quality. The protection of wetlands is also difficult. It complicates attempts to establish strategies to reduce risks and preserve these important resources. Promoting the wetland wise use programme is crucial to human livelihoods and survival.

1.4.1 Conservation Measures and Future Strategies

Human activities have had a different effect on the wetland environment and in the hydrological cycle, and an impact on wetlands themselves. Prioritization for recognising the environmental implications of wetland management activities is crucial. Development of holistic approaches and strategies to maximise the protection and conservation of the wetland ecosystem is essential to maintain the ecological balance and regulating earth system. Wetland management project plans need to be incorporated into economic development planning and with a socio-cultural and location-specific priority—establishing a sound ecosystem through the combination of land use planning and economic development. Local community participation in the implementation of management practises and priority needs to be given to the indigenous knowledge framework for the conservation of the wetland Ecosystem. Regular monitoring and scientific studies to determine the importance of wetlands contribute to quantifying ecological, economic and social values. It can be beneficial to ensure successful conservation and restoration.

Ramsar Convention focused on wetlands, recognizing their significance, particularly as waterfowl habitats (Matthews 1993). On realizing their worth, people and governments started laying stress on the conservation of the wetlands. An unusual spurt of interest in wetlands was seen at all levels among scholars, technical and management persons, researchers working in the field of society and economics, different social organisations, and Governments. The Ramsar Convention has been instrumental and highly successful in mobilizing most nations to pledge for wetland conservation. The Convention also talked about the wise and optimum use of wetlands and not the conservation only (Ramsar Convention Bureau 1998, 2001 2018). The signatories have to formulate laws that ensure the conservation of listed wetlands and the optimum use of wetlands in their domain.

The participation of local stakeholders, in particular the communities, is an important determinant of the effectiveness of conservation initiatives. The Ramsar Convention on Wetlands defines participatory management as a learning mechanism that seeks to enhance the collective analysis and action ability of all those interested in wetland conservation. Maintaining and preserving wetlands that are inclusive has a far greater chance of success. The Ramsar Convention indicates that local people should be encouraged to appreciate the principles of wetlands as champions of wetland protection and wise use and should also be involved in inappropriate policy formulation, planning and management.

Implementation of a performance appraisal accountability system among the stakeholders directly and indirectly involved in the wetlands conservation programme. Legal action needs to be taken against those who are responsible for wetlands degradation. Good governance and implementation of wetlands conservation legislation and law at the national and global level can improve the wetland ecosystem's health. Capacity building programme including creating a reliable institutional mechanism for imparting training to wetlands handlers, policymakers, and other key stakeholders. Developing conservation framework and modules in online and offline mode, creating groups of wetlands managers for collaborative research and knowledge are important tools and approaches for conserving wetlands and mitigating the adverse impacts on the wetland ecosystem. Education and awareness are the most crucial components for the successful conservation of wetlands.

1.4.2 Conclusion and Recommendation

Wetlands are highly significant for well being of humankind and they safeguard the environment as well uphold the health of the planet earth. Wetlands provide sufficient habitat and breeding grounds for rare and important endangered species and migratory birds, depending on the environment and seasons of the world's various geographical locations. They have a range of hydrological and ecological services, apart from being a source of livelihood, sustenance and other socio-economic activities. Management and conservation of wetlands systems are directly connected with socio-ecological security and maintaining the ecological balance. Presently these wetland systems are under significant threat from anthropogenic pressure and climatic changes impacts. Numerous factors are directly or indirectly responsible for wetlands degradation. The majority of wetlands are either disappearing and degrading worldwide due to rapid population growth, unregulated and uncontrolled human activities, land use and land cover change, agricultural expansion, urbanisation, discharge of effluents from the terrestrial environment, global warming and climate change. These are not only environmental or ecological problems, but they have directly affected the livelihoods, culture and sustainability of human well-being and the maintenance of the natural ecological system. Freshwater biodiversity due to changes in the hydrology are rapidly declining due to intense floods in the floodplains region. Climate change, water shortages, diminishing freshwater supply effects on aquatic flora and fauna has increased the risk of biodiversity loss and climate change. Wetland habitats lead, directly or indirectly to vital measures of the Sustainable Development Goal (SDG). Wetlands conservation are very important to achieving sustainable development targets 2030.

Globally many scholars have done intensive work on wetlands conservation, but presently more attention needs to focus to understand the intergradations of different wetlands composition, structure, biological, physio-chemical and socio-cultural as well as economic importance. Wetlands ecosystems are ecologically sensitive and fragile and would be incorporated in the priority area for conservation at national and international level. Presently some of the wetlands at the local and regional level are totally neglected from research and conservation points of view. Only selected wetlands which are internationally or nationally in the list of importance are given priority for conservation and management. More priority needs to be given for the documentation, monitoring, restoration and conservation to reduce the pressure on wetlands. Effective implementation of the wetlands regulation, policy, legislation and strengthening the existing regulation with the participation of all stakeholder needs to be implemented at ground level.

At present, many challenges are existing for wetland conservation and the development of future conservation strategies not only at local level but globally. Some of the challenges are implementation of regulation, legislation, lack of governance, uncontrolled and unregulated anthropogenic pressure, climate change, lack of cooperation from the local community and different stakeholders, public participation and insufficient funding for management of wetlands. The Nutrient cycle affects species composition and richness. Ecological integrity and participation of all stakeholders for protection of wetlands and preservation of wetlands ecosystem would contribute to conserving the extinction of endangered and critically endangered and threatened species of aquatic flora and fauna. People participation for wetlands conservation and strengthening awareness and communication strategies from bottom to top level approach are effective measures for strengthening wetlands conservation participation. Healthy wetlands systems are vital to maintaining the ecological sustainability and socio-ecological relationship. Community-level participatory wetlands conservation model based on participatory learning action programme need to develop by using multidisciplinary and interdisciplinary approaches. Management and partition of resource allocation within and in the vicinity of wetlands are required to be given priority in the future. Local communities' pressures on these wetlands for using the ecosystem services can be monitored regularly to reduce pressure and explore new livelihood options and alternative ways of livelihood. Strengthening education, awareness and national and international cooperation may strengthen the new attitude and sound policy. More applied research on wetlands restoration, monitoring and management need to conduct covering different spatial unit. Strict legislation, science based participatory regulation and effective mechanism for wetlands conservation and cooperation among different organisations and institutions. More research funding and opportunities to conduct wetland assessment conservation related study are the major requirements for wetlands conservation and the development of future wetland conservation strategies.

Acknowledgements

The authors are grateful and acknowledge all the distinguished scholars whose publications were referred or cited during preparation and finalization of this chapter. We would also acknowledge reviewers and publishers for reviewing the chapter and giving their valuable inputs and suggestions to improve the manuscript.

References

Bousquet, P., Ringeval, B., Pison, I. et al. (2011). Source attribution of the changes in atmospheric methane for 2006–2008, *Atmos. Chem. Phys.*, 11, 3689–3700, doi:10.5194/acp-11-3689-2011.

Briggs, S.V. (1981). Freshwater wetlands. In: Australian Vegetation (ed. R.H. Groves), 335–360. London: Cambridge University Press.

Cowardin, L.M., Carter, V., Golet, F.C., and LaRoe, E.T. (1979). *Classification of Wetlands and Deepwater Habitats of the United States.* Washington, DC: U.S. Fish and Wildlife Service, FWS/OBS-79/31.

Davidson, N.C. (2014). How much wetland has the World lost? Long-term and recent trends in global wetland area. *Marine and Freshwater Research*, 65:936–941.

Davidson, N.C. and Finlayson, C.M. (2018). Extent, regional distribution and changes in the area of different classes of wetland Marine and Freshwater Research, 2018, 69, 1525–1533, 2018 https://doi.org/10.1071/MF17377 available on https://www.publish.csiro.au/MF/acc/MF17377/MF17377_AC.pdf (assessed on 2 January 2021).

Davidson, N. C., Fluet-Chouinard, E., and Finlayson, C. M. (2018). Global extent and distribution of wetlands: trends and issues. *Marine and Freshwater Research*, *69*(4), 620–627.

Dixon, M. J. R., Loh, J., and Davidson, N. C. (2016). Tracking global change in ecosystem area: The Wetland Extent Trends index. *Biological Conservation*, *193*, 27–35.

Erwin, K.L. (2009). Wetlands and global climate change: the role of wetland restoration in a changing world. *Wetlands Ecology and Management* 17:71–84.

Fan, Y., and Miguez-Macho, G. (2010). Potential groundwater contribution to Amazon evapotranspiration. *Hydrology and Earth System Sciences*, *14*(10), 2039–2056.

Finlayson, C. M. and Valk, A. G. V. D. (1997). Classification and Inventory of the World's Wetlands. *Global Ecology and Biogeography Letters*, 6(5), pp. 407–408.

Finlayson, C. M., Davis, J.A., Gell, P.A. et al. (2013). The status of wetlands and the predicted effects of global climate change: the situation in Australia, *Aquatic Science.* 75:73–93.

Finlayson, M. and Moser, M. (eds). (1991). Wetlands. International Waterfowl and Wetlands Research Bureau, *Facts on File Ltd. Oxford*, Slimbridge, U.K.

Fisher, R. E., Sriskantharajah, S., Lowry, D. et al. (2011). Arctic methane sources: Isotopic evidence for atmospheric inputs. *Geophysical Research Letters*, *38* (21).

Gardner, R. C., and Finlayson, C. (2018). Global wetland outlook: State of the World's wetlands and their services to people. In *Ramsar Convention Secretariat.*

Gopal, B. (1977). Wetlands and their management. In: *Current Trends in Indian Environment* (eds. D. Bandhu and E. Chauhan), 193–198. New Delhi: Today & Tomorrow's Printers & Publishers.

Gopal, B. and Sah, M. (1995). Inventory and Classification of Wetlands in India. *Vegetatio.* 118: 39–8.

Gopal, B., Kvet, J., Loffler, H. et al. (1990). Definition and classification. In: *Wetlands and Continental Shallow Water Bodies* (eds. B.C. Patten, S.E. Jorgenson, H.J. Dumont, et al.), 9–15. The Hague: SPB Academic Publishing.

Hart, B. T., Bailey, P., Edwards, R. et al. (1990). Effects of salinity on river, stream and wetland ecosystems in Victoria, Australia. *Water Research, 24*(9), 1103–1117.

Hu, S., Niu, Z., and Chen, Y. (2017). Global Wetland Datasets: a Review. *Wetlands, 37*(5), 807–817. https://doi.org/10.1007/s13157-017-0927-z.

Johnson, P. and Gerbeaux, P. (2004). New Zealand wetland types. Wellington, New Zealand: Department of Conservation, Te Papa Atawhai.

Junk, W.J., An, S., Finlayson, C.M. et al. (2013). Current state of knowledge regarding the World's wetlands and their future under global climate change: a synthesis. *Aquatic Sciences* 75: 151–167.

Kandus, P., Minotti, P.G., Morandeira, N.S. et al. (2018). Remote sensing of wetlands in South America: status and challenges. *International Journal of Remote Sensing* 39 (4): 993–1016. Taylor and Francis Ltd. doi:https://doi.org/10.1080/01431161.2017.1395971.

Keddy, P.A. (2010). *Wetland Ecology: Principles and Conservation*, 2e, 2. New York: Cambridge University Press.

Kraus, M. L. (1995) Wetlands Worldwide. *BioScience*, 45(9), p. 640.

Kusler, J., Mitsch, W. and Lar, J. S. (1994). Wetlands. *Scientific American*, 270(1), pp. 64B–70.

Mao, D., Wang, Z., Wang, Y. et al. (2021). Remote Observations in China's Ramsar Sites: Wetland Dynamics, Anthropogenic Threats, and Implications for Sustainable Development Goals. *Journal of Remote Sensing* Volume 2021, Article ID 9849343, 13 pages https://doi.org/10.34133/2021/9849343.

Martin, A.C., Hutchkiss, N.F., Uhler, M. and Bourn, W.S. (1953). *Classification of wetlands of the United States.* Special Science Report–Wildlife 20. U.S. Fish and Wildlife Service, Washington, D.C. pp 14.

Matthews, G. V. T. (Geoffrey V. T., & International Union for Conservation of Nature and Natural Resources. Ramsar Convention Bureau. (1993). *The Ramsar Convention on Wetlands : its history and development.* Ramsar Convention Bureau.

Millennium Ecosystem Assessment (2005). Ecosystems and Human Well-Being: Wetlands and Water Synthesis, World Resources Institute Washington, DC.

Mitchell, S. A. (2013). The status of wetlands, threats and the predicted effect of global climate change: the situation in Sub-Saharan Africa. *Aquatic sciences, 75*(1), 95–112.

Mitra, S., Wassmann, R., and Vlek, P. (2003). Global inventory of wetlands and their role in the carbon cycle. Discus. Pap. 18771.

Mitsch, W.J. and Gosselink, J.G. (1986). *Wetlands*, 539. New York: Van Nostrand Reinhold.

Mitsch, W.J. and Gosselink, J.G. (1993). *Wetlands*, 2e, 722. New York: Van Nostrand Reinhold.

Mitsch, W.J. and Gossilink, J.G. (2000). The value of wetlands: Importance of scale and landscape setting. *Ecological Economics* 35: 25–33.

Mitsch, W.J. and Hernandez, M.E. (2013). Landscape and climate change threats to wetlands of North and Central America. *Aquatic Sciences* 75: 133–149.

Paijmans, K., Galloway, R.W., Faith, D.P., et al. (1985). *Aspects of Australian Wetlands.* CSIRO Division of Water and Land Resources Technical Paper No. 44. Canberra, p. 77.

Prigent, C., Matthews, E., Aires, F. and Rossow, W. (2001). Remote sensing of global wetland dynamics with multiple satellite data sets. *Geophysical Research letters*, 28: 4631–4634.

Qin, D., Plattner, G. K., Tignor, M. et al. (2014). Climate change 2013: the physical science basis. *Contribution of Working Group I to the Fifth Assessment Report of the Intergovernmental Panel on Climate Change (eds TF Stocker et al.)*, 5-14.

Ramsar Convention Bureau (1998). *Information Sheet on Ramsar Wetlands.* Gland, Switzerland.

Ramsar Convention Bureau (2001). *Wetlands Values and Functions.* Ramsar Convention Bureau. Gland, Switzerland.

Ramsar Convention Secretariat (2010). *Designating Ramsar Sites: Strategic Framework and Guidelines for the Future Development of the List of Wetlands of International Importance.* Gland: Ramsar Convention Secretariat. https://www.ramsar.org/sites/default/files/documents/pdf/lib/hbk4-17.pdf

Ramsar Convention on Wetlands (2018). *Global Wetland Outlook: State of the World's Wetlands and their Services to People.* Page 12. Gland, Switzerland: Ramsar Convention available https://www.ramsar.org/sites/default/files/documents/library/gwo_e.pdf

Reis, V., Hermoso, V., Hamilton, S. K. et al. (2017). A Global Assessment of Inland Wetland Conservation Status. In *BioScience* (Vol. 67, Issue 6, pp. 523–533). Oxford University Press. https://doi.org/10.1093/biosci/bix045

Reis, V., Hermoso, V., Hamilton, S. K. et al. (2019). Characterizing seasonal dynamics of Amazonian wetlands for conservation and decision making. *Aquatic Conservation: Marine and Freshwater Ecosystems*, *29*(7). https://doi.org/10.1002/aqc.3051

Russi, D., ten Brink, P., Farmer, A. et al. (2013). The economics of ecosystems and biodiversity for water and wetlands. *IEEP, London and Brussels, 78.*

Schuyt, K. and Brander, L. (2004). *Living waters- Conserving the source of life. The economic Values of the World's wetlands.*, Gland/Amsterdam: WWF.

Scott, D.A., (1989). *A Directory of Asian Wetlands.* IUCN, Gland, Switzerland and Cambridge, UK.

Semeniuk, C.A. and Semeniuk, V. (1995). A geomorphic approach to global classification for inland wetlands. *Vegetatio.* 118: 103–124.

Seneviratne, S.I., Corti, T., Davin, E.L. et al. (2010). Investigating soil moisture–climate interactions in a changing climate: a review. *Earth Science Reviews* 99: 125–161.

Shaw, S.P. and Fredine, C.G. (1956). *Wetlands of the United States, their extent, and their value for waterfowl and other wildlife.* Circular 39. U.S. Fish and Wildlife Service, Washington, D.C. Pp 67.

Shengjie, H.U., Niu, Z., and Yanfen, C. (2016). Global Wetland Datasets: a review. *Wetlands* https://doi.org/10.1007/s13157-017-0927-z.

Stanton, J.P. (1975). *A Preliminary Assessment of Wetlands in Queensland.* CSIRO Division of Land Use Research Technical Memorandum 75/10, Canberra, p. 42.

Tiner, R.W. (2005). *In Search of Swampland: A Wetland Sourcebook and Field Guide*, 3–5. Rutgers University Press.

Tockner, K., and Stanford, J. A. (2002). Riverine flood plains: present state and future trends. *Environmental conservation*, 308–330.

Vörösmarty, C.J., McIntyre, P.B., Gessner, M.O. et al. (2010). Global threats to human water security and river biodiversity. *Nature 467* (7315): 555–561.

Warner, B.G. and Rubec, C.D.A. (eds.) (1997). The Canadian wetland classification system. (National Wetlands Working Group) Second edition. Waterloo, Ontario, Canada: Research Centre, University of Waterloo.

Xu, T., Weng, B., Yan, D. et al. (2019). Wetlands of international importance: status, threats, and future protection. *International Journal of Environmental Research and Public Health 16* (10) https://doi.org/10.3390/ijerph16101818.

Zedler, J. B., and Kercher, S. (2005). Wetland resources: Status, treecosystem ystem services, and restorability. In: *Annual Review of Environment and Resources* (Vol. 30, pp. 39–74). https://doi.org/10.1146/annurev.energy.30.050504.144248.

Zoltai, S. C., and Vitt, D. H. (1995). Canadian wetlands: environmental gradients and classification. *Vegetatio*, *118*(1): 131–137.

2

Ramsar Convention

History, Structure, Operations, and Relevance

Rohit Rattan[*1], *Bharti Sharma*[2], *Rakesh Kumar*[3], *Vijay Saigal*[4], *and Sudeep Shukla*[5]

[1] *Western Himalayas Conservation Programme, World Wide Fund for Nature India, New Delhi, India*
[2] *School of Biosciences and Biotechnology, BGSB University, Rajouri, Jammu and Kashmir, India*
[3] *Department of Environmental Sciences, University of Jammu, Jammu, Jammu and Kashmir, India*
[4] *Department of Law, University of Jammu, Jammu, Jammu and Kashmir, India*
[5] *Environment Pollution Analysis Lab, Bhiwadi, Alwar, Rajasthan, India*
* *Rohit Rattan, Email: envrohit@gmail.com*

2.1 Background

Wetlands are among the most biodiverse ecosystems on the Earth. They occur where the water table is at or near the surface of the land or where the land is covered by shallow water (Ramsar Convention Secretariat 2011). In wetlands, water is the primary factor that controls the associated biotic and abiotic environment. Wetlands offer multiple benefits, including ecological as well as socioeconomic, to humans and other biota. They provide habitats to as many as 20% of the planet's various life-forms (Gopal 1977). The major civilizations have emerged and flourished along the fertile floodplains of some of the largest river systems of the world like Indus, Nile, Euphrates, and Tigris, and often wetlands are called "the cradle of human civilization" (Grist 1975). The wide range of ecosystem services that the diverse types of wetlands provide have been classified into four types, i.e. provisioning, regulating, cultural, and supporting services (Millennium Ecosystem Assessment 2005). The wetlands are often referred to as "biological supermarkets," owing to their high primary productivity, rich biodiversity and hosting of intricately woven food webs (Gawler 2000; Prasad et al. 2002). Despite the multiple intangible benefits they offer, the true worth of the wetland services has never really been accounted for. The overexploitation, degradation, and conversion of wetlands around the world have probably been going on ever since humans have found access to them. Unfortunately, of all-natural ecosystems, wetland ecosystems have suffered the most (Davidson et al. 2005). Studies have revealed that the wetlands around the world had degraded by about 87% since 1700 in data-existing regions, with a majority of this degradation occurring in the twentieth and early twenty-first centuries (Gawler 2000). In Europe, their thoughtless conversion has been going on since the times of the Romans (Davidson et al. 1991); in North America and southern Africa

Wetlands Conservation: Current Challenges and Future Strategies, First Edition.
Edited by Sanjeev Sharma and Pardeep Singh.

since the seventeenth century (Dahl 1990; Kotze et al. 1995); and in China, for at least last 2000 years (An et al. 2007). It has been estimated that at least 50% of valuable wetland ecosystems have been lost since 1900 (Dugan 1993; OECD 1996; Mitsch and Gosselink 2015; and Ramsar Convention Bureau 2015). In fact, around 35% of the wetland area has been lost between 1970 and 2015 (Gardner and Finlayson 2018). In terms of economic value, Costanza et al. (2014) estimated an annual loss of 9.9 trillion dollars in the value of marsh wetland ecosystem services over a period of 14 years from 1997 to 2011 (Costanza et al. 2014).

Though the wetland conversion and their corresponding decline have been going on for centuries, it was only in the early part of the twentieth century that their significance was realized and, consequently, deliberations toward their conservation began to gain strength. In North America, many researchers started raising their concerns on waterfowl decline as a result of wetland drainage since the 1920s (Schmidt 2006). In Europe, the conservationists had been projecting the issues of wetland degradation from the early 1960s (Hoffmann 1964; Swift 1964). This prompted the IUCN to launch the MAR Project (from "MARshes," "MARécages," and "MARismas") for the conservation and management of wetlands in the early 1960s. The project MAR was conceived during the MAR conference held in French Camargue from 12 to 16 November 1962 (Hoffmann 1964; Ramsar Convention Secretariat 2011). The recommendations made during the MAR conference later paved the way for the establishment of an international convention on wetlands at the Ramsar Convention in 1971 (Matthews 1993; Ramsar Convention Secretariat 2011; Davidson 2014).

2.2 The Ramsar Convention

After the MAR conference in 1962, a series of meetings were held in different parts of the world such as St. Andrews (1963), Noordwijk (1966), Morges (1967), Leningrad (1968), Vienna (1969), Moscow (1969), Espoo (1970), and Knokke (1970). During these eight years of deliberations, the official document for "Convention on Wetlands of International Importance" was finalized. The Convention was initially scheduled to be organized at the city of Babolsar in Iran, but the venue was later shifted to the resort town of Ramsar located on the shores of the Caspian Sea and having better connectivity and access to logistics (Carp 1972; Matthews 1993). The Ramsar Convention was organized on 2 February 1971 by the Game and Fish Department of Iran. The Ramsar Convention was officially named "The Convention on Wetlands of International Importance especially as Waterfowl Habitat." The convention was attended by official delegates from 18 nations which included Germany, India, Iran, Ireland, Jordan, the Netherlands, Pakistan, South Africa, Spain, Sweden, Switzerland, Turkey, the USSR, and the United Kingdom. Observers from five other countries, namely Bulgaria, Greece, Hungary, Italy, and Romania, also attended the convention. In addition, delegates from various intergovernmental agencies (like FAO and UNESCO) and nongovernmental organizations (like CIC, IBP, ICBP, IUCN, IWRB, and WWF) were also among those who attended the convention (Matthews 1993; Ramsar Convention Secretariat 2011).

The treaty was agreed on 2 February 1971 and signed by the representatives of 18 nations on 3 February 1971. As per terms laid down in convention, the Ramsar Convention would come into force only after being ratified by at least seven countries, which was fulfilled in December 1975 upon ratification by Greece (Matthews 1993). In the subsequent years, as the concerns about wetland conservation gained momentum, more nations joined the Ramsar Convention. As of October 2019, there are 171 members of the Ramsar Convention with a global count of 2390 Ramsar sites spread over an area of more than 2.5 million km^2 (Ramsar Sites Information System 2020). The goal of Ramsar Convention emphasizes that "the loss of wetlands, any further, would be irreparable," to people, and aims to "stem the loss and degradation of wetlands now and in the future," through the wise-use of all wetlands; designation and management of Wetlands of International Importance ("Ramsar Sites") and international cooperation (Davidson 2014).

2.3 The Convention Text

The original official text of the Ramsar Convention, agreed upon and adopted by the contracting parties of the convention, was signed on 2 February 1971. Since its adoption, the convention has undergone modification on two occasions, first by the Paris Protocol and later by Regina Amendments. Paris protocol was adopted during an Extraordinary Conference of Parties (COP) held at UNESCO headquarters in Paris in December 1982 and came into force in 1986. The Paris Protocol is known for recognizing the need for and adopting a procedure for amending the convention. The protocol added Article 10bis which lays down the process to bring amendments to the convention. The Regina Amendments were a series of amendments to Articles 6 and 7 that were accepted at an Extraordinary Conference of the Contracting Parties held in Regina, Canada, on 28 May 1987 (Matthews 1993). These amendments did not alter the fundamental structure and guiding principles of the convention, rather these were related to the convention's operation. They defined the powers of the Conference of the Parties, established an intersessional Standing Committee, established a permanent secretariat, and set up a budget for the Convention. Regina amendments came into force in May 1994. The treaty's current text has been subdivided into 13 articles (i.e. articles 1–10, article 10bis, and articles 11 and 12).

2.4 Wetland Definition and Classification

The term "wetland" represents a host of ecosystems inundated by water for a certain period of the year and possessing characteristics unique to wetland ecosystems. While there have been many definitions proposed by many scientists/researchers but the most versatile and widely accepted definition has been given by the Ramsar Convention which has taken care of every aspect of the wetlands. Article 1.1 of the Ramsar convention defines wetlands as "the areas of marsh, fen, peatland or water, whether natural or artificial, permanent or temporary, with water that is static or flowing, fresh, brackish or salt, including areas of marine water the depth of which at low tide does not exceed 6 m." Article 2.1 further states that "the riparian and coastal zone adjacent to the wetlands and islands or bodies of marine water deeper than six meters at low tide" may also be considered as wetlands. The typology of wetland accepted by the Ramsar Convention (which also accompanies the Ramsar definition) incorporates the three attributes to describe 42 different types of habitats across coastal/marine and inland environments (Table 2.1). These typologies also include streams, rivers, caves, coral reefs, alpine meadows, springs, dams, wastewater treatment areas as well as the more traditionally considered marshes and swamps (Ramsar Convention Secretariat 2011).

The Ramsar Convention has adopted a Ramsar Classification for wetland types (Table 2.1) which includes 42 types, grouped into three categories: Marine and Coastal Wetlands, Inland Wetlands, and Human-made Wetlands. The marine wetlands are considered to be wetlands up to a depth of 6 m at low tide (and this depth is believed to be the maximum depth to which various sea ducks can dive during feeding). The treaty also includes waters deeper than 6 m, as well as islands, within the boundaries of protected wetlands. The lakes and rivers are understood to be covered by the Ramsar definition of wetlands in their entirety, regardless of their depth (Ramsar Convention Secretariat 2010, 2011).

Table 2.1 Ramsar classification system for wetland types.

	Marine/Coastal Wetlands
A	Permanent shallow marine waters; in most cases, less than 6 m deep at low tide; includes sea bays and straits
B	Marine subtidal aquatic beds; includes kelp beds, sea-grass beds, and tropical marine meadows
C	Coral reefs
D	Rocky marine shores; includes rocky offshore islands and sea cliffs
E	Sand, shingle, or pebble shores; includes sand bars, spits, and sandy islets; includes dune systems and humid dune slacks
F	Estuarine waters; permanent water of estuaries and estuarine systems of deltas
G	Intertidal mud, sand, or salt flats
H	Intertidal marshes; includes salt marshes, salt meadows, saltings, raised salt marshes; includes tidal brackish and freshwater marshes
I	Intertidal forested wetlands; includes mangrove swamps, Nipah swamps and tidal freshwater swamp forests
J	Coastal brackish/saline lagoons; brackish to saline lagoons with at least one relatively narrow connection to the sea
K	Coastal freshwater lagoons; includes freshwater delta lagoons
Zk (a)	Karst and other subterranean hydrological systems, marine/coastal

Table 2.1 (Continued)

	Inland Wetlands
L	Permanent inland deltas
M	Permanent rivers/streams/creeks; includes waterfalls
N	Seasonal/intermittent/irregular rivers/streams/creeks
O	Permanent freshwater lakes (over 8 ha); includes large oxbow lakes
P	Seasonal/intermittent freshwater lakes (over 8 ha); includes floodplain lakes
Q	Permanent saline/brackish/alkaline lakes
R	Seasonal/intermittent saline/brackish/alkaline lakes and flats
Sp	Permanent saline/brackish/alkaline marshes/pools
Ss	Seasonal/intermittent saline/brackish/alkaline marshes/pools
Tp	Permanent freshwater marshes/pools; ponds (below 8 ha), marshes, and swamps on inorganic soils; with emergent vegetation water-logged for at least most of the growing season
Ts	Seasonal/intermittent freshwater marshes/pools on inorganic soils; includes sloughs, potholes, seasonally flooded meadows, and sedge marshes
U	Non-forested peatlands; includes shrub or open bogs, swamps, and fens
Va	Alpine wetlands; includes alpine meadows and temporary waters from snowmelt
Vt	Tundra wetlands; includes tundra pools and temporary waters from snowmelt
W	Shrub-dominated wetlands; shrub swamps, shrub-dominated freshwater marshes, shrub-carr, and alder thicket on inorganic soils
Xf	Freshwater, tree-dominated wetlands; includes freshwater swamp forests, seasonally flooded forests, and wooded swamps on inorganic soils
Xp	Forested peatlands; peat-swamp forests
Y	Freshwater springs; oases
Zg	Geothermal wetlands
Zk (b)	Karst and other subterranean hydrological systems, inland
	Human-made Wetlands
1.	Aquaculture (e.g. fish/shrimp) ponds
2.	Ponds; includes farm ponds, stock ponds, and small tanks (generally below 8 ha)
3.	Irrigated land; includes irrigation channels and rice fields
4.	Seasonally flooded agricultural land (including intensively managed or grazed wet meadow or pasture)
5.	Salt exploitation sites; salt pans, salines, etc.
6.	Water storage areas; reservoirs/barrages/dams/impoundments (generally over 8 ha)
7.	Excavations; gravel/brick/clay pits; borrow pits, mining pools
8.	Wastewater treatment areas; sewage farms, settling ponds, oxidation basins, etc.
9.	Canals and drainage channels, ditches
Zk (c)	Karst and other subterranean hydrological systems, human-made

2.5 Mission of the Convention

The mission of the Ramsar Convention was adopted by the Conference of Parties in 1999 and was later redefined in 2002. The Convention's mission statement is: "the conservation and wise-use of all wetlands through local, regional, and national actions and international cooperation, as a contribution toward achieving sustainable development throughout the world" (Matthews 1993; Ramsar Convention Secretariat 2011, 2016).

2.6 Structural Framework of the Convention

The Ramsar Convention has four main organs including the Conference of Parties (COP), Standing Committee, Ramsar Secretariat and the Scientific and Technical Review Panel (STRP) (Figure 2.1). These organs have been discussed below in detail.

i) **Conference of Parties (COP):** The Conference of Parties constitutes "the policy-making organ of the Convention." The representatives from the governments of the various contracting parties meet every three years. Since the inception of the Ramsar Convention, 13 ordinary and 2 extraordinary meetings of the conference of contracting parties have been organized (Table 2.2). During the COP, the contracting parties receive their national reports for the last three years (previous triennium) and approve the new work-plan for the next triennium. The budget for the approved work-plan for the succeeding triennium is also discussed and approved. Further, discussions are held on a wide range of ongoing and emerging environmental issues and points of guidance from different contracting parties are summed up and given due consideration. As per the procedures laid down under Articles 6 and 7 of the Convention, broad duties are assigned to various contracting parties. There is also the provision for the representatives of nonmember States, intergovernmental institutions, and national and international nongovernmental organizations (NGOs) to attend or participate in the conference of parties but only as non-voting observers (Matthews 1993; Ramsar Convention Secretariat 2011, 2016).
ii) **Standing Committee:** The Standing Committee of the Ramsar Convention is a temporary body that has intersessional existence. It is an executive body elected by the conference of parties that represents the COP between its triennial meetings, within the framework of the decisions made by the COP. A new standing committee is elected by the COP in each ordinary meeting which serves for the next triennium and seizes to exist when a new standing committee is elected in the next meeting of the COP. The standing committee was, for the first time, established by Resolution 3.3 of the 1987 Conference of the Contracting Parties. The tasks of the standing committee were initially set out based on Resolution 5.1 (of 1993) in the Framework for Implementation of the Ramsar Convention but are currently defined by Resolution VII.1 (of 1999) (Matthews 1993; Ramsar Convention Secretariat 2011, 2016).

   The Standing Committee, normally, meets once every year, generally in the Convention Secretariat in Switzerland. In addition to this, it also meets before each meeting of the Conference of the Contracting Parties. During the meeting of the COP, it is transformed into the Conference Committee. During the meeting, COP selects a

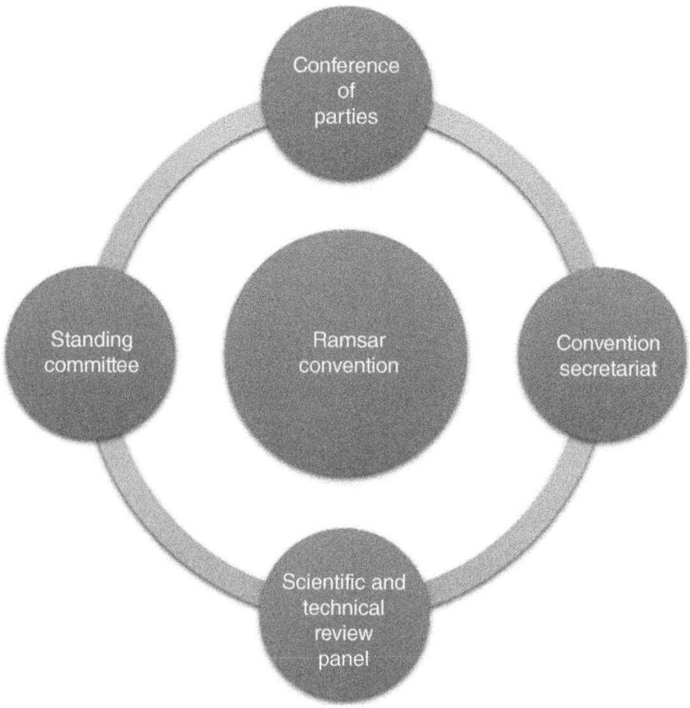

Figure 2.1 Ramsar Convention and its four organs.

Table 2.2 Chronology of meetings of the conference of contracting parties.

	Ordinary meetings of the Conference of Parties	
COP	**Location**	**Year**
1st	Cagliari, Italy	1980
2nd	Groningen, Netherlands	1984
3rd	Regina, Canada	1987
4th	Montreux, Switzerland	1990
5th	Kushiro, Japan	1993
6th	Brisbane, Australia	1996
7th	San José, Costa Rica	1999
8th	Valencia, Spain	2002
9th	Kampala, Uganda	2005
10th	Changwon, Republic of Korea	2008
11th	Bucharest, Romania	2012
12th	Punta del Este, Uruguay	2015
13th	Dubai, UAE	2018
	Extraordinary meetings of the Conference of Parties	
1st	Paris, France	1982
2nd	Regina, Canada	1987

new standing committee for the next three years and therefore the older standing committee seizes to exist. The composition of the standing committee currently includes 16 regional members who are representatives of different Ramsar regions (which are defined based on the regional distribution of various contracting parties) and 2 *ex-officio* members (representatives of the host countries of the most recent and the upcoming meetings of the COP). Moreover, there are seven permanent observers which include the host countries of the Ramsar Secretariat (Switzerland) and the Wetlands International (Netherlands), and the five International Partner Organizations which include BirdLife International, IUCN – The World Conservation Union, International Water Management Institute (IWMI), Wetlands International, WWF International (Matthews 1993; Ramsar Convention Secretariat 2011, 2016).

iii) **Convention Secretariat:** The Ramsar Convention Secretariat carries out the day-to-day coordination of the Convention's activities. It is located at the headquarters of the International Union for Conservation of Nature (IUCN) in Gland, Switzerland. The Secretariat staff are legally considered to be employees of the IUCN. The Secretariat is headed by a Secretary-General who is answerable to the Standing Committee (Matthews 1993; Ramsar Convention Secretariat 2011, 2016).
The main roles (among others) of the Ramsar Secretariat largely include:
   a) maintaining the list of Ramsar sites and all changes to the List and to the Ramsar Sites Database
   b) assisting in the convening of and organizing the meetings of the COP, the Standing Committee and the STRP
   c) providing the administrative, scientific, and technical support to Contracting Parties
   d) dissemination of the decisions, resolutions, and recommendations of the meeting of the COP and the Standing Committee
   e) keeping the Contracting Parties, the Ramsar community, and the public informed of developments related to the Convention
   f) organizing Ramsar Advisory Missions at the request of Contracting Parties
   g) developing avenues of cooperation with other conventions, intergovernmental institutions, and national and international NGOs

iv) **Scientific and Technical Review Panel (STRP):** The Scientific and Technical Review Panel (STRP) is a supplementary entity of the Ramsar Convention which was set up by Resolution 5.5 (of 1993). Its primary role is to provide scientific and technical guidance to the various policymaking and executive organs of the Convention. The panel guides the Conference of the Parties, the Standing Committee and the Convention Secretariat on the matters which require scientific and technical support. They are elected by the Standing Committee based on nominations from the Parties. They serve as the experts and not as the representatives of their countries. The total number of full-members of the STRP is 17, i.e. six regional representatives (one chosen from each of the six Ramsar regions), six thematic experts chosen for their expertise in the priority areas of work for the period; and representatives from the five International Organization Partners. Apart from the 17 full-members, the STRP also has a host of additional expert consultants and advisors and organizations which are invited to participate as and when required. The term of an STRP is for three years in unison with that of the standing committee and seizes to exist when a new standing committee takes the charge (Matthews 1993; Ramsar Convention Secretariat 2011, 2016).

2.7 Operational Framework of the Convention

The working of the Ramsar Convention is looked after jointly by the Contracting Parties, the Standing Committee, and the Convention Secretariat. The convention's operations are carried out with the active participation of these three organs but not without technical advice from the STRP as well as the support from the International Organization Partners (IOPs). The important decisions related to operations of the Convention are taken by means of voting after thorough discussions by the representatives of the Contracting Parties on the matters under consideration. Various important decisions related to the convention including resolutions and recommendations are taken during the meeting of the conference of parties which have been categorized into two groups, ordinary and extraordinary meetings. An ordinary meeting of the Conference of the Contracting Parties (COP) is organized every three years whereas the necessity for an extraordinary meeting of the Conference of Parties may arise if an amendment to the convention is to be discussed. Table 2.2 enlists the various ordinary and extraordinary meetings of COP held till the year 2018 (Matthews 1993; Ramsar Convention Secretariat 2011, 2016).

2.7.1 Convention Membership

As per Article 9.2 of the Convention, "Any member of the United Nations or of one of the Specialized Agencies or of the International Atomic Energy Agency or Party to the Statute of the International Court of Justice may become a Party to Ramsar Convention." The convention is open for membership to any such nonmember nation or agency which satisfies the eligibility criteria laid under Article 9.2 (Matthews 1993; Ramsar Convention Secretariat 2011, 2016).

As per the guidelines, the instrument of ratification/accession has to be submitted to the depository of the convention rather than to any of the organs of the convention directly. Depository receives, reviews, and accepts the instrument of ratification/accession and does not have any other role in administration or implementation. The United Nations Educational, Scientific and Cultural Organization (UNESCO) serves as the Depository of the Ramsar Convention 2018 (Matthews 1993).

In order to join the Ramsar Convention, the concerned nation requires agreeing to and, subsequently, working toward the fulfillment of three requirements mentioned below:

i) Submit a model instrument of ratification/accession, duly signed by the Head of State or Government or by the Minister of Foreign Affairs and forwarded through proper diplomatic channels to the depository of the convention. The depository formally notifies Ramsar Secretariat and COP about it.
ii) Identify and designate at least one of its wetlands to be included in the list of "wetlands of international importance." Afterward, the party needs to designate suitable wetlands within its territory for inclusion in the list.
iii) Must agree to contribute its share of (a percentage based on UN's scale of assessments) to the triennial budget of the convention approved by the conference of parties during their ordinary meetings.

India joined the convention in the year 1981 and as per the mandate, India designated two of its wetlands, Chilka Lake (in Orissa) and Keoladeo Ghana NP (in Rajasthan) as

Ramsar sites. Later, India expanded its network of Ramsar sites which currently has 42 wetlands with a cumulative area of 10814 km^2 (Ramsar Sites Information Service 2020). The Ramsar sites designated by India have been provided in Table 2.3 and graphically shown in Figure 2.2.

Article 2.2 of the Ramsar Convention acts as the guiding light for the declaration of a Ramsar site. Any nation, desirous of declaring a wetland as a Ramsar Site has to adhere to the criteria (Table 2.4) laid down for this purpose. The process starts with the compilation of baseline information related to the area, altitude, wetland type, legal jurisdiction as well as important aspects related to hydrology, biology, land-use, sociocultural aspects, conservation measures, and threats. The information is compiled in prescribed formats called the Ramsar Information Sheets (RIS). The RIS are submitted to Ramsar Secretariat. After the information provided in the RIS is verified and is believed to be correct and complete, the process for designation of the Ramsar site is moved forward. The RIS of the wetland is added further to the Ramsar Sites Database. As per Resolution VI.13 (1996), Ramsar Information Sheets for all the Ramsar Sites must be updated and resubmitted to the Secretariat, at least every six years (Matthews 1993).

Out of all the world's wetlands designated as Ramsar sites, 20.3% have been designated under criterion 2, i.e. such wetlands which support vulnerable, endangered, or critically endangered species or threatened ecological communities. About 17.7% of the Ramsar sites have been designated under criterion 1 which has been reserved for those wetlands which "contains a representative, rare, or unique example of a natural or near-natural wetland type found within the appropriate biogeographic region" (Figure 2.3). Lake Tsomoriri in Ladakh (India) has also been categorized under criterion 1. Only a miniscule percentage of 0.6 of the total Ramsar sites has been designated under criterion 9, which recognizes such wetlands which "regularly support 1% of the individuals in a population of one species or subspecies of wetland-dependent non-avian animal species" (Ramsar Sites Information Service 2020).

2.7.2 Ramsar Regions

For ease of management and implementation, the convention has divided the world into six Ramsar regions. This system of regionalization was adopted under Resolution 3.3 (in 1987). It has significant implications for effective operations, especially in terms of the structure of the Standing Committee and for the Contracting Parties to cooperate through regional meetings. The six regions under Ramsar Convention are:

I) Africa
II) Asia
III) Europe
IV) Neotropics (Central and South America and the Caribbean)
V) North America (Canada, Mexico, and the US)
VI) Oceania

Out of all the Ramsar regions, Europe has the most number of Ramsar sites (1116) followed by Africa (413), Asia (352), North America (219), Neotropics, (208) and Oceania (82)

Table 2.3 List of Ramsar sites of India (as mentioned in the numerical order in Figure 2.2).

S. No.	Name	State/UT	Area (km^2)	Date of designation
1.	Wular Lake	Jammu and Kashmir	189	23.03.1990
2.	Hokera Wetland	Jammu and Kashmir	13.75	08.11.2005
3.	Surinsar-Mansar Lakes	Jammu and Kashmir	3.5	08.11.2005
4.	Tsomoriri Lake	Ladakh	120	19.08.2002
5.	Tso Kar	Ladakh	95.77	17.11.2020
6.	Chandertal Wetland	Himachal Pradesh	0.49	08.11.2005
7.	Keshopur-Miani Community Reserve	Punjab	3.4	26.09.2019
8.	Pong Dam Lake	Himachal Pradesh	156.62	19.08.2002
9.	Beas Conservation Reserve	Punjab	64.2	26.09.2019
10.	Kanjli Lake	Punjab	1.83	22.01.2002
11.	Harike Lake	Punjab	41	23.03.1990
12.	Nangal Wildlife Sanctuary	Punjab	1.2	26.09.2019
13.	Ropar Lake	Punjab	13.65	22.01.2002
14.	Renuka Wetland	Himachal Pradesh	0.2	08.11.2005
15.	Asan Barrage	Uttarakhand	4.44	21.07.2020
16.	Upper Ganga River	Uttar Pradesh	265.9	08.11.2005
17.	Sambhar Lake	Rajasthan	240	23.03.1990
18.	Keoladeo Ghana NP	Rajasthan	23.73	01.10.1981
19.	Saman Bird Sanctuary	Uttar Pradesh	5.3	02.12.2019
20.	Sandi Bird Sanctuary	Uttar Pradesh	3	26.09.2019
21.	Sarsai Nawar Jheel	Uttar Pradesh	1.6	19.09.2019
22.	Nawabganj Bird Sanctuary	Uttar Pradesh	2.2	19.09.2019
23.	Parvati Agra Bird Sanctuary	Uttar Pradesh	7.2	02.12.2019
24.	Samaspur Bird Sanctuary	Uttar Pradesh	7.99	03.10.2019
25.	Sur Sarowar	Uttar Pradesh	4.31	13.11.2020
26.	Kanwar Taal	Bihar	26.2	21.07.2020
27.	Deepor Beel	Assam	40	19.08.2002
28.	Loktak Lake	Manipur	266	23.03.1990
29.	Rudrasagar Lake	Tripura	2.4	08.11.2005
30.	Bhoj Wetlands	Madhya Pradesh	32.01	19.08.2002
31.	Nalsarovar Bird Sanctuary	Gujarat	120	24.09.2012
32.	East Calcutta Wetlands	West Bengal	125	19.08.2002
33.	Sunderbans Wetland	West Bengal	4230	30.01.2019
34.	Bhitarkanika Mangroves	Orissa	650	19.08.2002
35.	Nandur Madhameshwar	Maharashtra	14.37	21.06.2019

(*Continued*)

Table 2.3 (Continued)

S. No.	Name	State/UT	Area (km^2)	Date of designation
36.	Lonar Lake	Maharashtra	4.21	13.11.2020
37.	Chilka Lake	Orissa	1165	01.10.1981
38.	Kolleru Lake	Andhra Pradesh	901	19.08.2002
39.	Point Calimere	Tamil Nadu	385	19.08.2002
40.	Vembanad Kol Wetland	Kerala	1512.5	19.08.2002
41.	Sasthamkotta Lake	Kerala	3.73	19.08.2002
42.	Asthamudi Wetland	Kerala	614	19.08.2002

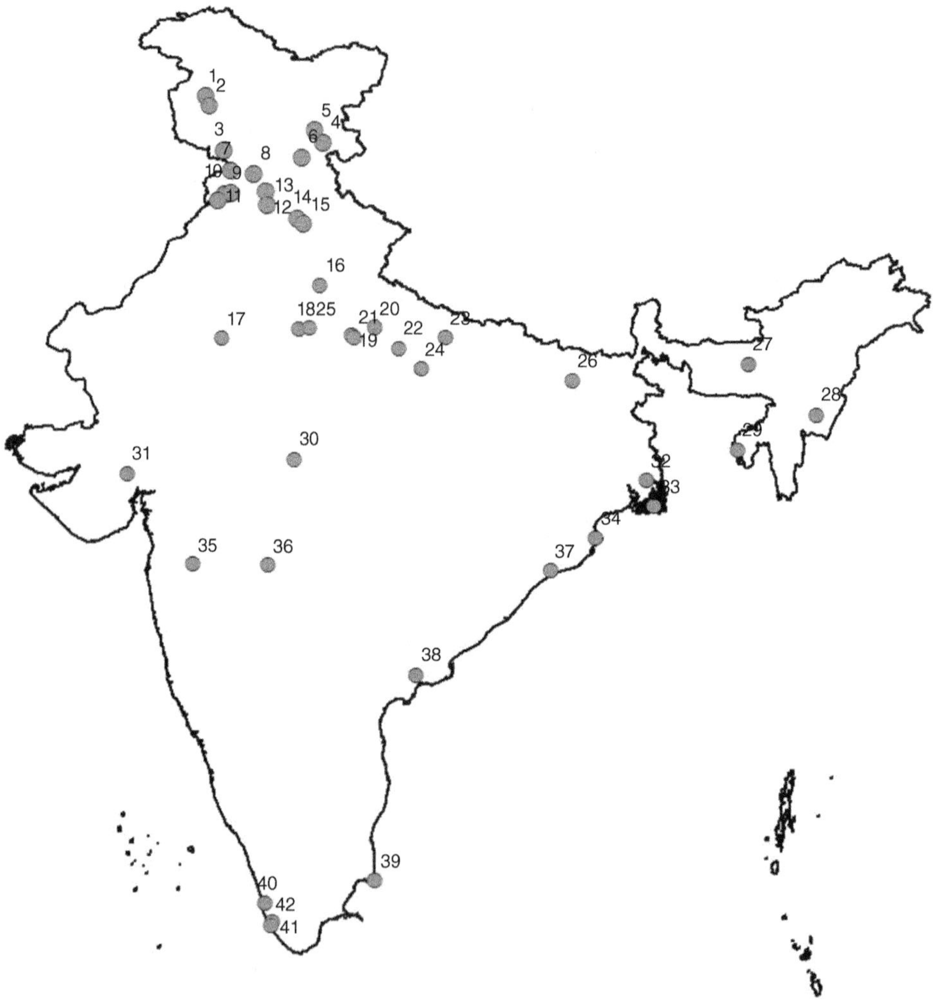

Figure 2.2 Ramsar sites in India.

Table 2.4 Criteria for the designation of Wetlands of International Importance.

Group A of the criteria Sites containing representative, rare, or unique wetland types		Criterion 1: A wetland should be considered internationally important if it contains a representative, rare, or unique example of a natural or near-natural wetland type found within the appropriate biogeographic region.
Group B of the criteria Sites of international importance for conserving biodiversity	Criteria based on species and ecological communities	Criterion 2: A wetland should be considered internationally important if it supports vulnerable, endangered, or critically endangered species or threatened ecological communities.
		Criterion 3: A wetland should be considered internationally important if it supports populations of plant and/or animal species important for maintaining the biological diversity of a particular biogeographic region.
		Criterion 4: A wetland should be considered internationally important if it supports plant and/or animal species at a critical stage in their life cycles or provides refuge during adverse conditions.
	Specific criteria based on waterbirds	Criterion 5: A wetland should be considered internationally important if it regularly supports 20 000 or more waterbirds.
		Criterion 6: A wetland should be considered internationally important if it regularly supports 1% of the individuals in a population of one species or subspecies of waterbirds.
	Specific criteria based on fish	Criterion 7: A wetland should be considered internationally important if it supports a significant proportion of indigenous fish subspecies, species, or families, life-history stages, species interactions, and/or populations that are representative of wetland benefits and/or values and, thereby, contribute to global biological diversity.
		Criterion 8: A wetland should be considered internationally important if it is an important source of food for fishes, spawning ground, nursery and/or migration path on which fish stocks, either within the wetland or elsewhere, depend.
	Specific criteria based on other taxa	Criterion 9: A wetland should be considered internationally important if it regularly supports 1% of the individuals in a population of one species or subspecies of wetland-dependent non-avian animal species.

(Figure 2.4). The Ramsar Secretariat has four 2-member "Regional Advisory Teams," each consisting of a Senior Advisor and an Intern/Assistant Advisor, for the following regions: Africa, Asia-Pacific, Europe, and the Americas (Neotropics and North America). Before every meeting of the Conference of Parties, regional or sub-regional meetings are organized to prepare for the COP meeting (Ramsar Sites Information Service 2020).

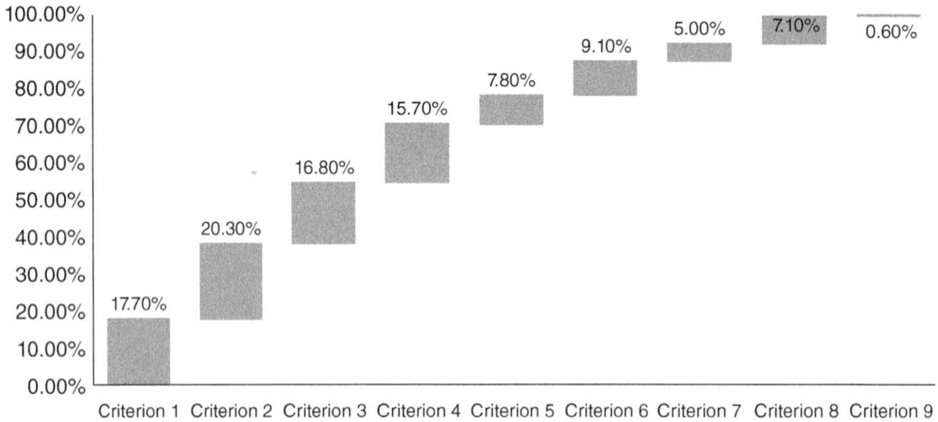

Figure 2.3 Percentage of Ramsar sites designated under different criteria. *Source:* Data from Ramsar Sites Information Service https://rsis.ramsar.org/?pagetab=2. © John Wiley & Sons.

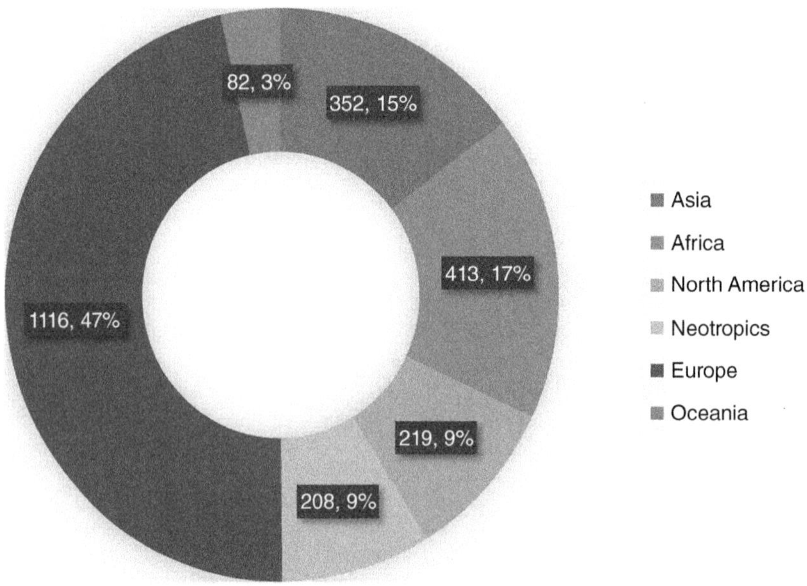

Figure 2.4 Number and percentage of Ramsar sites in Ramsar Regions. *Source:* Data from Ramsar Sites Information Service https://rsis.ramsar.org/?pagetab=2. © John Wiley & Sons.

2.7.3 National Ramsar Committees

Based on recommendation 5.7 of the Conference of Parties as well as Ramsar Strategic Plan, every member nation has to establish its National Ramsar Committee (often known as National Wetland Committee). The National Ramsar Committee of a Ramsar member provides help in the implementation of the Convention at the national level. A national Ramsar Committee particularly helps the different agencies, scientific and technical institutions, regional and local authorities, local communities, NGOs, and the private sector with regard to issues pertaining to the national wetland-related policies; management of Ramsar Sites; application of the Montreux Record; designation of new Ramsar sites; and with those related to the Ramsar Small Grants Fund (Matthews 1993; Ramsar Convention Secretariat 2011, 2016).

2.7.4 The Montreux Record

Those wetlands on the list of Ramsar sites which are "either facing or likely to face large-scale changes in their ecological character as a result of unsustainable technological developments, pollution or other human interference" are registered in the Montreux Record. The Montreux Record was established based on Recommendation 4.8 of the Conference of Parties in 1990 and is duly maintained as part of the Ramsar List. The Ramsar Advisory Mission, a technical assistance mechanism adopted by the Contracting Parties in 1990, has its role to provide assistance to the developed and developing countries in solving the problems or threats that make inclusion in the Montreux Record necessary (Matthews 1993; Ramsar Convention Secretariat 2011, 2016).

2.7.5 Ramsar Strategic Plan

The Ramsar Convention has set up mechanisms for developing and adopting innovative strategic plans of the treaty. The first strategic plan was adopted in the year 1997 for a period of three years. So far, there have been four strategic plans and the one under implementation is the fourth strategic plan which was adopted during the twelfth meeting of the Conference of Parties in Uruguay (2015). It shall be implemented over a span of eight years from 2016 to 2024. The fourth strategic plan lays out a new vision under the Convention's mission; four overall goals and 19 specific targets which are designed to support the wetland conservation efforts of the Contracting Parties, their partners, and other stakeholders. It envisions preventing, stopping, and reversing the global decline of wetlands. Overall, the Ramsar Strategic Plans have been a successful initiative and other conventions have borrowed from their success (Matthews 1993; Ramsar Convention Secretariat 2011, 2016).

2.7.6 Three Pillars of Ramsar Convention

The idea of "three pillars" holds great significance for the strategic plans of the Ramsar Convention as the majority of the convention's work is organized around these pillars (Figure 2.5). The idea of three pillars was introduced during the first strategic plan in 1997 and further recognized in the third Strategic Plan (2009–2015). The three pillars of the Ramsar Convention are:

1) The "Wise-use of Wetlands" has been identified as the first pillar of the Ramsar Strategic Plan. Its aim is to promote the conservation of wetlands through sustainable use of their services through actions like establishing national wetland policies; harmonizing the framework of laws and financial instruments affecting wetlands; undertaking inventory and assessment; ensuring public participation in wetland management and the

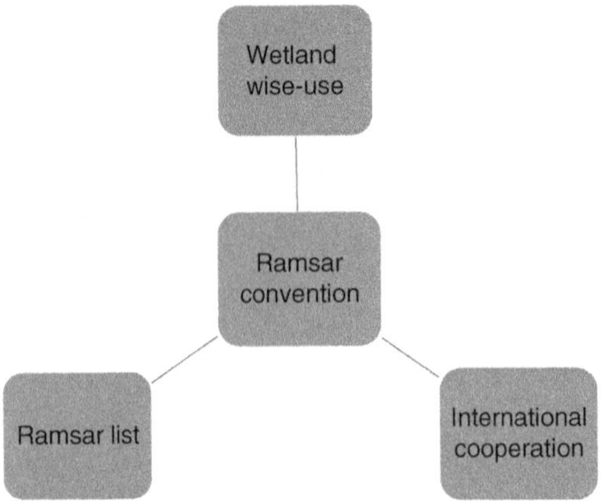

Figure 2.5 Three pillars of the Ramsar Convention.

maintenance of cultural values by local communities and indigenous people; promoting communication, education, participation, and awareness; and increasing private sector involvement (Matthews 1993; Ramsar Convention Secretariat 2011, 2016).

2) The second pillar is the identification of wetlands for their inclusion in the "List of Wetlands of International Importance." This helps in the identification, designation, and management of a comprehensive global network of Wetlands of International Importance (the Ramsar List) and ensures their effective conservation and management (Matthews 1993; Ramsar Convention Secretariat 2011, 2016).
3) "International Cooperation" for conservation and wise-use of wetlands has been identified as the third pillar of the strategic plan. This is particularly useful in the case of conservation and management of transboundary wetlands and the species harbored by them. This highlights the need for collaboration between different member nations, conventions as well as international organizations for easy flow of information and expertise, and sharing of financial resources, etc. (Matthews 1993; Ramsar Convention Secretariat 2011, 2016).

2.7.7 The Convention Budget

The convention budget is agreed upon for the applicable triennium. The decisions related to the financial regulations and the core budget are taken during the ordinary meetings of the Conference of Parties. The COP reviews the budget for the next triennium at each of its ordinary meetings. The Swiss Franc has been agreed upon as the official currency of the convention. The budget is prepared by the Ramsar Secretariat and is submitted to the standing committee for its endorsement. The standing committee further reviews the budget and submits it before the COP during ordinary meetings.

For the formal approval, there has to be an agreement among the contracting parties regarding the triennial budget. The final decision is taken by way of voting among the contracting parties which are present at the ordinary meeting. As per Article 6.5 of the Convention, if a formal vote is required, a two-thirds majority is needed for approval and adoption of the budget. Every contracting party has to contribute to the core budget of the Convention. The contribution is primarily made to the budget of the United Nations and the UN takes a decision on the percentage of the contribution which would be provided to the Ramsar budget and this is decided based on a scale approved by the UN. In addition to the contributions made by the member countries to the core budget, the Ramsar Secretariat also receives voluntary contributions from the contracting parties, NGOs, and other donors for special projects or other contractual agreements (Matthews 1993; Ramsar Convention Secretariat 2011, 2016).

2.8 External Partnerships and Synergies

Cooperation with other organizations: The Ramsar Convention, through the Secretariat and its other bodies, maintains close working links with other international, intergovernmental, and nongovernmental organizations to achieve a strategic alliance for wetland conservation. There are collaborations at multiples levels between the Ramsar Convention and various other organizations such as the UN Food and Agriculture Organization (FAO), the World Health Organization, the World Tourism Organization, and the Organization of American States (OAS). The Ramsar Convention also serves as an advisor on wetland-related project proposals submitted to the Global Environment Facility. The Secretariat participates regularly as an observer at meetings of the UN Commission on Sustainable Development and has been involved in collaborative work with UN-Habitat (Matthews 1993; Ramsar Convention Secretariat 2011, 2016).

International Organization Partners (IOPs): The Convention has close working relations with a group of five global nongovernmental organizations (NGOs). Out of the five, four NGOs have been associated with the Ramsar Convention since its origin and have been recognized as International Organization Partners (Resolution VII.3 1999). The International Water Management Institute was accorded the status of an IOP in 2005 (Resolution IX.16 2005), thereby taking the total number of IOPs to five (Ramsar Convention Secretariat 2011, 2016). The five IOPs of the Ramsar Convention (Figure 2.6) are:

A) BirdLife International (formerly ICBP)
B) International Union for Conservation of Nature (IUCN)
C) International Water Management Institute (IWMI)
D) Wetlands International
E) WWF International

The IOPs are important conservation partners for the Ramsar Convention at global, regional, national, and local levels. They provide technical advice, help in field-level implementation and also provide financial support. The IOPs also participate regularly as observers in all meetings of the Conference of the Parties and the Standing Committee, and as full members of the Scientific and Technical Review Panel.

Linkages with other international and regional conventions/treaties: There are a host of benefits of collaboration amongst international conventions with overlapping missions. The Ramsar Secretariat has put commendable efforts to develop synergies with other conventions. It strives to ensure close working relationships with different conventions at various levels. Within the United Nations system, the Ramsar Secretariat takes part in coordinating meetings of some conventions. It participates as an observer in the Joint Liaison

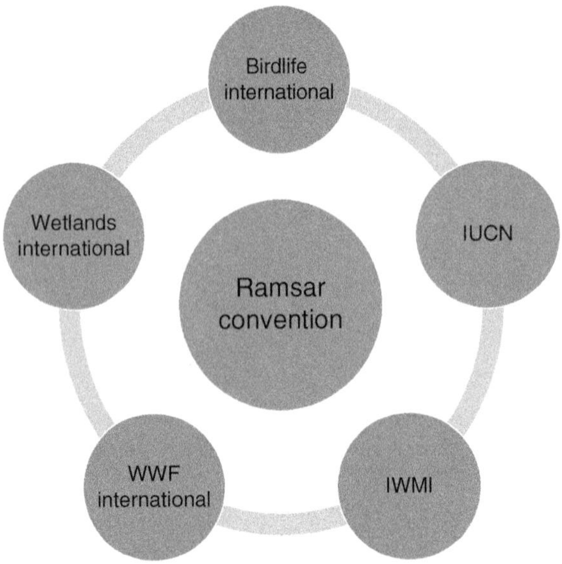

Figure 2.6 Ramsar Convention and International Organization Partners.

Group (JLG) of the "Rio Conventions," UNFCCC (the UN Framework Convention on Climate Change), the CBD, and UNCCD (the UN Convention to Combat Desertification). It also participates as a full member of the Biodiversity Liaison Group (BLG), which comprises the five biodiversity-related conventions – the CBD, CITES (the Convention on the International Trade in Endangered Species), CMS, Ramsar, and World Heritage (Matthews 1993; Ramsar Convention Secretariat 2011, 2016).

Brief details of the Ramsar Convention's cooperation with some of the major international conventions have been discussed below:

a) **The Convention on Biological Diversity (CBD):** The Ramsar Convention signed a Memorandum of Cooperation with the CBD in the year 1996 and followed by this, Ramsar participated in the COP-3 of the CBD as a "lead partner." Later, an innovative Joint Work Plan was put in place for the year 1998–1999. The relationship between these two has evolved further leading to a series of Joint Work Plans (Ramsar Convention Secretariat 2011, 2016).
b) **The Convention on Conservation of Migratory Species of Wild Animals (CMS):** Just like the Memorandum of Cooperation between Ramsar and CBD, a Memorandum of Understanding was signed between the Ramsar and CMS Secretariats in the year 1997. The MoU seeks to strengthen the joint promotion of the two conventions; joint conservation action; data collection, storage and analysis; and new agreements on migratory species, including endangered migratory species and species with an unfavorable conservation status (Ramsar Convention Secretariat 2011, 2016).
c) **UNESCO World Heritage Convention and Man and Biosphere Programme:** A Memorandum of Understanding was signed between the Ramsar Secretariat and the UNESCO World Heritage Convention in the year 1999. The MoU focused on having a partnership with regard to the issues which were common to both conventions. This

included promoting the nominations of wetland sites; coordinating the reporting about common wetland sites; collaboration on advisory missions to certain selected sites. Apart from this, the Ramsar Secretariat also works closely with the UNESCO Man and the Biosphere Programme (UNESCO-MAB) under the terms of a joint programme of work (Ramsar Convention Secretariat 2011, 2016).

d) **United Nations Environment Programme (UNEP):** An official collaboration has also been going on between Ramsar and UNEP. A collaboration was developed with the UNEP World Conservation Monitoring Centre (UNEP-WCMC) in 2010 with regard to harmonizing reporting requirements under the different instruments and on developing indicators of effectiveness, among other projects. Ramsar Secretariat also participates in UNEP's Environmental Management Group (EMG). An agreement was also signed between Ramsar and UNEP's Global Programme of Action for the Protection of the Marine Environment from Land-based Activities (UNEP-GPA) in 2006. The Ramsar Secretariat has also signed the Memorandum of Cooperation with UNEP's Convention for the Protection and Development of the Marine Environment of the Wider Caribbean Region (Cartagena Convention) (Ramsar Convention Secretariat 2011, 2016).

e) **UNECE "Water Convention"**: The Ramsar Convention also works closely with UNECE "Water Convention" (the Convention on the Protection and Use of Transboundary Watercourses and International Lakes) at the pan-European level (Ramsar Convention Secretariat 2011, 2016).

f) **European Environment Agency:** The Ramsar convention signed an agreement with the European Environment Agency in 2006 and for a collaborative project named the GlobWetland project. The project works on monitoring and management tools based on earth observation data related to different Ramsar Sites around the world (Ramsar Convention Secretariat 2011, 2016).

2.9 Education and Outreach

2.9.1 Communication, Education, Participation, and Awareness (CEPA)

The Communication, Education, Participation, and Awareness popularly known by the acronym CEPA is a collective term for the set of different activities meant for sensitizing the wide range of stakeholders. The Ramsar Convention's vision of the CEPA Programme is to "prevent, stop, and reverse the degradation of wetlands and use them wisely" with the overarching goal as "People taking action for the wise-use of wetlands."

The first programme of actions for CEPA was adopted by the Conference of Contracting Parties under Resolution VII.9 during the seventh meeting. Each contracting party has to name governmental and nongovernmental focal points for CEPA who become a part of a global network of experts to share information, promote the dissemination of resource material, and support the development or expansion of programmes. The CEPA's national focal person has to reach out to the relevant authorities and ensure practical implementation of CEPA at the national level. The activities under CEPA play a very vital role in the dissemination of information related to wetland wise-use and conservation (Ramsar Convention Secretariat 2011, 2016).

2.9.2 World Wetlands Day

The World Wetlands Day, celebrated every year on 2 February, marks the date of the adoption of the Convention on 2 February 1971. For the first time, it was celebrated in the year 1997 and since then, it has come a long way in popularizing the wetlands and highlighting the need for their wise-use and conservation. A host of governmental and nongovernmental organizations bring together a diverse group of stakeholders to celebrate the wetlands and appreciate the services provided by them. World Wetlands Day provides an excellent opportunity for conservationists to shift focus on wetlands and spread awareness about these immensely important ecosystems. Every year since 1997, the convention's website has been posting the resource materials and yearly themes for the World Wetlands Day. The resource materials largely include decorative and instructional posters, stickers, brochures and leaflets, bookmarks, pocket calendars, screen savers, quotable background papers, flash animations, and videos, etc. These resource materials are also available in their design files so that they can be customized to local languages and priorities. The WWD page on the Ramsar website leads to a listing of what promotional materials are available at any point in time and reports on WWD activities around the world (Ramsar Convention Secretariat 2011, 2016).

2.10 Legal Status

In the international context, the term "soft law" is applied to declarations, recommendations, and resolutions, etc., which are not legally binding for the signatories but often considered as sources of moral as well as political forces to help the application of allied, country-specific laws when the need arises. Such soft laws might be nonjusticiable but they certainly can help in invoking the applicable hard laws or the legal instruments available within the legal framework of the signatory nations against the actions that are in violation of the convention (Verschuuren 2008).

The Ramsar convention, despite being a soft law, is supported at the national and local levels, by the legal framework meant for the protection and conservation of the environment, wildlife, and water resources, etc. In India, the Ramsar Convention draws strength from a host of laws that can be invoked in case of violations of the Ramsar Convention. Although India lacks a comprehensive National Wetland Policy, it does have Wetlands (Conservation and Management) Rules, 2010 (amended in 2017) which empower the country's legal system against wetland conversion and degradation. Some of the laws which strengthen the implementation of the Ramsar Convention in the India include Indian Forest Act, 1927; Wildlife (Protection) Act, 1972; Environment (Protection) Act, 1986; Water (Prevention and Control of Pollution) Act, 1974; Forest (Conservation Act), 1980; Coastal Zone Regulation Notification, 1991; Wildlife (Protection) Amendment Act, 1991; Biodiversity Act, 2002; and National Environment Policy, 2006 (Pritchard 2009).

Similarly, various other countries have enacted laws the enable the conservation and wise-use of wetlands as mandated under the Ramsar Convention. Although a soft law, the convention may, at times, act as a guiding light in making the right decisions with regard to large developmental projects. This may avoid unnecessary hassle and wastage of time and financial resources on such projects which might have severe implications for wetland ecosystems. There have also been instances where development projects challenged in the light of the Ramsar Convention were ultimately rejected. One such instance which has been widely written about is the construction of a resort on an island in the Netherlands. A private company, Crown Court Estate, planned to build a resort named "Mangrove Village" on the Island of Bonaire (one of the islands of the Netherlands Antilles). The project would be spread over an area of 44 150 m^2 and some of its area falling within the boundary of a Ramsar site, the Lac/Sorobon wetland. The Governor of the island realized that the project was causing an infringement of Article 3 of the Ramsar Convention and the guidelines adopted in the Annex of Resolution VIII.9. The project was rejected by the Governor based on the argument that it breached the provisions of the Ramsar Convention (Verschuuren 2008).

2.11 Effectiveness of the Convention

Evaluating the effectiveness and success of an International Convention like Ramsar is a task not as easy and straightforward as it seems yet a fair level of the effectiveness of the convention can be gauged based on the indicators mentioned as below:

1) Number of signatories: In the year 1971, only a meagre number of 18 nations realized the significance of wetlands and attended the Ramsar Convention but by the year 2019, the total number of signatories have risen to 171 which could be considered a big success of the convention and the way it has been able to lobby for the conservation of wetlands (Ramsar Sites Information Service 2020).
2) Number/area of Ramsar sites: The number of Ramsar sites has been on a constant rise since the Ramsar Convention. Currently, the total number of Ramsar sites around the world is 2390 with their cumulative area of 2.5 million km^2 (Ramsar Sites Information Service 2020).
3) Increase in protection: It has been observed that the Ramsar-designated sites have experienced an increase in protection efforts over the years. The increase in protection has particularly been noticeable in the developing world (Castro et al. 2002).
4) Reduction in threat levels: It has been noticed that the Ramsar-designated sites with greater levels of threats have witnessed a significant reduction in the levels of threats they experienced in the past. (Castro et al. 2002).
5) Decline in degradation: The degradation of wetlands is still going on but at a lesser rate. As per a case study, the Mangrove wetlands are still getting degraded but the rate of their loss has reduced around the world except in Asia (where it has increased) (Pritchard 2009).

While the above-mentioned indicators point toward a healthy trend, there is a need to be wary of the fact that wetland degradation is still continuing. In some of the cases it has been realized that the wetlands which earlier had lower levels of threats are now facing greater levels of threats. As per some assessments, there has also been a decline of around

35% in the spatial extent of wetlands globally between the years 1970 and 2015 (Ramsar Convention on Wetlands 2018). This points to the fact that while the Ramsar Convention has successfully proved itself to be a great tool for wetland conservation across the globe, there still are some serious gaps that will take time to fill.

References

An, S., Li, H., Guan, B. et al. (2007). China's Natural Wetlands: past problems, current status, and future challenges. *Ambio* **36** (4): 335–342.

Carp, E. (ed.). (1972). *Proceedings of the International Conference on the Conservation of wetlands and Waterfowl*, 30 January to 3 February 1971, Ramsar, Iran.

Castro, G., Chomitz, K., and Thomas, T.S. (2002). *The Ramsar Convention: Measuring its Effectiveness for Conserving Wetlands of International Importance*. Ramsar COP8 DOC.

Costanza, R., de Groot, R., Sutton, P. et al. (2014). Changes in the global value of ecosystem services. *Global Environmental Change* **26** (1): 152–158.

Dahl, T.E. (1990). *Wetland Losses in the United States 1780's to 1980's*. Washington, DC: U.S. Department of the Interior, Fish and Wildlife Service.

Davidson, N.C. (2014). How much wetland has the world lost? Long-term and recent trends in global wetland area. *Marine and Freshwater Research* **65** (10): 936–941.

Davidson, N.C., Laffoley, D.'.A., Doody, J.P. et al. (1991). *Nature Conservation and Estuaries in Great Britain*. Peterborough, UK: Nature Conservancy Council.

Davidson, N.C., D'Cruz, R., and Finlayson, C.M. (2005). *Ecosystems and Human Well-being: Wetlands and Water Synthesis: A Report of the Millennium Ecosystem Assessment*. Washington, DC: World Resources Institute.

Dugan, P. (1993). *Wetlands in Danger – A World Conservation Atlas*. New York, NY: Oxford University Press.

Gardner, R. and Finlayson, M. (2018). *Global Wetland Outlook: State of the World's Wetlands and Their Services to People*. Gland: Ramsar Convention.

Gawler, M. (2000). *Strategies for Wise Use of Wetlands: Best Practices in Participatory Management*. IUCN, Wetlands International and WWF.

Gopal, B. (1977). Wetlands and their management. In: *Current Trends in Indian Environment* (eds. D. Bandhu and E. Chauhan), 177–181. New Delhi: Today & Tomorrow's Printers and Publishers.

Grist, D.H. (1975). *Rice*, 5e, 601. London, UK: Longman Group.

Hoffmann, L. (1964). *Proceedings of the MAR Conference Organized by IUCN, ICBP and IWRB*. Stes-Maries-de-la-Mer, France, 12–16 November 1962. Morges: IUCN Publications New Series 3.

Kotze, D.C., Breen, C.M., and Quinn, N. (1995). Wetland losses in South Africa. In: *Wetlands of South Africa* (ed. G.I. Cowan), 263–272. Pretoria: Department of Environmental Affairs and Tourism.

Matthews, G.V.T. (1993). *The Ramsar Convention on Wetlands: Its History and Development*. Gland: Ramsar Convention Bureau.

Millennium Ecosystem Assessment (2005). *Ecosystems and Human Well-being: Synthesis*. Washington, DC: Island Press.

Mitsch, W.J. and Gosselink, J.G. (2015). *Wetlands*. New York, NY: Wiley.

OECD Development Assistance Committee (1996). *Guidelines for Aid Agencies for Improved Conservation and Sustainable Use of Tropical and Subtropical Wetlands*. Paris: Organization for Economic Co-operation and Development.

Prasad, S.N., Ramachandra, T.V., Ahalya, N. et al. (2002). Conservation of wetlands of India – a review. *Tropical Ecology* **43** (1): 173–186.

Pritchard, D. (2009). The Ramsar Convention on Wetlands and its indicators of effectiveness. *International Expert Workshop on the 2010 Biodiversity Indicators and Post-2010 Indicator Development*. UNEP World Conservation Monitoring Centre (UNEP-WCMC).

Ramsar Convention Bureau (2015). *Wetlands for Our Future: Act Now to Prevent, Stop, and Reserve Wetland Loss*. Gland: Ramsar Convention Bureau.

Ramsar Convention on Wetlands (2018). *Global Wetland Outlook: State of the World's Wetlands and Their Services to People*. Gland: Ramsar Convention Secretariat.

Ramsar Convention Secretariat (2010). *Designating Ramsar Sites: Strategic Framework and Guidelines for the Future Development of the List of Wetlands of International Importance*. Gland: Ramsar Convention Secretariat.

Ramsar Convention Secretariat (2011). *The Ramsar Convention Manual: A Guide to the Convention on Wetlands*, 5e. Gland: Ramsar Convention Secretariat.

Ramsar Convention Secretariat (2016). *An Introduction to the Convention on Wetlands*. Gland: Ramsar Convention Secretariat.

Ramsar Sites Information Service. (2020). https://rsis.ramsar.org/?pagetab=2 (accessed 30 April 2020).

Schmidt, P.R. (2006). North American Flyway Management: a century of experience in the United States. In: *Waterbirds Around the World* (eds. G.C. Boere, C.A. Galbraith and D.A. Stroud), 60–62. Edinburgh, UK: The Stationery Office.

Swift, J.J. (1964). *Proceedings of the First European Meeting on Wildfowl Conservation*. St. Andrews, Scotland, 16–18 October 1963. London: The Nature Conservancy.

Verschuuren, J. (2008). Ramsar soft law is not soft at all. *Milieu en Recht* **35** (1): 28–34.

3

Ecological Importance of Wetland Systems

K.C. Jisha[1] *and Jos T. Puthur*[2]

[1] *Department of Botany, MES Asmabi College, P. Vemballur, Kerala, India*
[2] *Plant Physiology and Biochemistry Division, Department of Botany, University of Calicut, Malappuram, Kerala, India*

3.1 Introduction

There are many forms of wetlands like tidal zones, rice fields, marshes, bogs, or swamps, etc., and all of them have characteristics that enable them to be classified as wetlands. These are the areas on earth, above which or close to the soil surface, water can be seen for at least a major portion of the year. The Ramsar Convention (1971) broadly defines wetlands as: "areas of marsh, fen, peatland or water, whether natural or artificial, permanent or temporary, with water that is static or flowing, fresh, brackish, or salt, including areas of marine water, the depth of which at low tide does not exceed six metres."

Wetlands are the fundamental constituents of the regional hydrological cycle and, moreover, they are extremely productive, hold large floral and faunal diversity, and also offer many ecosystem services like food, flood control, purification of water, assimilation of wastes, recharging of groundwater, climate regulation, erosion control, increasing the aesthetic value of the land, and providing several recreational, sociocultural activities. Wetlands, like lakes, rivers, marshes, and coastal estuaries, provide enumerable services that help human life. The major ecosystem services include fish, fiber, purification of water, protection of shorelines, regulation of climate change, flood control, recreations, tourism, etc. (Millennium Ecosystem Assessment 2005). The major ecosystem services of wetlands are discussed here.

3.2 Importance of Wetlands in Flood Control

The heavy or long-lasting rain leads to the reduction in the infiltration rate of water and also increases the surface runoff on the land surface into surface waters, thus leading to flooding. Flooding not only slows down the economic growth of countries, but also damages the ecological balance (Li and Cai 2002). Ecosystems can control the floods by lowering the surface runoff and river discharge; or reducing the flooding effects through

Wetlands Conservation: Current Challenges and Future Strategies, First Edition.
Edited by Sanjeev Sharma and Pardeep Singh.

retaining space for surplus water (Nedkov and Burkhard 2012). The action of wetlands can be compared to that of a sponge, i.e. like sponges, these habitats can hold water in the rainy season and can release it in the dry periods (Bucher et al. 1993). Conservation of wetlands has been encouraged as an effective method for controlling flood by organizations like IUCN (Dugan 1990), Wetlands International (Davies and Claridge 1993) and the Ramsar Convention on wetlands of international importance (Davis 1993).

According to Depietri et al. (2012), for maintaining sufficient flood control and water purification, 3–7% of the watershed area in temperate regions must be retained as the wetlands. The wetland system acts as the center of the flood control system. Wetlands decrease the flow of water, allowing the replenishment of groundwater supply, that most of the organisms depend on. When a flood occurs, wetlands decrease the flood impact in the areas downstream and around them very effectively. Wetlands can mitigate the harmful effects of flood by several methods like storing the excess water or by holding the water or allowing the water to infiltrate into the soil (Jia et al. 2011; Yu et al. 2012). The rate of flood reduction by wetlands depends on their physical capacity (Narayan et al. 2017). Moreover, the success of flood reduction by the wetlands depends on many other factors like the type of wetland flora, topography, field saturation before the flooding, etc. But, usually, because of the less storage capacity of wetlands, the effectiveness of flood reduction is less (Leon et al. 2018). According to Galloway report, wetlands in the uplands could hold minor floods, and, at the same time, are unable to prevent bigger floods. In such cases, one strategy is that if we drain the water in the wetland before the flood, the wetland can hold more water at the time of the flood (Leon et al. 2018). In countries like India, water bodies such as ponds and lakes are usually made to collect rainwater for general purpose. These wetlands actually reduce the flood impact by holding the water during the rainy season (Neelakantan and Ramakrishnan 2017).

Many papers report that wetlands play a major role in reducing floods, while some conclude that wetlands play no roles in controlling floods, while others provide proofs of wetlands increasing floods. According to Acreman and Holden (2013), this variation may be due to the diversity of ecosystems which are treated as wetlands and, moreover, all these wetlands never show similar hydrological behavior. Besides this, the same kind of wetland can serve as a sink or source for flooding based on the specific features of the individual wetlands, its hydrological situations, its location, or on how it is maintained.

3.3 Role of Wetlands in Groundwater Replenishment

Groundwater replenishment may occur naturally or by anthropogenic activities like the construction of basins or wells. The natural methods include localized or diffuse-type recharges. In diffuse recharging, the water-level rise occurs as a result of precipitation, while in localized recharging, the water-level rise occurs as a result of drainage from surrounding wetlands (Alley 2009). In order to maintain a high water table in wetlands, groundwater discharge is necessary. On the other hand, groundwater recharge to the underlying aquifers replenishes groundwater supplies and this wetland discharge and recharge function is an important and, at the same time, complex part of wetland hydrology. Wetlands are said to be natural sponges, which hold water until it can percolate deep into the ground. The water that fails to reach the water table is gradually released into close by streams or rivers and this gradual discharge of water prevents flooding during heavy precipitation. The flora of the wetlands also plays a very important function in reducing the speed of water when it flows over the landscape. In this manner, wetlands provide plenty of water storage services and thus reduce the flood and erosion rates (Alley 2009).

Many freshwater wetlands can be seen associated with the areas where the surface water enters an underground aquifer, thus effectively recharging the groundwater supplies. The groundwater recharge is very significant because it forms the major source of drinking water for the human population or is important for producing stream flows for plants, fishes, animals, and other organisms that thrive near the wetlands throughout severe dry summer periods. The recharge of aquifers by wetlands mainly depends on the availability of water within the wetland and the presence of small depressions in the surrounding lands and also on the hydraulic conductivity of the aquifers (Van der Kamp and Hayashi 1998). Many studies on tiny wetlands of semiarid northern prairies revealed that the wetlands act as the centers of groundwater recharge (Winter 1989; La Baugh et al. 1998) and these services of prairie wetlands are accepted as a key role in the ecosystem (Fuller 1988; Leitch and Hovde 1996). Rehm et al. (1982) suggested that recharge of groundwater also occurs from outside the wetlands and it may be from the upland shallow ephemeral puddles or from sandy areas with low moisture-retention capacity. As far as the shallow wells are concerned, wetlands play an important role in maintaining the water level through recharging them. But the changes in land use may increase or decrease the surface runoff from nearby lands to the wetlands which affect the water level and groundwater recharge property of the wetlands (Van der Kamp and Hayashi 1998). In aquifers, the water level changes in accordance with the seasonal variations, the water level will be high during the winter or rainy season and the level will be falling during warm summer periods. Many human activities like deforestation increase the surface runoff and thereby reduce the water-recharging properties of wetlands. At the same time, certain other human activities like agricultural tillage and creating artificial water bodies increase the ground water recharge (Alley 2009).

3.4 Role of Wetlands in Stabilization and Storm Protection of Shorelines

Seashores always experience a wide range of natural calamities like tsunamis, storms, and hurricanes. Such calamities are a normal occurrence that usually upset the coastal zones and get amplified by environmental deviations. The effects of climate change like the rise in the sea level and ocean warming surely enhance the possibility of these coastal hazards (Nicholls et al. 2007). It is a practice to construct sea walls for protection from storms, but such constructions may not be successful; moreover. it may cause adverse effects on the environment also. In this context, coastal wetlands can be viewed as an ecofriendly, cost-effective method of protection from natural calamities like storms, tsunamis, etc. Besides this major protective function, wetlands collect sediments which also offer safety from sea level rise (Suna and Carsona 2020).

Coastal marshes have a key role in mitigating these hazardous effects like wave attenuation, which cause a decrease in wave height or energy due to the passage of waves across the marsh vegetation. The attenuation of wave energy is achieved by the frictional drag posed by the flora in the marshes (Tsihrintzis and Madiedo 2000; Boesch et al. 2006; Leonard et al. 2006). According to Barbier (2007), it is very important to know the shoreline protection abilities of wetlands to relate it with the wetland degradation and the worth of restoration of wetlands. Being a low-cost alternative for barrier creations, coastal wetland preservation and restoration could be adopted by the rural communities for reducing the natural calamities like a storm (Walton et al. 2006; Halpern et al. 2007; Costanza et al. 2008). In addition to the above-said benefits, coastal wetlands offer many other services like wood and non-wood products, fisheries and ecotourism opportunities for local coastal communities (Barbier 2007).

Shoreline stabilization is the process through which the saline marsh vegetation enhances the deposition of sediments, increases the elevation of the marsh through root production, and thus leads to the stabilization of these marsh sediments. In the case of lakes and ponds also, wetland flora can prevent soil erosion during floods by binding with the soil particles and also by causing sedimentation around their roots. Because of the ecological importance of this erosion protection function of wetlands, at certain places, wetlands are being restored especially in seashores for the protection from hurricanes, tsunamis, and storms. There should be a minimum elevation of the salt marsh to avoid the drowning of marsh plants and loss of marsh edge. Thus the mechanisms which keep up the marsh elevation surely sustain the marsh shorelines and thus decrease the erosion rates (Shepard et al. 2011).

3.5 Role of Wetlands in Sediment and Nutrient Retention

Sedimentation is the preservation of particulate substances which get deposited on the wetland soil plane. The flora of wetlands absorbs these particulate substances and through their decomposition, they will again get deposited in the wetland soil. The plant litter present in the wetland floor acts as a good precursor for microbial action. Thus wetland soils not only absorb different types of nutrients, but also offer suitable conditions for microbial action of nutrients (Johnston 1991).

The soils of wetlands show a vast diversity in the texture as well as in the organic matter content. In wetlands, the water forms the major medium for the transportation of materials into and out of the system and the water which comes from different sources contributes different types of sediments and nutrients to the wetlands in different quantities. Generally, the high sediment and nutrient contributions come from anthropogenic sources of water like household discharges, agriculture runoff, etc., and the least contribution of sediments and nutrients occurs from groundwater and precipitation (Reed 1995). The quality of water in the wetlands becomes better through the sedimentation process since it reduces the turbidity and other suspended solid particles in the water and, at the same time, retains the phosphorus in the water. The elevation of the marsh surface is mainly due to the sediment deposition together with the production of roots by the marsh vegetation (Penland and Ramsey 1990; Cahoon et al. 1999). Moreover, the subsidence and compaction of the sediments also interfere with the marsh surface elevation, especially in the quickly subsiding marshes (Khurana and Pritpal 2012). When any substance enters into a wetland, it may be stored or modified either biologically or by chemical action or it may be released via water or atmospheric currents. Thus the wetlands are the storage ecosystems, which could trap the sediment and nutrients to the downstream waters and the capacity of the wetland to do these functions depends on various factors like hydrology, distribution of sediments, and the nutrient inputs, etc. Usually, the terrestrial ecosystems get inputs from the atmosphere only, but in the case of wetlands, the wetlands can get inputs from different sources like surface, runoff waters, atmosphere, and groundwater. These diverse sources account for the nutrient-retention capacity of wetlands (Johnston 1991).

According to Silvan et al. (2004), wetlands can effectively retain nitrogen and phosphorus which can be released into these wetlands from forestry. The wetland plants play a significant role in retaining the nutrients in wetlands. The nitrogen and phosphorus in the wetlands can be absorbed by the wetland plants through various forms like ammonia, nitrate, etc., by the roots and they utilize these nutrients for the production of organic compounds within their body. These nutrients are again released back into the wetland water when the plants are decomposed by microbial action. Moreover, the shoot system of wetland flora causes the sedimentation of particles by decreasing the water current and thereby favors the settlement of particles.

3.6 Role of Wetlands in Water Purification

Wetlands serve as major agents of water purification. The rainfall carries various pollutants from the urban and agricultural lands to the water bodies. The major pollutants consist of fertilizers, oil, and grease from trucks and cars, soil particles, pesticides, and road salts. Wetlands can remove these pollutants from the surface waters and thus can improve water quality. This is mainly achieved through various processes like trapping of sediments, detoxifying chemicals, and by removing various nutrients from the water.

We can also use wetlands for wastewater treatment. Ample recycling of wastewater is recommended for protecting our water bodies, environment, and human health. The straight release of untreated contaminated water to the soil and different wetlands imparts detrimental effects on human health (Scholz 2010) and aquatic organisms (FAO 2003). Due to these harmful effects, the proper treatment of wastewater and the recycling methods are very much significant to supply enough pure water in the coming decades since the water resources are less (Vymazal 2014). An alternative method for this purpose includes the creation of artificial wetlands and it involves interconnections of soil, different wetland plants, and microorganisms which together can function as a single unit to treat the wastewater (Vacca et al. 2005). The success of such an artificial wetland wastewater treatment system mainly depends on the design, nature, plant type, microbes, and the local weather conditions (Picek et al. 2007; Ström and Christensen 2007; Weishampel et al. 2009). Moreover, Almuktar et al. (2017) studied the chances of household wastewater recycling through artificial wetlands for crop irrigation purposes and according to their opinion, artificial wetlands are highly effective in removing most of the contaminants and thus satisfied the general standards of wastewater reused for irrigation purposes. Moreover, wetlands can also remove various heavy metals from the water and the removal was found to be 55% for chromium (Cr) (Arroyo et al. 2010), for nickel (Ni), it was 25–35%, 25–87% for zinc (Zn), and for copper (Cu), it was about 9% (Galletti et al. 2010), for cadmium (Cd), it was found to be 33%, and for cobalt (Co), it was 75% (Pedrero et al. 2010), and bacterial removal between one and six log units (Feigin et al. 2012). These results remind us that wetland technology is the most attractive one for the treatment of wastewater and further reuse for irrigation. All these investigations imply that if all the artificially created wetlands are well designed and properly maintained, they can be effectively used for local secondary and tertiary wastewater treatments.

3.7 Biodiversity of Wetlands

The species diversity in the wetlands was found to be highly significant as that of tropical rain forests and coral reefs. The vegetation of wetlands belongs to almost all taxonomic groups like algae, bryophytes, pteridophytes, and angiosperms. Plant genus such as *Typha*, *Phragmites*, *Scirpus*, *Cyperus*, etc., are usually found to be dominated in the wetlands. Different types of aquatic plants including submerged, rooted, and free-floating were usually found in wetlands. Moreover, lots of planktonic and filamentous algae along with other organisms like fungi and bacteria occur in different wetlands. Wetlands also serve as the habitat for different types of animals and the diversity of fauna comprises members from protozoa to mammals (Gopal 2015). Mainly two types of fauna are found in the wetlands – one group entirely depends on the wetlands for completing its whole life cycle but another group needs wetlands only for a period in its life. Gopal and Junk (2000) identified mainly six classes of wetland fauna: (i) those living in the proper wetlands; (ii) habitual deep water migrants; (iii) habitual terrestrial migrants; (iv) habitual migrants from some other wetlands; (v) irregular visitors; and (vi) those which indirectly depend on wetland biota. Different types of fish, reptiles, amphibia, insects, and birds use wetlands for various purposes like feeding, breeding, and for making shelters (Gopal 2015).

Since the water in estuarine wetlands contains many different types of nutrients, it acts as a feeding and spawning ground for many species of shellfish and fish. Like amphibians, insect and plants, different types of mammals like otter, mink, etc., live in the North American wetlands. These coastal wetlands are also considered as an important nesting habitat for different types of birds. In the US, among the endangered plant and animal species, one-third depend on wetlands for their survival (USEPA). Different types of migratory birds visit wetlands and without these visits, they are unable to complete their life cycle. Similarly, many economically important fish species, oysters, shrimps, crabs, etc., are unable to survive without wetlands (Cronk and Fennessy 2009). As far as the wetland plants are concerned, they form an inevitable part of the wetland ecosystem and include mainly hydrophytes. Among the 250 000 angiosperms, only 3–5% is found to be distributed in wetlands. In addition to the angiosperm members, some lower group members like Azolla and *Ceratopteris* (Pteridophyta), *Taxodium distichum*, *Pinus elliotti* (Gymnosperms), *Sphagnum* (Bryophyta), etc., were found to be adapted to wetlands. They perform many important functions in wetlands and they form the foundation of the food chain, thus forming the base of energy circulation in the ecosystem. The different types of plants differ in their primary productivity level, but it was proven that most herbaceous plants dominating wetlands possess the maximum primary productivity and it was comparable to that of tropical rain forests. Moreover, these wetland florae form the major habitat for different types of microbiota, thus the wetland flora significantly influence the species diversity of the wetland ecosystem (Cronk and Fennessy 2009).

3.8 Wetland Products

Wetlands are the source of many economically important natural resources on which the economy of the rural community mainly depends. They offer a lot of significant benefits to the local society as well as to the individuals who live distant from them. Wetlands are widely recognized for their essential role in nourishing the vast biodiversity and also for giving various goods and services (Ramachandra et al. 2011). Moreover, wetlands provide many raw materials which are effectively used by the rural community (Malabika et al. 2012). The simplest and the most common example of the wetland product includes drinking water for humans. Wetlands form the main source of water supply for the wells, dams, springs, etc. It offers a variety of tangible and non-tangible services to different communities (Karanja et al. 2001; Wetlands Management Department 2009). Water for the household purposes and for livestock, irrigation, food resources, handicrafts, and building materials include the tangible benefits of wetlands, while the non-tangible benefits are water purification, flood control, water table maintenance, climate regulation, and protection from storms (Kakuru et al. 2013). The major wetland products include different types of fishes, shrimps, crabs, different types of berries, wild rice, and timber. Certain important medicinal plants were also found in wetlands. Many of the fishing industries, especially in the south-east countries, are focused on wetland products. In addition to this, many wetland plants are employed for construction purposes like thatching, timber works, and others are used as fuel, fibers, in the paper industry, for making dyes, etc. These also serve as the habitats for diverse flora and fauna, possess a heritage and aesthetic importance, and contain different medicinally important flora and fauna (Karanja et al. 2001; Wetlands Management Department 2009). Besides these significances, wetlands provide food security by indirectly contributing toward food production through nutrient retention and modification of weather.

3.9 Sociocultural Values of Wetlands

Cultural services of ecosystems occur by means of interaction between people and nature (Fish et al. 2016), implying that cultural ecosystem services and their values are context-specific (Bryce et al. 2016). Wetlands are usually linked with the cultural practices that make the humans flourish, adjust to the changing environmental conditions, and utilize nature in a sustainable way. Humans utilize the wetlands for many purposes like transportation, fishing, energy production, farming, and drinking water. In addition to these tangible benefits, wetlands also have many important intangible and cultural services. According to Millennium Ecosystem Assessment (2005), different wetlands like lakes, ponds, rivers, paddy fields, estuaries, etc., offer many important functions which directly influence the social well-being of humans, thereby alleviating poverty. The people who are in close association with these wetlands greatly depend on the benefits from wetlands and thus they are more prone to be hurt by the degradation of these wetlands. Wetlands usually act as the sites of pilgrimage and religious fulfillment and the water collected from such wetlands is often used in rituals and for curative purposes. These spiritual values of wetlands demand the proper maintenance and protection of wetlands. According to Sighn (2013), in India, about 250 million pilgrimages visit every year and among these, the majority of pilgrimage centres are concentrated around wetlands. Wetlands also form a major area of education and serious research by providing opportunities for proper education and various pieces of training. Much of the cultural, ecological, and historical wealth, which is plentiful in most of the wetlands, offers indefinite chances for ecological education and community awareness classes. The wetlands constitute the exceptional study areas for understanding the vegetative organization of the ecosystem and many environmental functions like nutrient cycling, plant succession, community structure, ecological interactions, etc. Verschuuren (2007) provided a non-exhaustive classification of sociocultural values describing spiritual services as "the qualities of wetlands that inspire humans to relate with reverence to the sacredness of wetlands."

3.10 Wetlands in Relation to Recreation and Tourism

Nowadays, environment-based recreation is growing faster and forms the base of tourism. The majority of these environment-based recreational activities, including bird-watching, are associated with wetlands. Due to the aesthetic values of these water bodies, many nature lovers, especially artists and writers, copy the scenic beauty of these ecosystems on their paper or canvas or by using any other digital methods. Wetlands offer infinite possibilities for many recreational activities, like boating, fishing, hiking, bird watching, trapping, etc. The aesthetic beauty of this ecosystem attracts everyone to its vicinity. Moreover, the different types of flora and fauna present in the wetlands make it an enjoyable place. Globally, wetlands offer significant opportunities to tourism and recreation and, thus, in turn, provide great monetary profits to the local communities, tourism industry, and to governments, and generally the money generated by these means is used for their conservation (Ramsar-UNWTO 2012). Due to the recreational values of the well-conserved and maintained wetlands, they can attract visitors and can provide income for the empowerment of the local people. Tourism is a very important economic part in which seaside wetlands contribute a lot in many countries and the incorporation of tourism along with the protection of the wetlands can be seen worldwide. Camargue wetland in France (Beltrame et al. 2013); Nabugabo wetland in Uganda (Bikangaga et al. 2007); Junam and Upo wetlands in South Korea (Do et al. 2015); Ondiri and Manguo wetlands in Kenya (Macharia et al. 2010); and Nariva Swamp in Trinidad (Pemberton and Mader-Charles 2005) are examples for such a type of conservation. Tourism is the basis of monetary thrust for conserving various natural habitats like wetlands and biosphere reserves since they constantly draw visitors (Brandl et al. 2011). According to Millennium Ecosystem Assessment (2005), in the United States itself, recreational fishing could earn a comparable income and about 35–45 million public got actively involved in this process by utilizing a sum of $24–37 billion per year on this hobby.

Research work on ecotourism in the Ramsar-listed Tsomoriri wetland of Ladakh, India concluded that the local families earned US $700–1200 during the summer season itself from the wetlands (Anand et al. 2012). Moreover, Do et al. (2015) described a heavy increase in the wetland-tourism with greater than 21 million visitors regularly visiting the four wetlands in South Korea between 2007 and 2012. On the basis of a survey in 11 protected areas in India, Karanth and DeFries (2011) concluded that the average growth rate of wetland tourism remained at nearly 15%, and of which 80% were domestic visitors, thus indicating that eco-tourism is a developing industry in India. The aesthetic value of wetlands can be improved by various methods and it mainly includes the protection of wetlands through changing the landscape, conserving the wildlife habitat, conserving the flora and fauna and all these efforts surely improve the tourist attraction of the water bodies (Pueyo-Ros et al. 2016).

3.11 Wetland and Climate Change

Nowadays, changing climate has become the key risk factor for the species' survival and ecosystem's stabilization worldwide (Hulme 2005). Climate changes associated with wetlands include variations in hydrology, land-use changes, temperature changes, etc. (Ferrati et al. 2005). According to Ramsar-STRP (2002), the major impacts of extreme climate events are deviations in the base flows, altered hydrology including the depth and the hydro-period, enhanced temperature stress in wildlife, extensive range, and action of disease and pest vectors, increased flooding, landslide, mudslide damage and avalanche, greater soil erosion rate, increased flood runoff resulting in the decline of recharge in some floodplain aquifers, reduction in the quantity and quality of water resources, greater risk of fires, enhanced coastal erosion and degradation to coastal buildings and infrastructure, increased damage to coastal habitats like mangroves and coral reefs, and enhanced tropical cyclone actions. Moreover, exotic species will flourish under changing climates and it will impart pressure on wetlands and other ecosystems (Root et al. 2003). Wetlands are vulnerable and most susceptible to the changes occurring in the water quality and quantity. The most significant effects of climate change on wetlands are the alterations in the hydrological regimes. Similarly, other effects of climate change like increased temperature, evaporation, transpiration, fire, etc., play important roles in deciding the regional and local impacts on wetlands (IPCC (International Panel on Climate Change) 1998; Burkett and Kusler 2000; USGCRP (US Global Change Research Program) 2000). For conserving the wetlands properly, hydrology of the wetlands should be maintained, should reduce the rate of pollution, should control the entry of exotic species and also should protect the wetland biodiversity and integrity. If we conserve in this way, the wetlands will continuously provide the major ecosystem services under the changed climatic conditions also (Kusler et al. 1999).

Climate change will surely cause the degradation and loss of several wetlands, thereby reducing their species' richness, causing a direct decline in the services for human beings who depend on the wetlands for their livelihood. It was projected that global climate change will result in a high precipitation rate that will surely provide more water to the ecosystem and the humans as well. But this increased precipitation will not be uniform in all the portions of the earth. At certain regions, we have to face scarcity of water due to decreased precipitation. Due to climate changes, the sea level rise causes many unusual occurrences of storms, tidal surges, tsunamis, etc., which surely will damage the coastal wetlands, thereby creating a threat to the flora and fauna of such coastal wetlands. Climate change has already damaged a major portion of coral reef which is the very vulnerable ecosystem to climate change. Moreover, it was already reported that climate change will bring about the reduction of population size in high-Arctic breeding waterbird due to the loss of habitat and the distribution pattern of many fishes changes according to the variation in climate (Hulme 2005).

3.12 Summary

Wetlands form a significant part of our nature. They provide protection to our shores from wave action, decrease the flood impact, purify the water, and increase the water quality. Moreover, they offer habitats for many animals and plants and many wetlands contain unique floral and faunal diversity that is found nowhere else. Thus wetlands supply a very important range of environmental, social, and economic services to the world. The major ecological roles of wetlands are summarized in Figure 3.1.

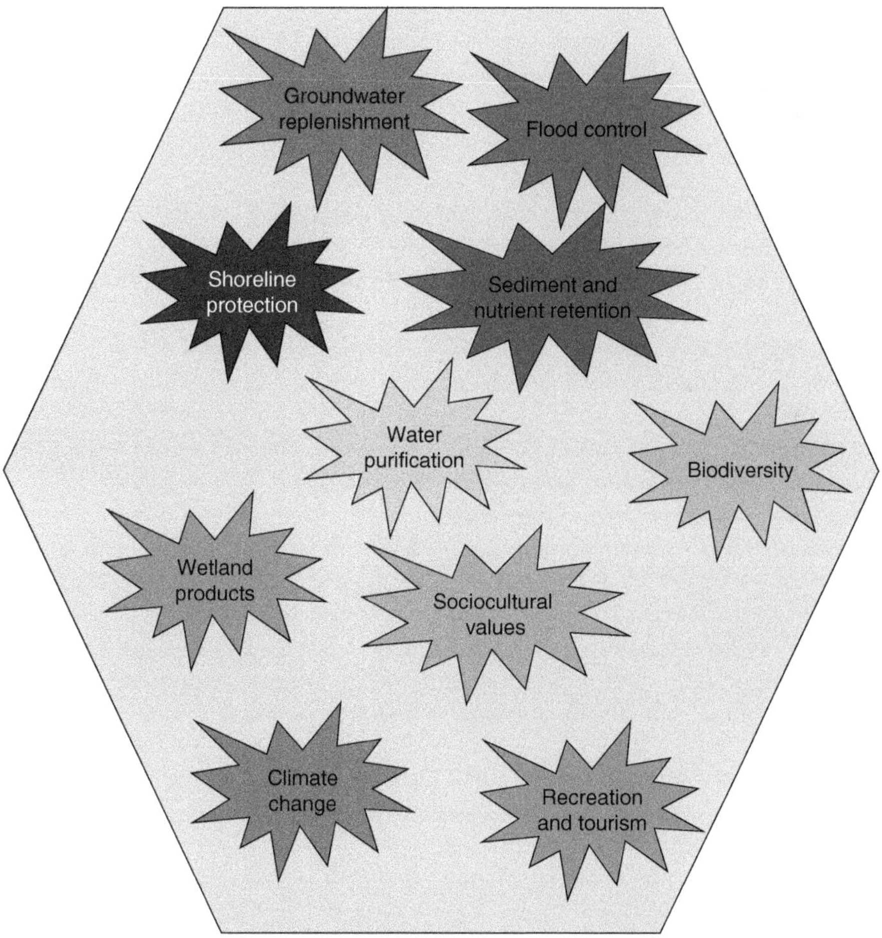

Figure 3.1 Major ecological roles of wetlands.

Acknowledgments

JTP acknowledges the financial assistance provided by the KSCSTE through research grant (KSCSTE/5179/2017-SRSLS).

References

Acreman, M. and Holden, J. (2013). How wetlands affect floods. *Wetlands* 33: 773–786.

Alley, W.M. (2009). Ground water. In: *Encyclopedia of Inland Waters* (ed. G.E. Likens), 684–690. Academic Press.

Almuktar, S.A.A.N., Abed, S.N., and Scholz, M. (2017). Recycling of domestic wastewater treated by vertical-flow wetlands for irrigation of two consecutive *Capsicum annuum* generations. *Ecological Engineering* 107: 82–98.

Anand, A., Chandan, P., and Singh, R.B. (2012). Homestays at Korzok: supplementing rural livelihoods and supporting green tourism in the Indian Himalayas. *Mountain Research and Development* 32: 126–136.

Arroyo, P., Ansola, G., and de Luis, E. (2010). Effectiveness of a full-scale constructed wetland for the removal of metals from domestic waste water. *Water, Air, and Soil Pollution* 210: 473–481.

Barbier, E.B. (2007). Valuing ecosystem services as productive inputs. *Economic Policy* 22: 177–229.

Beltrame, C., Cohen-Shacham, E., Trouillet, M., and Guillet, F. (2013). Exploring the links between local management and conservation applying the ecosystem services concept: conservation and tourism service in Camargue, France. *International Journal of Biodiversity Science, Ecosystem Services & Management* 9: 166–177.

Bikangaga, S., Picchi, M.P., Focardi, S., and Rossi, C. (2007). Perceived benefits of littoral wetlands in Uganda: a focus on the Nabugabo wetlands. *Wetlands Ecology and Management* 15: 529–535.

Boesch, D.F., Shabman, L., Antle, L.G. et al. (2006). *A New Framework for Planning the Future of Coastal Louisiana After the Hurricanes of 2005*, 48. Cambridge, MD: *University of Maryland Center for Environmental Science.*

Brandl, K., Baumgartner, C., Birkett, M., et al. (2011). Sustainable tourism and nature conservation. A report, SURF Nature Project. *European Regional Development Fund*, Brussels.

Bryce, R., Irvine, K.N., Church, A. et al. (2016). Subjective wellbeing indicators for large-scale assessment of cultural ecosystem services. *Ecosystem Services* 21 (Part B): 258–269.

Bucher, E.H., Bonetto, A., Boyle, T., et al. (1993). Hidrovia – an initial environmental examination of the Paraguay – Parana waterway. *Wetlands for the Americas Publication No. 10*, Manomet, MA.

Burkett, V. and Kusler, J. (2000). Climate change: potential impacts and interactions in wetlands of the United States. *Journal of the American Water Resources Association* 36: 313–320.

Cahoon, D., Day, J., and Reed, D. (1999). The influence of surface and shallow subsurface soil processes on wetland elevation: a synthesis. *Current Topics in Wetland Biogeochemistry* 3: 72–88.

Costanza, R., Perez-Maqueo, O., Martinez, M.L. et al. (2008). The value of coastal wetlands for hurricane protection. *Ambio* 37: 241–248.

Cronk, J.K. and Fennessy, M.S. (2009). Wetland plants. In: *Encyclopedia of Inland Waters* (ed. G.E. Likens), 684–690. Academic Press.

Davies, J. and Claridge, C.F. (1993). *Wetland Benefits: The Potential for Wetlands to Support and Maintain Development.* Asian Wetland Bureau Publication No. 87: IWRB Special Publication 27: Wetlands for the America Publication No. 11. Kuala Lumpur: Asian Wetland Bureau.

Davis, T.J. (1993). *Towards the Wise Use of Wetlands.* Gland: Ramsar Convention Bureau.

Depietri, Y., Renaud, F.G., and Kallis, G. (2012). Heat waves and floods in urban areas: a policy oriented review of ecosystem services. *Sustainability Science* 7: 95–107.

Do, Y., Ji, S.K., Kim, Y., and Joo, G. (2015). Wetland-based tourism in South Korea: who, when, and why. *Wetlands Ecology and Management* 23: 779–787.

Dugan, P.J. (1990). Wetland conservation – a review of current issues and required action. *IUCN – The World Conservation Union*, Gland.

FAO (2003). *Users Manual for Irrigation with Treated Wastewater.* Food and Agriculture Organization (FAO) of the United Nations. Cairo: FAO Regional Office of the Near East.

Feigin, A., Ravina, I., and Shalhevet, J. (2012). *Irrigation with Treated Sewage Effluent: Management for Environmental Protection*, vol. 17. Berlin: *Springer-Verlag.*

Ferrati, R., Canziani, G.A., and Moreno, D.R. (2005). Estero del Ibera: hydrometeorological and hydrological characterization. *Ecological Modelling* 186: 3–15.

Fish, R., Church, A., and Winter, M. (2016). Conceptualising cultural ecosystem services: a novel framework for research and critical engagement. *Ecosystem Services* 21: 208–217.

Fuller, G.A. (1988). The value of prairie sloughs. *Canadian Water Resources Journal* 13: 52–58.

Galletti, A., Verlicchi, P., and Ranieri, E. (2010). Removal and accumulation of Cu, Ni and Zn in horizontal subsurface flow constructed wetlands: contribution of vegetation and filling medium. *The Science of the Total Environment* 408: 5097–5105.

Gopal, B. (2015). Guidelines for rapid assessment of biodiversity and ecosystem services of wetlands, Version 1.0. *Asia-Pacific Network for Global Change Research (APN-GCR), Kobe, Japan, and National Institute of Ecology*, New Delhi, 134 p.

Gopal, B. and Junk, W.J. (2000). Biodiversity in wetlands: an introduction pages 1–10. In: *Biodiversity in Wetlands: Assessment, Function and Conservation*, vol. 1 (eds. B. Gopal, W.J. Junk and J.A. Davis). Leiden: Backhuys Publishers.

Halpern, B.S., Silliman, B.R., Olden, J.D. et al. (2007). Incorporating positive interactions in aquatic restoration and conservation. *Frontiers in Ecology and the Environment* 5: 153–160.

Hulme, P.E. (2005). Adapting to climate change: is there scope for ecological management in the face of a global threat? *Journal of Applied Ecology* 42: 784–794.

IPCC (International Panel on Climate Change) (1998). The regional impacts of climate change: an assessment of vulnerability. In: *A Special Report of IPCC Working Group II* (eds. R.T. Watson, M.C. Zinyowera and R.H. Moss). Cambridge: Cambridge University Press.

Jia, H., Ma, H., and Wei, M. (2011). Calculation of the minimum ecological water requirement of an urban river system and its deployment: a case study in Baijing central region. *Ecological Modelling* 222: 3271–3276.

Johnston, C.A. (1991). Sediment and nutrient retention by freshwater wetlands: effects on surface water quality. *Critical Reviews in Environmental Control* 21: 491–565.

Kakuru, W., Turyahabwe, N., and Mugisha, J. (2013). Total economic value of wetlands products and services in Uganda. *The Scientific World Journal* 2013: 1–13.

Karanja, F., Emerton, L., Mafumbo, J., and Kakuru, W. (2001). Assessment of the economic value of Pallisa District Wetlands, biodiversity economics programme for Eastern Africa. *IUCN-The World Conservation Union and Uganda National Wetlands Programme*, Kampala, Uganda.

Karanth, K.K. and DeFries, R. (2011). Nature-based tourism in Indian protected areas: new challenges for park management. *Conservation Letters* 4: 137–149.

Khurana, M.P.S. and Pritpal, S. (2012). Waste water use in crop production: a review. *Resources and Environment* 2: 116–131.

Kusler, J., Brinson, M., Niering, W., et al. (1999). Wetlands and climate change: scientific knowledge and management options. *White Paper Institute for Wetland Science and Public Policy*, Association of Wetland Managers/Wetland International, Berne.

La Baugh, J.W., Winter, T.C., and Rosenberry, D.O. (1998). Hydrologic functions of prairie wetlands. *Great Plains Research* 8: 17–37.

Leitch, J.A. and Hovde, B. (1996). Empirical valuation of prairie potholes: five case studies. *Great Plains Research* 6: 25–39.

Leon, A.S., Tang, Y., Chen, D. et al. (2018). Dynamic management of water storage for flood control in wetland system: a case study in Texas. *Watermark* 10: 325.

Leonard, L., Croft, A., Childers, D. et al. (2006). Characteristics of surface-water flows in the ridge and slough landscape of Everglades National Park: implications for particulate transport. *Hydrobiologia* 569: 5–22.

Li, H. and Cai, Y. (2002). Ecological risk assessment of flood disaster in major cities in Taihu basin. *Journal of Catastrophology* 17: 91–95.

Macharia, J.M., Thenya, T., and Ndiritu, G.G. (2010). Management of highland wetlands in central Kenya: the importance of community education, awareness and eco-tourism in biodiversity conservation. *Biodiversity* 11: 85–90.

Malabika, B.R., Pankaj, K.R., Nihar, R.S., and Asis, M. (2012). Socioeconomic calculations of wetland based occupations of lower gangetic basin through participatory approach. *Environment and Natural Resources Research* 2: 30–44.

Millennium Ecosystem Assessment (2005). *Ecosystems and Human Well-being: Wetlands and Water Synthesis*. Washington, DC: World Resources Institute.

Narayan, S., Beck, M.W., Wilson, P. et al. (2017). The value of coastal wetlands for flood damage reduction in the North eastern USA. *Scientific Reports* 7: 9463.

Nedkov, S. and Burkhard, B. (2012). Flood regulating ecosystem services – mapping supply and demand, in the Etropole municipality, Bulgaria. *Ecological Indicators* 21: 67–79.

Neelakantan, T.R., and Ramakrishnan, K. (2017). Protection of urban water body infrastructure – policy requirements. *Conference Series; Earth and Environmental Science*, 80.

Nicholls, R.J., Wong, P.P., Burkett, V. et al. (2007). Coastal systems and low-lying areas. In: *Climate Change 2007: Impacts, Adaptation and Vulnerability Contribution of Working Group II to the Fourth Assessment Report of the Intergovernmental Panel on Climate Change* (eds. M.L. Parry, O.F. Canziani, J.P. Palutikof, et al.), 315–356. Cambridge, UK: Cambridge University Press.

Pedrero, F., Kalavrouziotis, I., Alarcón, J.J. et al. (2010). Use of treated municipal wastewater in irrigated agriculture—review of some practices in Spain and Greece. *Agricultural Water Management* 97: 1233–1241.

Pemberton, C.A. and Mader-charles, K. (2005). Ecotourism as a means of conserving wetlands. *Journal of Agricultural and Applied Economics* 37: 463–474.

Penland, S. and Ramsey, K.E. (1990). Relative sea-level rise in Louisiana and the Gulf of Mexico: 1908–1988. *Journal of Coastal Research* 6: 323–342.

Picek, T., Čížková, H., and Dušek, J. (2007). Greenhouse gas emissions from a constructed wetland—plants as important sources of carbon. *Ecological Engineering* 31: 98–106.

Pueyo-Ros, J., Ribas, A., and Fraguell, R.M. (2016). Recreational uses and motivations of visitors in sea side wetlands of Costa brava (Spain). *International Journal of Sustainable Development and Planning* 11: 385–395.

Ramachandra, T.V., Alakananda, A., Rani, A., and Khan, M.A. (2011). Ecological and socio-economic assessment of Varthur wetland, Bengaluru (India). *Journal of Environmental Science & Engineering* 53: 101–108.

Ramsar Convention on Wetlands of International Importance Especially as Water fowl Habitat. Ramsar (Iran), 2 February 1971. UN Treaty Series No. 14583.

Ramsar-STRP (Scientific and Technical Review Panel of the Ramsar Convention on Wetlands). (2002). New guidelines for management planning for Ramsar sites and other wetlands. "Wetlands: Water. Life, and Culture." *8th Meeting of the Conference of the Contracting Parties to the Convention on Wetlands (Ramsar, Iran, 1971) Valencia*, 18–26 November 2002, Spain.

Ramsar-UNWTO. (2012). Destination wetlands: supporting sustainable tourism. *A Report Published Jointly by Secretariat of the Ramsar Convention on Wetlands (Ramsar, Iran, 1971) and the World Tourism Organization (UNWTO).*

Reed, D.J. (1995). The response of coastal marshes to sea-level rise: survival or submergence? *Earth Surface Processes and Landforms* 20: 39–48.

Rehm, B.W., Moran, S.R., and Groenewald, G.H. (1982). Natural groundwater recharge in an upland area of central North Dakota, U.S.A. *Journal of Hydrology* 59: 293–314.

Root, T.L., Price, J.T., Hall, K.R. et al. (2003). Fingerprints of global warming on wild animals and plants. *Nature* 421: 57–60.

Scholz, M. (2010). *Wetland Systems—Storm Water Management Control*. Berlin: *Springer Verlag*.

Shepard, C.C., Crain, C.M., and Beck, M.W. (2011). The protective role of coastal marshes: a systematic review and meta-analysis. *PLoS One* 6: 1–11.

Sighn, R. (2013). Studies of Hindu pilgrimage: emerging trends and bibliography. In: *Hindu Tradition of Pilgrimage: Sacred Space and System* (ed. R. Sighn), 7–14. New Dehi: Dev Publishers and Distributers.

Silvan, N., Vasander, H., and Laine, J. (2004). Vegetation is the main factor in nutrient retention in a constructed wetland buffer. *Plant and Soil* 258: 179–187.

Ström, L. and Christensen, T.R. (2007). Below ground carbon turnover and greenhouse gas exchanges in a sub-arctic wetland. *Soil Biology and Biochemistry* 39: 1689–1698.

Suna, F. and Carsona, R.T. (2020). Coastal wetlands reduce property damage during tropical cyclones. *Proceedings of the National Academy of Sciences of the United States of America* 117: 5719–5725.

Tsihrintzis, V.A. and Madiedo, E.E. (2000). Hydraulic resistance determination in marsh wetlands. *Water Resources Management* 14: 285–309.

USGCRP (US Global Change Research Program) (2000). *Climate Change and America: Overview Document. A Report of the National Assessment Synthesis Team.* Washington, DC: *US Global Change Research Program.*

Vacca, G., Wand, H., Nikolausz, M. et al. (2005). Effect of plants and filter materials on bacteria removal in pilot-scale constructed wetlands. *Water Research* 39: 1361–1373.

Van der Kamp, G. and Hayashi, M. (1998). The ground water recharge functions of small wetland in the semi arid Northern prairies. *Great Plains Research: A Journal of Natural and Social Sciences* 366: 39–56.

Verschuuren, B. (2007). Sociocultural importance of wetlands in northern Australia. In: *Conserving Biological and Cultural Diversity: The Role of Sacred Natural Sites and Landscapes* (eds. T. Schaaf and C. Lee), 141–150. Paris: *United Nations Educational Scientific and Cultural Organization (UNESCO).*

Vymazal, J. (2014). Constructed wetlands for treatment of industrial wastewaters: a review. *Ecological Engineering* 73: 724–751.

Walton, M.E.M., Samonte-Tan, G.P.B., Primavera, J.H. et al. (2006). Are mangroves worth replanting? The direct economic benefits of a community-based reforestation project. *Environmental Conservation* 33: 335–343.

Weishampel, P., Kolka, R., and King, J.Y. (2009). Carbon pools and productivity in a 1-km^2 heterogeneous forest and peatland mosaic in Minnesota, USA. *Forest Ecology and Management* 257: 747–754.

Wetlands Management Department. (2009). Ministry of water and environment, Uganda Bureau of Statistics, International Livestock Research Institute, and World Resources Institute. *Mapping a Better Future: How Spatial Analysis Can Benefit Wetlands and Reduce Poverty in Uganda*, World Resources Institute, Washington, DC.

Winter, T.C. (1989). Hydrologic studies of wetlands in the Northern prairies. In: *Northern Prairie Wetlands* (ed. G. Van der Valk), 17–54. Ames: Iowa State University Press.

Yu, D.Y., Xun, B., Shi, P.J. et al. (2012). Ecological restoration planning based on connectivity in an urban area. *Ecological Engineering* 46: 24–33.

4

Ecological and Societal Importance of Wetlands

A Case Study of North Bihar (India)

Alvia Aslam[1], *Purushothaman Parthasarathy*[2], *and Rajesh Kumar Ranjan*[1]

[1] *Department of Environmental Science, Central University of South Bihar, Gaya, Bihar, India*
[2] *Department of Civil Engineering, College of Engineering and Technology, Faculty of Engineering and Technology, SRM Institute of Science and Technology, Chengalpattu, Tamil Nadu, India*

4.1 Introduction

Wetlands have been known to be transitional ecosystems interlinking the terrestrial systems to aquatic systems. They possess the unique character of self-sustenance due to frequent interaction with the surroundings (Cowardin et al. 1979). These unique wetland systems, apart from performing the balancing functions, also house huge diversity of flora and fauna (Yim and Ni 1998). The major services provided by them include recharging groundwater, nutrient absorption, acting as a sink for pollutants, maintaining hydrological balance, flood control, and maintaining the trophic structure of various species of flora and fauna (Ward et al. 2002) (Figure 4.1).

India, marked by diverse topographical and climatological features, has given rise to different types of water bodies – some fresh and others saline (Prasad et al. 2002). Wetlands account for 1.5% of the total land area in India which accounts to 15.26 million hectares (ha). Out of this, Bihar accounts for about 4.03 million hectares of wetland area (Kumar et al. 2013). The Gangetic plains of North Bihar are drained by the interconnected confluence of rivers originating from the Himalayas and, thereby, give rise to a number of floodplain wetlands and oxbow lakes. The intricately linked mesh of rivers, floodplain wetlands, and oxbow lakes supports thriving of various species of flora and fauna due to frequent exchange of water, sediments, and nutrients and promotes greater stability (Kumar and Ambastha 2016). The geomorphology of North Bihar created with complex fluvial exchanges over the course of time has also led to occurrences of shallow depressions and meanders known locally as chaurs, taals, mauns, etc. (Table 4.1). These are heavily deluged during the monsoon and dry up during the months from March to June. The seasonal nature of these wetlands increases their importance in the context of water availability and ground water recharge with a bonus of livelihood opportunities they provide (Kumar and Ambastha 2016). About 4.96% of the geographical area spanning over 21 districts of North Bihar accounts for 269 418 ha of the wetland region (Jha et al. 2014). Some prominent

Wetlands Conservation: Current Challenges and Future Strategies, First Edition.
Edited by Sanjeev Sharma and Pardeep Singh.

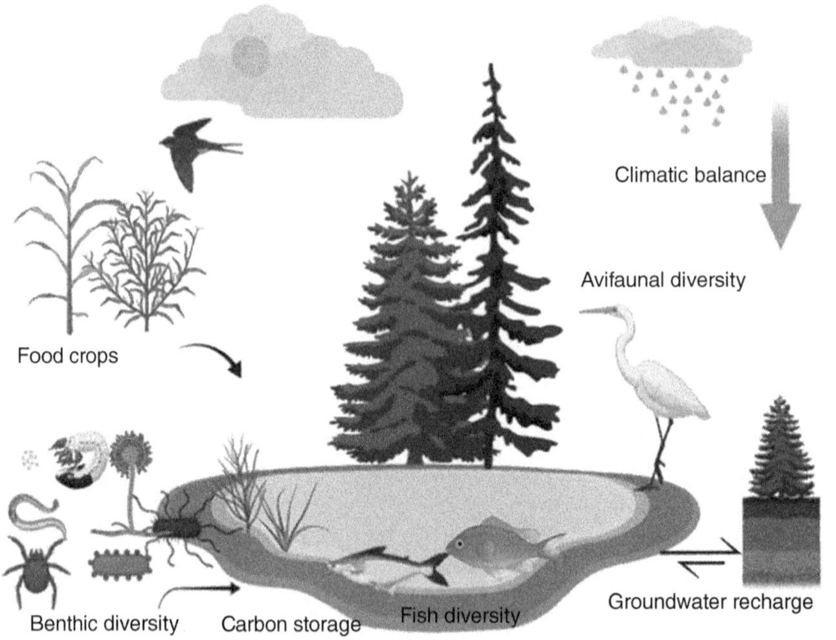

Figure 4.1 Major services provided by the wetlands. *Source:* Based on Ward et al. (2002). © John Wiley & Sons.

regions in North Bihar harboring various wetlands are East and West Champaran, Darbhanga, Vaishali, Begusarai, Muzaffarpur, Samastipur, etc. (Prasad et al. 2020). The major wetlands of North Bihar are Kabartal (Begusarai), Kusheshwar Asthan (Darbhanga), Baraila (Vaishali), Moti jheel (East Champaran), and Gogabeel (Katihar). These wetlands will be discussed further in this review as they are significantly associated with the livelihood, ecosystem services, sustainability of groundwater resources, biodiversity, and flood control. The hydrological regimes of North Bihar are extremely important to assess in order to come up with future restoration policies, regulating frameworks, and ensuring water security for more than six crore human population of the North Bihar region (Bihar State Profile 2016).

The bowl-shaped physiography of North Bihar has led to the development of many permanently waterlogged areas and causes frequent flooding episodes (ZSI 2011). Even though many embankments have been built throughout the postindependent era (3500 km), still there has been a 2.5 times increase in the flood occurrences compared to pre-embankment times. This has posed a greater threat on the sustainability and long-term prosperity of this region which is constantly undergoing fast-track changes induced naturally or unnaturally altering the overall hydrological regime of this region (ZSI 2011). Many natural and manmade ponds called pokharis exist throughout the Bihar floodplain which caters to the irrigational needs of the farmers and is advantageous to fish-rearing groups of fishermen/farmers. However, with rapid urbanization, population load, and excessive pollution, farmers are facing tough times securing their livelihood and covering

Table 4.1 Lithographic and geographic details of the North Bihar plains.

Parts of North Bihar	Age	Basin interfluves	Districts	Geography	Geology and soil types
North-eastern area	Quaternary	Gandak-Kosi interfluve and Gandak-Kosi-Mahananda interfluve	Purnea, Katihar, Saharsa, Madhepura, Araria, Kishanganj, Supaul, Khagaria, Begusarai	Kosi fan and in the Ganga-Kosi interfluve flood plain	Soil alluvium deposits (sand, clay-silt fragments) Light yellow fine sand with clay/silty clay; course to medium textured; micaceous
Mid-northern part	Quaternary	Gandak-Kosi interfluve	Madhubani, Darbhanga, Samastipur, Muzaffarpur, Sitamarhi, Vaishali	Bhangar floodplain (older alluvium group) and khaddar (newer alluvium group)	Unoxidized to feebly oxidized grey to pale yellowish-grey silty clay (Aeolian) Soft ferruginous concretions and rare kankars (Kosi) and pedocal Palaeosol (Gandak); noncalcareous and nonsaline; medium to heavy textured
North-western part	Tertiary	Ganga-Gandak-Kosi interfluve	Bettiah, Motihari, Gopalganj, Siwan, Seohar, east Champaran, west Champaran	North Champaran Hills (upper Siwaliks); Tarai belt and floodplains	Sandstones and clay stones, hilly zone and sandy to loamy in texture, and brown to yellow and grey-olive green in colour.

Source: Adapted from State of Environment Report (2007). © John Wiley & Sons.

drinking water needs. Despite being a water-surplus region, potable drinking water has been a big issue. Prolonged land-use practices, agricultural exploitation, dredging, sedimentation, and logging are also some of the human activities which have profoundly disturbed the ecological character of the region and destroyed the genetic pool of organisms (Ranjan and Kumari 2018). This leads to disruption of smooth functioning of wetlands and loss of ecological value. The whole Ganga and Brahmaputra floodplain has been the recipient of massive human interference which has created a huge anomaly in the environmental balance and climatic changes are rampantly being witnessed in today's scenario.

4.2 Geographical and District-Wise Distribution of Wetlands in North Bihar

Bihar is the twelfth largest state as far as the geographical extent is concerned (94 163 km^2) but is the third largest from the perspective of the population (10.38 crores) as per the census of 2011. The geographical position of North Bihar (31°55′ to 27°3′ latitude and 83°20′ to 88°17′ longitude) is advantageous for the people of the region as it receives an adequate quantity of rainfall and surface water from the channels of major rivers and its tributaries flowing through it (Jha and Chandra 1997). The north-western region receives an average rainfall of 1234.7 mm whereas the north-eastern floodplains receive 1382.2 mm of rainfall (SAPCC 2015). The total number of wetlands in the districts falling in the Ganga basin sum up to be 5183 including all types of lakes, ponds, and streams and water-logged areas. Rivers and streams constitute the largest and most dynamic components of such water bodies. In another major finding, the National Wetland Atlas of Bihar (2010) reported that 4416 wetlands cover about 46 828 ha of the area with 130 of them having area greater than 100 ha. A significant number of wetlands (~17 582), mostly including chaurs, ponds, and waterlogged regions were less than 2.25 ha in size (The Wetland Report 2012) (Table 4.2). Most of them are seasonal, being fed by overflow from rivers and monsoon rains (Sinha 2011).

The region is deluged by diverse rivers like Ganga, Kosi, Gandak, Burhi Gandak, Kamla-Balan, Baghmati, and Kareh to name a few among others.

Northern Plain is a major physiographic unit of the Indian landmass, covering about 56 980 km^2 of area. The Ganga basin in Northern Bihar is characterized by a flat terrain comparatively with slope tilting toward the southern and South-East direction. It has been reported that the basin was formed during the late paleogene-neogene period. The quadrilateral structure restrained by the piedmont belt is then followed by water-surplus surface, swamps, depressions, and floodplains, which, ultimately, cause the alluvial fans to be formed through fluvial deposition and sedimentation (Ghosh et al. 2004). The region's fluvial geomorphology from west to east is markedly defined by the Ghagra-Gandak interfluve, the Gandak-Kosi interfluve, and the Kosi Fan Belt (Ghosh et al. 2004).

The North Bihar plain falls under the tropical monsoon-type climatic zone with long hot summers and short cool and pleasant winters with few months of monsoon during late June to September (Singh and Sinha 2020). The summers commence from late March and

Table 4.2 District-wise distribution of waterlogged areas in north Bihar.

S. No.	District	Waterlogged area (thousand hectares)	Wetland area (ha)	% of total wetland area in north Bihar
1.	Araria	26.885	4157	1.60
2.	Begusarai	2.908	20 365	7.85
3.	Bhagalpur	12.033	24 171	5.99
4.	Darbhanga	20.251	8709	3.36
5.	East Champaran	34.394	12 477	4.81
6.	Gopalganj	21.692	7122	2.75
7.	Katihar	38.404	31 011	11.95
8.	Kishanganj	12.98	10 954	4.22
9.	Madhepura	14.222	3539	1.36
10.	Madhubani	11.486	8958	3.45
11.	Muzaffarpur	32.692	10 490	4.04
12.	Purnia	36.3	12 401	4.78
13.	Saharsa	9.277	12 086	4.66
14.	Samastipur	18.074	15 022	5.79
15.	Saran	32.34	21 170	8.16
16.	Sheohar	2.004	1476	0.57
17.	Sitamarhi	4.947	2601	1.00
18.	Siwan	23.048	7105	2.74
19.	Supaul	24.707	19 285	7.43
20.	Vaishali	34.495	17 148	6.61
21.	W. Champaran	22.127	21 697	8.36
	Total	**435.266**	**271 944**	**101.48**[a]

[a] Bhagalpur (south of Ganga) has been added to the list. So, there is an increase in the % area.
Source: Adapted from National Wetland Atlas of Bihar (2010) and Jha et al. (2014). © John Wiley & Sons.

find a peak in May and June where the temperature reaches up to 45 °C. The cold months set in from November and stay till February with the minimum temperature dipping to around 3–4 °C (ZSI 2011). The region also witnesses dust and thunderstorms and heat waves (8–16 km/h) locally called "loo" during hot months (ZSI 2011).

The areas under wetland vary from one district to another. The area varies from 2.71% (East Champaran) to 11.43% (Vaishali). Bihar saw a total of 34% reduction in water levels during the post-monsoon to pre-monsoon lapse in Bihar wetlands (Bassi et al. 2014). The decline in water level during summer also varies with the maximum recession seen for Bhagalpur while lowest for Muzaffarpur (The Wetland Profile 2012). There are many important waterlogged regions in North Bihar which have not been able to catch the eye of the government. According to the report of Chowdhury et al. (2018), Saran district has the largest span of waterlogged area (37 489 ha) including inland wetlands, rivers/

streams/canals, reservoirs/lakes/ponds, followed by Vaishali, West Champaran, Supaul, Bhagalpur, and Katihar with 37 198, 32 968, 32 307, 30 543, and 23 556 ha, respectively. Vaishali (21 401 ha) has the highest area under inland wetlands followed by Purnea (19 153 ha), Siwan (16 264), and East Champaran (15 792 ha) districts. Some districts have large areas under reservoirs/lakes and ponds like Darbhanga (1595 ha), Madhubani (931 ha), East Champaran (917 ha), and Muzaffarpur districts (833 ha) (Chowdhury et al. 2018). Some important wetlands of North Bihar are briefly discussed below (Table 4.3).

4.2.1 Kabartal

It is one of the most important wetlands of Bihar famous for its waterfowl. It is located between 25° 37′N and 86° 08′E in the Begusarai district of Bihar (Figure 4.2) (Box 4.1). The lake is surrounded by the river Burhi Gandak and spreads over 6000+ ha of area (Kumar 2013). It serves as a major habitat for many resident and migratory waterfowls and other avifauna (Box 4.1). Kabartal has long been providing recreation, tourism, fisheries, irrigation, and domestic water supply services (Jain et al. 2007a,b). Due to its diverse importance recently Kabartal Wetland has been designated as one of the Ramsar sites (no. 2436) making it to the list of "Wetlands of international importance" (Ramsar 2020). Traditionally, the wetland has been used for water supply for irrigation and domestic purposes, fishing, netting of migratory waterfowl for sale, harvesting of wild rice, and gathering of the edible mollusk, *Pila globosa*. The net sown area is very high (Kumar 2013). It is almost 60% of the total land resource. The major crops grown in the villages are rice, maize, and wheat. Paddy, wetland paddy, and sugarcane are cultivated during the Kharif season in the villages near the wetland.

4.2.2 Baraila Jheel

Baraila wetland is a prominent floodplain marsh spanning about 1625.34 ha of surface area. It is located from 25°57′N to 25° 36′N latitudes and 87° 53′E to 87°25′E longitudes in the Vaishali district (Figure 4.2). According to the official gazette notification of the government, it has been declared as a protected area under the National Wetland Conservation and Management Programme (NWCMP).

Baraila jheel, or taal as commonly called, is a haven for a variety of fishes, crabs, snails, gastropods, and amphibians and thus plays a great role in defining the biodiversity of the area and balancing the trophic structure. The eight villages on the periphery of the wetland area also come under the eco-sensitive zone. The recent management plan developed for Baraila jheel aims at improving the lives of the local population and developing the area for ecotourism apart from restoration (WII 2017). The few patches of the area (about 2 km^2) exclusively belonging to the state government have been declared as the bird sanctuary named Salim Ali Jubba Sahni Bird Sanctuary. This wetland too provides food, fuel, and fodder requirements of the local population apart from working as a major flood mitigation system.

Table 4.3 General physiography and land-use patterns of the important wetlands of North Bihar.

Name of wetland and district	Associated rivers	Location	Area (ha)	Degradation activities	Effects on the environment
Kabartal (Begusarai)	Burhi Gandak	25° 37′ 00″ N, 86° 08′ 00″ E	6311	River erosion	Infertility of soil, habitat loss
Baraila jheel (Vaishali)	Baya, Noon	25° 44′ 57″ N, 85° 35′ 22″ E	1625.34	Siltation	Wetland extinction, flooding, loss of biodiversity
Moti jheel (East Champaran)	Gandak	26° 38′00″ N, 84° 55′00″ E	130		Nutrient imbalance, loss of biodiversity, loss of livelihoods, water imbalance
Gogabeel (Katihar)	Gogabil	25° 24′ 02″ N, 87° 45′ 11″ E	88	Overexploitation of wetland resources	
Kusheshwar Asthan (Darbhanga)	Kamla-Balan, Bagmati and Karree	26° 10′ 00″ N, 86° 02′ 30″ E	2932	Urbanization	Ecological imbalance, global warming effect, pollution and loss of biodiversity
					Environmental pollution, land degradation
				Mining and industrialization	Toxicity and contamination of soil and water
					Biodiversity loss, diseases
				Agricultural expansion (overuse of fertilizers and pesticides)	
				Construction of dams and embankments	Extinction of wetlands, erratic floods, imbalance of hydrological regime

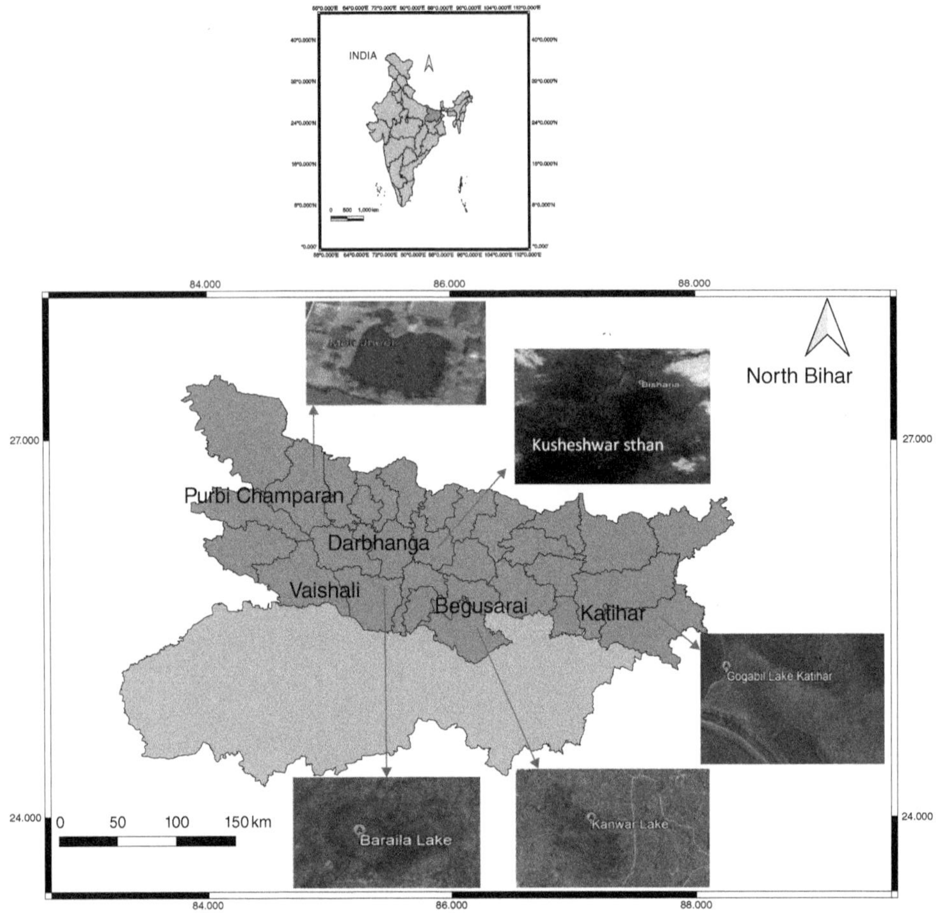

Figure 4.2 Some major wetlands of North Bihar.

4.2.3 Kusheshwar Asthan

Kusheshwar Asthan wetland is a confluence of many big and small lakes (16) fed by river water. The main area is located at 26° 10′ 00″ N and 86° 02′ 30″ E in the Darbhanga district of Bihar (Figure 4.2). The largest lake of this wetland system is Larail chaur (more than 600 ha) and the smallest one is Mahisath (40 ha). This wetland recognized as a bird sanctuary since the pre-independence period is supposed to be the second-largest protected area. The water from nearby rivers, namely Kamla, Bagmati, and Kareh are the main contributors to this low-lying mesh of lakes. The wetland becomes one whole system of the waterbody (more than 10 000 ha) when several chaurs overflow and join with Simri jheel and Kabartal (Yahya 1995). The main domestic usage of this wetland includes fishery and agriculture.

4.2.4 Jagatpur Wetland

Jagatpur wetland is an important bird sanctuary, 12 km from the Bhagalpur city in the district of Bhagalpur (Kumar and Choudhary 2010). Though this wetland is located at the

Box 4.1 Ramsar Site (Kabartal): The Lost Glory

Kabartal is one of the most important wetlands of North Bihar which is spread across 7400 hectares of area in the Begusarai district. It is reported to be one of the largest freshwater lakes of south Asia. Apart from being a dynamic habitat to a wide variety of flora and fauna, it provides life-sustaining services like irrigation water, food, and fodder. Kabartal was also declared as a protected area under the Act of protected area. It was also provided with the tag of an important bird area because of large avifaunal diversity. In addition, Kabartal is one of the most researched water bodies of Bihar also because it is rapidly losing its charm and urgently needs restorative actions. Despite being declared as a protected area by the Bihar government and having a bird sanctuary status, it is overzealously abused and exploited for food, fodder, fuel, and raw materials. It has become a recipient of domestic pollutants from nearby areas and the water level has been rapidly receding over the last few decades. The wetland has lost areas at an alarming rate. From having an area of 6786 ha in 1984 to 6043 ha in 2002 and then being reduced to 2032 ha in 2012, the area is undergoing shrinking at a rapid pace of 20 ha every year (Kumar 2013). The wetland is choked due to eutrophication and the water has turned acidic and turbid. The average DO reported in various studies is 7.6 mg/L. The poaching and expansion of the human population in and around the lake have caused loss of habitat and avifaunal diversity in recent past decades. Uncontrolled agricultural practices (paddy cultivation in protected areas), constructing structures like railway embankments and dykes around rivers have greatly turned some parts of land barren and vegetation less (Ghosh et al. 2004). Careless usage of pesticides like DDT, Aldrin, Endosulfan, etc., has rendered the water and sediments highly toxic. The quantity and quality of vulnerable benthic communities residing at the transitional zone of water and sediment bed have greatly been altered. This has greatly compromised the sustainability and resilience of the ecosystem. In the long run, people may have to pay huge ecological prices indirectly.

south of the main Ganga channel (25° 20′ 219″ N, 87° 02′ 623″E), it is a very important wetland owing to its proximity to the Ganga dolphin sanctuary (Kumar and Choudhary 2010). It is a perennial freshwater floodplain with an area of 0.4 km^2. The main source of water includes monsoon rains and underground seepage. The area has private as well as government rights (Kumar and Choudhary 2010).

4.2.5 Moti Jheel

Moti jheel is one of the important oxbow lakes in the district of East Champaran of Bihar. It is situated at 26° 38′ N, 84° 55′ E (central coordinates) (Figure 4.2). Moti jheel divides Motihari town into two parts. The formation appears to be the part of Dhanuati River passing very close to it which had meandered. The region has several oxbow lakes which were cut off from the main river due to channel migration and simultaneous fluvial processes. The area is about 130 ha in size and provides shelter to large varieties of migratory and resident bird species. The lake is filled by overflowing Gandak River during monsoon and rain water. However, domestic effluents from the town find their way into the wetland.

4.2.6 Gogabeel Pakshi Vihar

Gogabeel Pakshi Vihar in the district of Katihar (25°24′02″ N latitude and 87°45′11″ E longitude) is one of the important wetlands in North Bihar (Figure 4.2). This is an oxbow lake formed due to the meandering of Mahananda River and Kankhar River in the north and Ganga River in the south and east. It is an important bird-watching site. Recently, in November 2018, it has now entered the spot of the 15th protected areas of Bihar. It has been developed into the first-ever community reserve and conservation reserve of Bihar with the help of the state coordinator of the Indian Bird Conservation Network (IBCN) and the support of the local community. Gogabeel (88 ha) had initially been notified as a closed area in 1990 which was later revised till 2000. The lake is surrounded by many temporary chaurs and gives rise to two more important chaurs, namely Bhagar Beel and Baldia chaur which are seasonal and shallower than Gogabeel.

4.3 Wetlands: Promoters of Sustainable Livelihood and Services

Wetlands have had their own set of struggles in the past which were quite different from today's scenario. Before the 1970s, wetlands were considered as wastelands and their drainage and conversion for cultivation purposes were not only a common practice but also encouraged by certain sections of communities and even governments. The properties, services, and functions of the wetlands were not clearly understood and were considered a breeding ground for mosquitoes and unwanted plants. Especially, such thinking in the country, particularly in Bihar, is still engrained where wetlands are not recognized as important ecological systems. However, the wetlands provide numerous ecological and material goods and services including water for irrigation, fisheries, forest produce, and recreation. Some indirect services that have huge repercussions on the global level include carbon storage, flood control, biodiversity sustenance, groundwater recharge, and nutrient absorption and removal, and water filtration among others (Turner et al. 2000; Bassi et al. 2014).

The irrigation department under the Bihar Wasteland (Reclamation, Cultivation and Improvement) Act, 1946 has been responsible for draining, ditching, and filling the depression areas and this pattern has caused a significant amount of chaurs to vanish from the northern belt of Bihar (Kumar 2019). Despite few researchers highlighting the beneficial functions and properties of wetlands, the positive restorative steps taken are too miniscule to have a significant impact on the ecological state of these wetlands.

The northern plains of Bihar are endowed with highly sensitive ecosystems which are rich in floral and faunal diversity and, in turn, benefit hundreds of people with employment locally (Jha and Chandra 1997). The oxbow lake like Bhagar Beel alone helps more than 60 farmers to procure fishes. The daily average harvesting of fishes ranges from 200 kg in summer to 300 kg in monsoon (Prasad et al. 2020). However, with passing time and increased pressure and conflict between the demand for resources and conservation needs, the socioeconomic lives of the local population depending on local employment have been extremely pathetic (Choudhary et al. 2006). The region has been a hub of enviro-socio-political discords where feudal conflicts, environmental uncertainties, and human intervention have taken a toll on the ecological health of the Gangetic basin (Kelkar and Krishnaswamy 2010).

Most of the people dependent on water resources make their living through fishery, agriculture, navigation, irrigation, and cattle rearing (Tiwary et al. 2009; Sinha 2011). The wetlands, namely Kabartal and Gogabeel jheel are ecologically regulated by freshwater organisms. Crabs like *Paratelphusa spinigera* are most commonly available. They happen to hold an important position in the trophic status and are a promising source of sustainable livelihood to villagers where crabs can be advantageously produced, cultured, and marketed (Tiwary et al. 2009). The fish demand remains incessantly high despite high production. The annual production of 4 lakh metric tonnes of fishes is in deficit compared to the demand of 5.8 metric tonnes annually (Sinha 2011).

In an economic initiative taken in the district of Samastipur, a 110-acre complex was to be developed for the management and boosting up of fisheries, dairy, duckery, and horticulture components. 44 tanks spanning an area of 75 acres has been deemed operational under this strategy. Similarly, another integrated project including hatchery, dairy, horticulture, biogas, and vermicompost components has been given a green signal and is being developed in the districts of Sitamarhi (211 acres) and Madhubani (83 acres) (Sinha 2011). Cultivation of gorgon/foxnut (*Euryale ferox Salisb.*) and similar nutritious water crops hold promising sustainable employment opportunities for locals of Darbhanga, Kosi, and Purnea divisions of North Bihar (Sinha 2011).

The district of East Champaran has been a rich source of Oyster shells of *Parreysia* sp. collected from the nearby Sikarahana River which has played a prominent role in reinforcing the pearl button industry (Kumar and Singh 2013). About 67% of people from the vicinity depend on commercially important raw materials and subsistence goods like food, fodder, firewood, fiber, medicine, construction items, etc. (Cavendish 2000; Bluffstone et al. 2001; Kumar 2013).

The survey around Kabartal area has revealed that 20% of the wetland is put to direct consumptive use in the form of fishing. Out of 160 homes in the villages, 31 opt for fishing activity. Others harvest macrophytes (shoot of *Commelina benghanlensis*, *Cynodon dactylon spp.*) for fodder and grazing purposes (Kumar 2013). Overall, these wetland systems have created a dynamic trophic structure that promotes and sustains the livelihood of more than 4.9 million fishermen of the state (Kumar and Ambastha 2016).

4.4 North Bihar Wetland Biodiversity: Status and Role

The Gangetic plains of northern Bihar and the associated river-wetland system are quite vulnerable to climatic changes and also extremely diverse. The variations in the water level and pulsing of flood waters in and out create a unique structural mesh that supports all kinds of flora and fauna (Singh et al. 2007; Dey et al. 2014). Also the soil and sediment of the North Bihar basin and surrounding regions are extremely fertile and thus are a rich source of plant- and animal-based resources. The wetlands primarily promote biodiversity in the form of various phytoplanktons, zooplanktons, gastropods, macrophytes and fishes. In this brief review, we will be discussing some important diversity of fauna of Gangetic plains, macrophytes, phytoplanktons, zooplanktons, fishes, and birds found in the wetlands of North Bihar (Table 4.4).

Table 4.4 Major biodiversity (species distribution) in North Bihar wetlands.

Macrophytes		**No. of species**	**% composition**
Groups	Dicotyledons	91	66.42
	Monocotyledons	42	30.66
	Thallophyta	2	1.46
	Algae and Pteridophyta	2	1.46
	Total	**137**	
Zooplanktons			
Groups	Rotifera	25	49.00
	Cladocera	18	35.30
	Copepods	8	15.70
	Total	**51**	
Molluskans			
Groups	Gastropods	15	68.2
	Pelecypoda	7	31.8
	Total	22	
Fishes	**Total**	**48**	
Avifauna			
Groups	Waterbirds	39	54.2
	Terrestrial	33	45.8
	Total	**72**	

Critically endangered	**Scientific name**
Oriental white-backed vulture	*Gyps bengalensis*
Long-billed vulture	*Gyps indicus*
Endangered	
Greater adjutant	*Leptoptilos dubius*
Vulnerable	
Lesser adjutant	*Leptoptilos javanicus*
Pallas' fish eagle	*Haliaeetus leucoryphus*
Greater spotted eagle	*Aquila clanga*
Lesser kestrel	*Falco naumanni*
Swamp francolin	*Francolinus gularis*
Sarus crane	*Grus antigone*
Sociable lapwing	*Vanellus gregarius*
Indian skimmer	*Rynchops albicollis*

Source: Adapted from ZSI (2011) and Adhishwar and Choudhary (2014). © John Wiley & Sons.

The water bodies of North Bihar have overall been infested with a diverse variety of macrophytes. Some of them have been of immense advantage as far as services and employment are concerned while others have been triggering nuisance and cause of increasing eutrophication (Sinha 2011; Jha 2012). Macrophytes have been suggested to indicate the extent of organic and chemical pollution in a water body (Tripathi and Shukla 1991). Its structure and spread hint about the species' richness, climate, and quality of ecosystem sustaining in an area (Sharma et al. 2007; Adhishwar and Choudhary 2014). The Gogabeel lake of Katihar is found to be rich in species' diversity documenting 137 macrophyte species grouped under marginal (74), submerged (11), floating (13), and emergent (39) categories (Adhishwar and Choudhary 2014). Further, it was found that dicots were the most dominant category with 91 species and covered about 66.42% of the total species enlisted. Monocots followed with 30.66% with 42 species. Pteridophyceae lagged far behind with only 2 species being notified. Poaceae family was the most dominant (14 species) followed by Asteraceae and Cyperaceae (10 species each). The wetland is dominant in emergent and marginal species like *Eichhornia crassipes* (free-floating), *Hydrilla verticillata* (submerged), *Cyperus rotundus* (emergent), and *Cynodon dactylon* (marginal). In this wetland, the dispersion of macrophytic communities is not segregated by any clear boundary or division but is interspersed with no clear territorial demarcation (Devi and Sharma 2007). Some submerged varieties dependent on light and dominant in shallow waters were found in decent numbers. In addition, some noticeable species like *Ceratophyllum demersum*, *Utricularia stellaris*, *Urena* sp., *Ottelia alismoides*, etc., were also reported (Adhishwar and Choudhary 2014). Rooted emergent varieties were also witnessed in shallow waters, e.g., *Nelumbo nucifera*, *Lemna minor*, *Spirodela polyrrhiza*, and *Azolla pinnata*. The dry upland region of the wetland was populated by some emergent species like *Eclipta alba*, *Phyla nodiflora, etc.* (Adhishwar and Choudhary 2014). In Kabartal, sedges like *Eleocharis plantaginea* are frequently noticed all along the peripheries of the wetland among the dominant form of emergent vegetation. Among submerged varieties, *Vallisneria spiralis* and *Potamogeton pectinatus* have also been documented. Dominant floating species include *Pistia stratiotes* and *Lemna minor* among others (Shardendu et al. 2012; Irfan 2015). Some macrophytic species are of great consumptive values and are used as food, like *Ipomoea aquatica* (as young leaves), *Euryale ferox* (as seeds), and *Nelumbo nucifera* (as seeds). Some fodder-relevant species are *Commelina benghalensis*, *Cynodon doctylon*, *Oryza rufipogon*, etc. Species used and marketed as fuel are *I. aqatica* and *S. spontaneum* and important medicines are derived from the leaves of *Centella asiatica*, rhizomes/flowers of *N. nouchali* (Kumar 2013). A particular type of plant known as Chikota, found about three feet under the wetland, is good for the eyes. Other wetlands like Jagatpur Pakshi Vihar of Bhagalpur have been found with similar patterns of vegetation. The tropical trees throng the nearby wetland areas with common sightings of *Mangifera indica*, *Ficus religiosa*, *F. benghalensis*, *F. glomerata*, *F. infectoria*, *Dalbergia sissoo*, *Acacia nilotica*, *Eugenia jambolana*, *Borassus flabellifer*, *Phoenix dactylifera*, etc. The upland areas are hugely cultivated for food crops (Kumar and Choudhary 2010). In Kusheshwar Asthan wetlands, 50–90% of the area of some *chaurs* is covered by invasive species, namely *Eichhornia* and *Hydrilla verticillata*. Among marginal species, *chara* was the most commonly sighted (Jha and Chandra 1997). Baraila jheel of the Vaishali district in North Bihar also struggles with periodic infestations of *Eichhornia crassipes*, *Typha angustifolia*, *Ipomoea aquatica*, *and Ipomoea fistulosa*, etc. *Sesbania bispinosa* is cultivated to provide for the fodder requirement. The highly fertile floodplain of Baraila jheel is ideal for growing large varieties of food crops like wheat, maize, carrot, vegetables, potato, rice, tobacco, etc., among others.

Phytoplanktons play an important role in determining and balancing the ecological health of water bodies. A slight variation in their community structure can trigger a cause-and-effect cycle which leads to eutrophication and imbalance in the trophic structure. In the study done by Shekhar et al. (2008) on the Ganga River stretch from Munger to Manihari, 57 genera and 158 species of phytoplankton have been identified. Although this stretch falls toward the periphery of the southern part of Bihar, the river channel is still significant for the study of phytoplanktons. The prominent families witnessed in the region were *Chlorophyceae*, *Bacillariophyceae*, *Cyanophyceae*, and *Euglenophyceae*, with *Bacillariophyceae* outnumbering the other two families (69 species, 20 genera). The water of Kusheshwar Asthan too is defined by the algal distribution of *Cyanophyceae*, *Chlorophyceae*, and *Bacillariophyceae* (Krishna and Sinha 2014). However, the population and dominance varied depending upon the seasons and chemical composition at that time. Few researchers found the predominance of *Chlorophyceae* at a certain time of the year while others observed dominance of *Cyanophyceae* at other times (Vyas 1968; Munawar 1974; Kant and Anand 1978). *Cyanophyceae* dominance hints toward high nitrate content in the water. Variations in the population dominance can indicate variation in locations of sampling owing to changing pollution status (Kant and Kachroo 1975; Rai 1978; Kumar 2013). The samples from this site indicate occurence of organic pollution and presence of tolerant genera like *Oscillatoria*, *Scenedesmus*, *Chlorella*, *Spirulina*, and *Anabaena*. Some of the useful taxa frequently used for bioremediation experiments and purposes, such as *Anabaena*, *Chlorella*, *Fragilaria*, *Nostoc*, etc., were also noticed (Krishna and Sinha 2014).

In another study undertaken by Jha and Chandra (1997), the phytoplankton community was found to be rich in diversity and population at Mahisath (3317–4725 μ/l), Dahbhadi (3525–3841 μ/l), and Amaldaha (2930–3262 μ/l) chaurs of Kusheshwar Asthan. Larail chaur was however less loaded (1071–1927 μ/l). The phytoplankton distribution documents the dominance of green algae in all chaurs as *Chlorophyceae* > *Bacillariophyceae* > *Mysophyceae* > *Dinophyceae*. Except Kamaldaha, where blue-green algae outnumbered the green algae community with *Myxophyceae* > *Chlorophyceae* > *Bacillariophyceae* > *Dinophyceae*.

The study taken up by Zoological Survey of India (ZSI 2011) reveals that there is a good diversity of zooplanktons in the wetland water of North Bihar. Three prominent groups, namely *Rotifera*, *Cladocera*, and *Copepoda* have been identified. Out of the 51 species studied, 25 were rotifers, 18 cladocerans, and 8 copepods. The species identified during all the sampling periods of the study were *Brachionus caudatus*, *Brachionus rubens*, *Asplanchna brightwelli*, *Filinia longiseta*, *Filinia opoliensis*, *and Keratella tropica*. The percentage distribution analysis signifies the rotifers as the dominant group (55%) followed by cladocera (31%) and copepods (14%). The maximum number of species' diversity was witnessed in Baraila jheel (Vaishali) (ZSI 2011). In waters of Kusheshwar Asthan wetland, the study on zooplankton abundance indicates predominance of *copepods* in the order of *copepods* > *rotifers* > *cladocerans* > *protozoans* (Jha and Chandra 1997).

The fishery is an important livelihood option and closely associated with lives of the wetland-dependent people and the diversity of fishes also signifies the ecological health of the wetland system. Wetlands of North Bihar have been known to be a haven for diverse varieties of freshwater fishes. Although the production has seriously declined over the decades, demand has been constantly on the rise. The Kabartal has been identified with the presence of 41 commercially significant fish species which support hundreds of fishermen (Anon 2004). In the chaurs of Kusheshwar Asthan, more than a decade back, the fishery was restricted to the harvesting of carnivores and greater dominance of species like *Wallago attu* and *Channa punctatus* (murrel) was seen. The fishes mostly caught by fishermen include *Wallago attu*, *Channa punctatus*, *Ompok pabda*, *Mystus seenghala*, *Mystus aor*, *Notopterus chitala*, *N. notopterus*, *Clarius batrachus*, and *Mastacembelus armatus*. However, a point worth noticing is that the fishing pattern has considerably shifted from harvesting of major carp to large catfishes and minnows in recent years in Kusheshwar Asthan (Jha and Chandra 1997). Fish diversity in North Bihar wetlands is known for representing 50 species spanning over 19 families (ZSI 2011). The area also houses several turtle species, of which 10 are liable for conservation concerns (Das 1998). *Catla catla* (Catla), *Labeo bata* (Bata fish), *Labeo calbasu* (Calbasu), *Cirrhinus mrigala*, *Sperata aor*, *Mystus cavasius*, *and Labeo rohita* (Rohu), *Notopterus notopterus* (Moi) are some commonly sourced fish species extracted from the Baraila wetland (WII 2017). During monsoon, due to flooding pulse, the species diversity is more pronounced. However, *Wallago attu* and *Channa* species are rampantly getting depleted as complained by the local farmers (Prasad et al. 2020). The State Government has recently declared Mangur (*Clarias batrachus*) as the state fish.

Organic matter compounded with important minerals and nutrients in sediment-water matrix creates a viable environment for macrobenthos of wetlands. Benthos forms an important community of wetlands. They play a significant role in nutrient cycling and hold importance in ensuring the balanced trophic structure of the benthic environment. The northern basin in Bihar is marked by the presence of a plethora of benthic species, some of which are of direct consumptive value and others play an ecological role primarily. One of the most commonly found and consumed species are of crabs. Freshwater varieties of crabs generally found in the wetlands of Bihar include *Paratelphusa spinigera* (Tiwary et al. 2009). The crabs' abundance in wetland systems greatly depends on the fauna of lower trophic status like crustaceans and mollusks. Other benthic communities vary in quantity and quality throughout the wetlands of North Bihar. The Kusheshwar Asthan chaurs have benthic populations ranging from 835 numbers /m^2 (Larail) and 4217 numbers/m^2 (Kamaldaha). Among the identified communities of macrobenthos, gastropods and bivalves were the most dominant. However, in few chaurs, *Chironomids* were found to be more pronounced. The quality and type of gastropods are largely affected by the density of macrophytes. This was suggested by the dominance of *Lymnaea ovata*, *Guyrarulus* sp., and *Indoplanor bisexustus* in water bodies infested with weeds in Kamaldaha and Mahisath. On the contrary, less-loaded chaur like Larail saw the dominance of *Pila globosa* (Jha and Chandra 1997). On the basis of few studies performed on the detrital load balance, it has been deciphered that some active grazers like *Cirrhinus mrigala* or *Cyprinus carpio* are missing from the action in Kusheshwar Asthan wetland leading to inadequate utilization of energy and nutrients (Jha and Chandra 1997). In the Bhagar lake of Buxar, communities of gastropods and *Macrobrachium* were witnessed all year round (Prasad et al. 2020). The tribals and locals of the Kosi region have long been using and promoting the consumption of mollusks like *Pila*, *Bellamya*, *Lamellidens*, and *Parreysia* for the cure of varied internal aberrations of the body (Prabhakar and Roy 2009).

As far as avifauna is concerned, North Bihar has been a hub for the sighting of various varieties of local and migratory bird species. The study taken up by ZSI (2011) on avifauna enlisted 72 bird species, out of which, 39 were waterbirds and are reported to belong to 11 families and 30 genera. Out of these, 14 were migratory and 4 local migratory and the remaining 21 were residents. The terrestrial and riparian types were thronging the wetlands of North Bihar in quite big numbers and 33 species of them were sighted belonging to 22 families and 26 genera. Out of all the 72 recorded, 28 were winter visitors. Of the 33 riparian species, 12 were of migratory type. The Kabartal has been home to the maximum number of avifauna due to feasible habitat and availability of food (Pandey and Prakash 2019). Winter visitors are found to be concentrated during the months of January and February mostly (ZSI 2011). Dr. Salim Ali made frequent visits to this wetland and documented the presence of 166 bird species, of which 106 were resident and the remaining 60 had migrated from central India during winter months (Kumar 2013). Some commonly reported bird species included Little Grebe (*Tachybaptus ruficollis*), Great cormorant (*Phalacrocorax carbo)*, Indian cormorant (*P. fuscicollis*), Little cormorant (*P. niger*), and Grey heron (*Fulica atra*) among many (Kumar 2013). Jagatpur jheel has been studied and 34 bird species spread across 12 families and 8 orders have been recorded (Kumar and Choudhary 2010). Most were resident in nature (21); however, 7 and 6 of them were locally migrant and migrant, respectively. Jagatpur wetland has been significant as storks have been sighted in the region. Greater Adjutant Stork (*Leptoptilos dubius*) has been enlisted under Endangered (En) category, whereas Lesser Adjutant Stork (*L. javanicus*) has been notified as Vulnerable (Vu). Black-necked Stork (*Ephippiorhynchus asiaticus*) has been categorized as Near Threatened (NT)(IUCN 1996; BirdLife International 2001). Greater adjutant storks are of locally migrant nature and are generally restricted to the Asia-Pacific region. They have not been sighted outside Assam in recent decades and their reporting from Jagatpur bird sanctuary that too in decent numbers (20) has been of significant ecological value (Kumar and Choudhary 2010). It also qualifies the wetland to be proposed as a Ramsar site owing to the fulfillment of the criteria.

Bhagalpur city is also significant because of its closeness to the dolphin sanctuary where close to 198 bird species have been sighted and enlisted (del Hoyo and Collar 2014). The area was significant also because of the sighting of a single bird species called Marbled Teal *Marmaronetta angustirostris* (Vu) once during December 1999 by the banks of River Ganga near the Bhagalpur town. Some other bird species worth mentioning are Ferruginous Duck *Aythya nyroca (*NT*)*, Falcated Duck *Mareca falcata (*NT), Painted Stork *Mycteria leucocephala* (NT), and Asian Woolly necked *Ciconia episcopus* (Vu) (Dey et al. 2014). Some eminent ornithologists who have recorded these sightings include Choudhary and Mishra (2006), Choudhary and Ghosh (2014), and Mishra and Mandal (2010).

North Bihar plains are also reported to have a rich diversity of animals including that of blackbucks (*Antelope cervicapra*) and blue bulls or nilgai (*Boselaphus tragocamelus*). They can easily be seen in the agricultural fields and wetland beds during dry months and in the vicinity of Ganga dolphin sanctuary (Prasad et al. 2020). The most remarkable freshwater wildlife includes the endangered, the Ganga dolphins (*Platanista gangetica*) (Dey et al. 2014). Asia's first Dolphin Research Centre has been set up at Patna University, Bihar. Out of approximately 2500 Gangetic dolphins found in the Gangetic basin, 60% are exclusively located in the state of Bihar (Sinha 2011). Other important wildlife of great biodiversity significance includes Gharial (*Gavialis gangeticus*), Marsh Crocodile (*Crocodylus palustris*), and Smooth-coated Otter (*Lutrogale perspicillata*) (Dudgeon 2000). The dynamic Ganga basin known to be the lifeline of the state of Bihar is clearly one of the richest biodiversity hotspots in India (Islam and Rahmani 2004). The North Bihar wetlands boast of having 4 species of amphibians and 15 species of reptiles too (Daniel 2002).

4.5 Urbanization, Pollution, and Climate Change Impacts

No matter how diverse and dynamic the North Bihar wetlands seem, with the passage of time, it is certainly losing the battle to urbanization, pollution, and climate change. The primary reason for it is constant negligence and avoidance of the problems at sight (Sinha 2007; Dey et al. 2014). Very few researchers have helped with documenting the ecological health status of Bihar wetlands. Even fewer studies are there to provide an insight into the nutrient dynamics and biogeochemistry of these dynamic water bodies. Without proper research studies, these wetlands are staring at a dark future. It is also worth noticing that constant protection and monitoring protocols are difficult to execute in such dynamic terrain where a frequent change in the flow patterns, precipitation, and pulsing of rainwater occur (Singh et al. 2007). Open access wetlands and river channels often see some unwanted entities staking claims causing discords that further complicate the protection and management process (Kelkar and Krishnaswamy 2010).

Floodplain wetlands hold immense stature due to their unique fluvial and hydrological regime. Understanding their geomorphic evolution and sedimentary dynamics can provide a huge insight into the complex nature of these wetlands (Singh and Sinha 2020). Additional efforts of identifying the primary factors characterizing origin and present conditions will also be hugely helpful (McCarthy et al. 2011). An important study in this regard was undertaken by Singh and Sinha (2020) where he has investigated the Kosi-Ganga interfluve and lent an understanding of the distribution and diversity of the floodplain regime. The study has led to the identification of 6 prominent geomorphic units, namely active channel belt, active floodplain belt, inactive floodplain belt, and multiple local interfluves for Ganga-Burhi Gandak, Burhi Gandak-Baghmati, and Kosi-Baghmati (Singh and Sinha 2020).

A very alarming situation with North Bihar wetlands' face is that they may be turning into sources of carbon rather than sink due to improper management and excessive land-use changes. Alteration in the hydrological course and habitat loss at a faster pace has exacerbated the issue. Wetlands forever have been a cheap source of regulating and mitigating environmental pollution by acting as a sink of nutrients, especially carbon, nitrogen, and phosphorus (Bystrom et al. 2000). They absorb, transform, and cycle the incoming effluents, filter them, and rechannel them to purer forms which are a great blessing for us indirectly. However, people still fail to understand its importance and are converting the waterlogged areas to agricultural lands and establishing human settlements rampantly without any proper planning and vision for future course of action.

The wetlands are reported to sequester 200 times more carbon than present in the vegetation and thus keep the nutrients intact, recycle them when required, keep the biodiversity flourishing, and perform ecological functions which are crucial for the survival of every species (Mitsch et al. 2010). However, draining the wetlands results in net emission of carbon dioxide (inducing climate change), which, in turn, further alters the temperature and hydrological regimes (Lal 2008). As per the IPCC 2000 report, the wetland has the capacity to sequester 0.4 tonnes C/ha/year over a period of 50 years in a restored wetland. Wetlands have this potential because of a faster rate of deposition of organic matter than their breakdown and release (Mitsch et al. 2010).

With the growing population, the Indian wetlands are already under tremendous anthropogenic pressure. The North Bihar basin has all the more to worry about as the population load from 1951 to 2011 has increased from 29 million in 1951 to 104 million in 2011 as per the last census (NWA 2011). Such huge population density has triggered incessant urbanization due to the requirement of space, food, and resources for human beings (Zhao et al. 2006). This is leading to the conversion of fertile floodplains into agricultural fields aided by the drainage of water. The climate change regulation bodies throughout the world have already taken note of the vulnerabilities the wetlands face which may have a huge cause-and-effect scenario altering the rainfall, hydrological balance, and climate regimes (MEA 2005; UNESCO 2007).

The Ganga basin will be at the receiving end in such climatic scenarios where the change in temperature and rainfall pattern will follow periodic episodes of algal blooms, fish kills, global warming, and intense climatic natural disasters like floods and droughts (Sinha 2011). The smaller-sized wetlands face this threat even more as they get extinct faster and the deterioration of their water becomes very difficult to reverse. The wetlands are constantly under pressure due to excessive agricultural activities and runoff laden with fertilizers and pesticides being forced to get assimilated into the system (Verma 2001). The wetlands are left with an unused load of nutrients causing invasive macrophytes to encroach the water bodies. This leads to occurence of eutrophication and sedimentation is promoted furthering the species' loss (Verhoeven et al. 2006). All these problems slowly lead to the extinction of the wetlands. The lack of will power of governments, mismanaged policies, and inefficient execution has led to further deterioration of water quality of wetlands (Turner et al. 2000; Bassi et al. 2014). The academicians, experts, and scientific community of Bihar are not empowered enough to have a complete perspective of the level of pollution that has occurred especially from the context of altering nutrient dynamics and climate change in the wetlands of Bihar due to the knowledge gap in many fields of scientific importance. No full-fledged research has taken place in these fields which may provide us with a wholesome insight into the geochemistry, nutrient dynamics, and complex transformation reactions that might be taking place at the water-sediment continuum. However, there are few studies done that have assessed the pollution status of wetlands of Bihar.

Overall estimation and assessment of the ecological condition of wetlands have been researched by many scientists with the help of remote-sensing techniques (Manju et al. 2005). This has been proved to be helpful in ascertaining the health of wetlands spanning large areas easily. The characterization of wetlands is done by assessing the turbidity levels and aquatic vegetation spread through digital image-processing techniques. Manju et al. (2005) studied Moti jheel, one of the important wetlands situated in the East Champaran district of North Bihar. The district spans across an area of 3969 km^2. This study is extremely important and paves the way for a further detailed study on the major wetlands of Bihar. The study revealed that the district is comprised of 2.7% of wetland area (107.35 km^2), out of which 78 km^2 is waterlogged and others are oxbow lakes. The seasonal water recession of about 14 km^2 does occur in the summer months. Intense agricultural practices are noticed from the study in and around wetlands which has caused increased runoff episodes and eutrophication in some areas. The turbidity range has been higher as revealed from the post-monsoonal data. More area was categorized under medium-to-high turbidity range (Manju et al. 2005). Post monsoonal runoff in association with heavy sedimentation load is the primary reason for high-level turbidity. Stands of intense productivity were visible from the study which hinted at macrophytic infestation and in the areas which might have been eutrophic. The land use/change pattern indicates that the wetland area under medium vegetation has reduced and higher vegetation-covered area has increased. Most of the North Bihar wetlands have been witnessing a similar succession trajectory causing progressive deterioration of the waterbodies.

Kumar and Ambastha (2016) have synthesized land-use/land-cover change studies done over the Gangetic plains which highlight the rise in intensive agricultural practices over the last century. There has been a sharp decline of the area under forest (47%) and wetland (27%) during this period and is still witnessing similar patterns of extinction with the growing population and need for modernization (Flint 2002). Some macrophytic communities have overtaken large tracts of wetland area over the decades and have caused constant nutrient enrichment. Excessive usage of chemical and synthetic fertilizers has further exacerbated the problem (Kumar and Ambastha 2016). The primary invasive species commonly sighted in and around most of the wetlands of North Bihar are *Eichhornia crassipes* and *Ipomoea aquatica*.

Researchers have maintained that reduction in river flows due to natural and man-made reasons such as rainfall deficits and construction of irrigation structures have led to a decline in fish harvesting and productivity (Banerjee 1999; Choudhary et al. 2006; Kelkar and Krishnaswamy 2010). The major loophole arises when the river loses connectivity with the surrounding floodplains and much ecological and economic usage is jeopardized (Vass et al. 2011).

The study undertaken by Sinha (2011) has provided important details on the climate change impacts on the wetlands of North Bihar. Because of the change in river courses in this belt and the simultaneous extinction of big and small wetlands (floodplains, ponds, and oxbow lakes), the flood remediation capacity of the region has greatly been compromised. Natural as well as anthropogenic actions have led to increased flood upsurges during monsoon seasons and increased spell of droughts in dry seasons. The beauty and resilience of the North Bihar wetlands have been greatly impaired also because of threats of poaching and hunting of resident and migrant bird species. Loss of habitat has played a major role in the rapid decline of sightings of the bird species. Loss of benthic diversity has caused an imbalance in the feeding habits of the avifauna dependent on these wetlands. Increased infestation of weeds has worsened the situation (Islam and Rahmani 2008). Many background studies have revealed that similar circumstances have led to the complete extinction of many wetlands around the wetland, the prominent example being Lake Chad which was the world's sixth-largest lake (Gore 2006). The trend of climate change witnessed in North Bihar plains reveals changing temperature and rainfall patterns with a rising trend in temperature and declining trend in monsoon rainfall at a much faster rate compared to the 1970s. The period between 1901 and 1970 saw a minute change in average temperature per annum of 0.0078°C. However, the trend certainly switched to a more sudden change of 0.304°C per annum from 1971 to 2002. The rainfall deficit for both these studied time periods was 0.3477 mm and 13.01 mm, respectively, which is hugely worrisome. The rate of temperature change recorded for North Bihar is similar to the trend observed by the Intergovernmental Panel on Climate Change (IPCC). The simultaneous increase of temperature and decline in rainfall percentage were observed. This could be due to the fact that the saturation pressure of air increases with a rise in temperature. The rainfall trend is however not homogeneous for the whole of the Bihar state. Rising monsoonal episodes have been witnessed in Patna city from 1951 to 2001 (De and Prakas Rao 2004).

The quantity and quality of the algal population can be analyzed to infer the environmental factors and behaviors highlighting the ecological health of a wetland (Shekhar et al. 2008). Various species diversity indices can be utilized for this purpose. Shekhar et al. (2008) has studied an important Ganga channel near Bhagalpur town and Ganga dolphin sanctuary. Here, pH ranged from slightly acidic to alkaline both in pre- and post-monsoon durations. The turbidity values were found to be on the higher side during both pre- (5.6–17.7) and post-monsoon times (12.4–29.9). High-level runoff phenomena, with loads of silt, clay, and organic matter, lead to higher turbidity. As far as the phytoplankton community is concerned, the study area witnessed a higher phytoplankton abundance in the post-monsoon season. The presence of algal communities like *Scenedesmus*, *Anabaena*, *Oscillatoria*, and *Melosira* signify the presence of excessive nutrients, as these species are one of the most tolerant to high organic pollution (Zargar and Ghosh 2006). Similar findings have been reported by other researchers in the Ganga stretch (Nandan and Aher 2005; Zargar and Ghosh 2006; Kumar and Choudhary 2010). The pollution index of algal genera (Palmer 1969) was 32, which indicates the presence of organic pollution in the river channel. The river has been subjected to serious pollution while accumulating effluents from industries, agricultural lands, and farms in its course. Also the Shannon Weiner Index was in the range of 0.117–1.675 indicating heavy pollution in the waterbody (Kumar and Choudhary 2010).

Although wetlands throughout the world and even in India have been thoroughly studied for trace metal pollution assessment, the Bihar wetlands have still not been researched enough in this context except for few studies pioneered by Ghosh et al. (2004), Ranjan et al. (2016), Ranjan et al. (2017), Singh and Jayakumar (2016), Singh et al. (2020) among few. Ranjan et al. (2016) has documented significant findings regarding the trace metal concentration in the sediments of Kabartal. According to this study, the Kabartal is extremely contaminated with trace metals, especially cadmium. The trace metal abundance was in the order of Fe > Mn > Pb > Ni > Co > Cu > Cd. High Pollution Load Index (PLI) values also indicate that trace metal pollution exists in this wetland. Mostly metal concentration in a wetland is attributed to agricultural waste, storm water discharge, domestic sewage, and urbanization. Fe and Mn are found to be the most abundant metals present in the sediments of Kabartal. These metals are generally naturally abundant in the geosphere and become part of the soil strata on crustal weathering. However, anthropogenic sources like industries and factory discharge might have played an active role in their enrichment. Nickel too was found in greater quantity in Kabartal which might have been released from various anthropogenic activities and sources like sludge and storm water transport, and atmospheric deposition from Barauni oil refinery located at some distance (Abbasi et al. 1998). Other trace metals worth mentioning were copper and cobalt. Copper is a chalcophilic element and its concentration may have risen in the sediments due to the high affinity of organic matter (humic substances) to bind with copper. They also might have been released from the common anthropogenic sources in that area, like the storm water and agricultural runoff and natural enrichment from the existing anoxic conditions in submerged parts of the wetland. The trace metal which was found to be at a notoriously toxic level was cadmium. Cd might have been released in the water body from the organic matter breakdown under aerobic respiration processes and then would have settled back to the sediments. Anthropogenic sources from human settlements nearby must have played a major role in their accumulation over the years (Ranjan et al. 2016). Apart from quantifying heavy metals, the study has also focused on the distribution of phosphorus and its fractions in the sediments of Kabartal. It was reported that the distribution of phosphorus is heterogeneous in nature and the most dominant fraction was non-apatite inorganic phosphorus followed by organic phosphorus. Ranjan et al. (2016) was of the view that the phenomenon of low phosphorus content and high trace metal concentration is required to be further assessed and researched in the wetlands of Bihar.

Few researchers have also documented the chemical, hydrological, and fluvial status of Kabartal wetland (Singh and Jayakumar 2016). Kabartal, once a haven for migratory birds, is fast getting reduced to a drainage area (Ambastha et al. 2007). Ghosh et al. (2004) reported that the western part of the Kabartal has undergone heavy siltation and sedimentation leaving the raised platform devoid of surface water. On the eastern side, the rate of sedimentation is slower in comparison to the west. The intensive agricultural practices have altered the ecological regime to an extent that the biodiversity has been compromised extensively. The constant rate of increase of macronutrients (phosphorus and nitrogen) over the last several years indicates eutrophication and heavy anthropogenic pressure including nitrogenous organic matter from sewage discharge, farms, and agricultural runoffs, etc., in the wetlands (e.g. Ranjan and Kumari 2018). Ghosh et al. (2004) are of the view that the rapid deterioration of wetland has been superimposed due to heavy infestation of macrophytes like *Phragmites* and *Hydrilla spp.* Unsustainable agricultural practices and overexploitation of natural yields for food and fodder and fuel have stripped the wetland of natural vegetation causing invasive types to take over or rendering the land barren.

A major study related to hydrogeochemistry and trace elements was performed by Ranjan et al. (2017) on two very important wetlands of Bihar, namely Kabartal and Kusheshwar Asthan chaur. The study has found that Ca^{2+} and HCO_3^- are the most dominant ions in both wetlands. It was revealed that rock water interaction especially from the chemical dissolution of calcite and carbonate rocks promoted the high abundance of calcium species. Weathering process and interaction with water were the dominating processes with hardly any role of anthropogenic activity in the water chemistry of the two wetlands. However, the study indicates an influence of other sources apart from rock water interaction. The stability diagrams of the two wetlands reveal that the water of Kabartal is in equilibrium with gibbsite and kaolinite minerals whereas Kusheshwar Asthan chaur supports the kaolinite-based formations. The study also indicates that the sediments from both wetlands have originated from orthoquartzite-based minerals. Quartz (~36%) and clay (~21%) minerals were found to be the most dominant entities in both the study areas. The study performed on the macronutrient distribution revealed that organic matter and carbon were abundant and sufficient to support all types of productivities on site. Nitrogen and sulfur too were present in a good quantity necessary for ecological sustenance.

Another major problem that most of the North Bihar areas face every year is the flood. The wetlands perform a crucial service by abating the effect and intensity of floods by absorbing, diverting, and slowing down the rate of flow (Kumar 2013). The wetlands pose a cheaper alternative for the management and mitigation of floods and they also trap pollutants, silts, and sediments, which nullify a great percentage of possible flood damage altogether (Boyd and Banzhaf 2007). However, flooding activities have been on the rise because of the rapid extinction of wetlands and their shrinking and destruction (Prasad et al. 2002). It is a serious issue which leads to loss of lives as well as huge wastage and destruction of resources and goods. In the floods of 2010, in Bihar alone, 0.72 million population, and 3.24 million ha of the cultivated area were affected. According to Ganga Flood Control Commission (2012), 13.50 billion rupees was released for damage control and mitigation. The net economic loss is so high that Bihar is never fully able to sustain without external help and loans. The possible remedy and mitigation measure suggested by various geographical and hydrological experts is to connect multiple rivers so as to abate the overflowing of rivers and adjust the flow toward drier beds at a particular time. This is because different rivers flood at different intervals (ZSI 2011). Some prominent rivers facing periodic upsurge are Budhi Gandak, Harha, Massan, Pandai, Lalbakiya, Bakaiya, Bagamati, Dhaunsa, Adhawara Group of Rivers, Kamala, Jeevachh, Bhutahi, Balan, Kareh, Baya, Kosi, Paraman, Kankai, Mahananda, Naraini, Ghaghara, and Ganga.

As we discussed earlier, North Bihar wetlands used to be winter paradise for different varieties of migratory and resident bird species. However, in the last few decades, the birds' population has significantly declined. Bird populations of varieties like endangered Indian skimmer, river lapwings (*Vanellus duvaucelii*), river terns and Little Pratincole *Glareola lactea* have been severely affected due to human-based activities. The destruction of habitats of birds, their breeding, feeding, and nesting grounds has been altered for the cultivation of food crops to support rising population and demand. There has been an upsurge in the growth of the cultivation of cucurbits on previously uncultivated sand bars and mid-channel highlands which used to be the favourite breeding nests for birds. Expansion of roads and related construction activities on rivers has hastened the loss of biodiversity. Illegal poaching activities have constantly been on an increase in defiance of strict curbing policies. Thus, it should be the government's utmost prerogative to come up with a sound and efficient system of rules and regulations which may be successful in the restoration of these fragile ecosystems (Jha and Chandra 1997).

The well-known Bhagalpur Ganga dolphin sanctuary which has been enjoying a protected status for quite some time has been under incessant environmental and anthropogenic pressure. The growing population of village communities, need to manage the growing demand for food by accessing more lands for cultivation and farming, motorized river navigation, construction of dams and dykes altering the water regime, downstream construction of Farakka barrage have all caused an immeasurable decline in the number of fishes, dolphins, and bird population (Banerjee 1999; Kelkar and Krishnaswamy 2010). Initiative such as Vikramshila Biodiversity Research and Education Centre (VBREC) in Bhagalpur pledges to work on monitoring, conservation, and research of dolphin sanctuary (Choudhary et al. 2006).

The North Bihar population is also struggling with the issue of trace metals and pesticide contamination in water and thus fishes and crops. Some wetlands of North Bihar like Sukhaldari dam and Bhagar Beel have been reported to have the maximum concentrations of copper and lead, respectively. Among the fish varieties, species of *Puntius dorsalis* reported the highest bioconcentration of zinc and cadmium whereas *Channa punctatus* was said to accumulate the maximum concentration of lead. Further investigation by National Wetland Atlas (2011) has revealed that fishes belonging to Cheriya Bariyarpur of Begusarai district have accumulated the maximum concentration of organochlorine pesticides.

4.6 Legal Framework, Policies, and Challenges

Knowing the importance of the Northern Gangetic Plains and its biota, it should be the governments' and civic bodies' primary concern to restore and protect the dynamic and unique wetland systems. The North Bihar region which has been at the receiving end of habitat and species loss cycle driven by natural and anthropogenic activities calls for an extensive policy-driven action plan toward conservation and restoration of these wetlands. Developing baseline data for biotic diversity (avian, fish, and planktonic) is an essential step toward making a sustainable effort (Sinha 2007). A comprehensive bird checklist for the Bihar Ganga basin has been taken up as a major objective by researchers to fill in missing links and update the limited knowledge (Dey et al. 2014). Important wetland management studies have been taken up by many scientists in the last few decades. Important insights and suggestions on different aspects of North Bihar wetlands' resource management and their efficient utilization have been studied by many wetland experts (Roy et al. 2008; Kumar et al. 2013; Jha et al. 2014). On a broader level, the convention on wetland which has 168 nations as contracting parties as of December 2015 has developed a vision for the protection and conservation of wetlands of international importance. This vision is directed at maintaining a network of internationally important wetlands which have significant implications for global biodiversity and which play a major role in sustaining human life through performing crucial ecosystem services.

India is a contracting party to the environmental conventions related to wetlands and biodiversity, yet it has not come up with a solid action plan or framework regulating the wetlands and the activities related to them. However, as far as the Indian laws are concerned, many legal acts call for the protection, monitoring, and regulation of wetlands. Some of them are the Indian Fisheries Act, 1857; Wildlife (Protection) Act, 1972; Water (Prevention and Control of Pollution) Act; Territorial Water, Continental Shelf, Exclusive Economic Zone and other, Marine Zones Act, 1976; Coastal Zone Regulation Notification, 1991; Environmental (Protection) Act, 1986; etc. (MoEF 2007; Bassi et al. 2014). The National Environment Policy, 2006, however, for the first time, acknowledged the lack of wetland-regulating frameworks in India and stated the urgent need for the implementation of rules enforcing regulation and conservation of wetlands in the country (MoEF 2006; Dandekar et al. 2011). In 2009, the government had issued a set of formal guidelines for the conservation of about 122 wetlands. The guidelines had directed the state governments to carry out activities such as mapping and demarcation of different areas, regulation and mitigation of pollution-related damage, and protection of biodiversity encouraging sustainable livelihoods. In 2010, MoEF put a stamp on Wetlands (Conservation and Management) Rules, 2010 by notifying it thereby making it a law. Despite having several legal frameworks on board, the execution has always been full of loopholes and lacks political will.

The real on-ground challenge for the Bihar region is the per capita availability of clean water. It has been projected that there will be a major decline from about 1950 m^3/year in 2001 to a mere 1170 m^3/year by 2050 (Ghosh et al. 2004). The Ghagra-Gandak and Gandak-Kosi basins have already faced water deficits of about 43.4% and 37.8%, respectively. Sedimentation on a large scale has already wiped away many floodplain wetlands and flow channels, causing frequent episodes of floods, which is getting more alarming with each passing year (Ghosh et al. 2004).

Analyzing the varied problems we are facing currently, a massive programme aiming at sustainable usage of wetland needs to be promoted. The full-scale study of the hydrological as well as the ecological continuum across high altitude, plains, and deltaic regions needs to be updated. The emphasis should be given to protection and sustainable use of resources without compromising with its ecological values (Kumar and Choudhary 2010; Kumar and Ambastha 2016).

Very recently, in January 2020, the Ministry of Environment, Forest and Climate Change (MoEF&CC), Government of India has come out with a new rule [Wetlands (Conservation and Management) Rules, 2017] to identify and manage the wetlands. This rule also does not permit any major economic activity (industrial expansion, construction, and demolition) to take place within the wetlands. The ministry also directs the state to set up an authoritative body to chalk out strategies and conservation plans for wetlands. This new rule decentralizes wetland management by giving states more powers to not only identify and notify wetlands within their jurisdictions but also keep a watch on prohibited activities. The notification says, "The wetlands shall be conserved and managed in accordance with the principle of 'wise use' as determined by the Wetlands Authority." It also indirectly widens the ambit of permitted activities by inserting the "wise use" principle, giving powers to state-level wetland authorities to decide what can be allowed in the larger interest.

As is the case of other Indian states, Bihar too does not have a law pertaining to conservation, protection, and management of wetlands. However, the government of Bihar has taken steps in the field of promoting sustainable livelihoods by encouraging aqua culturing, organic farming, and ecotourism (Sinha 2011). The government of India has already drawn a conservation action plan for the Ganges river dolphins (Sinha et al. 2010). However, after the Wetlands (Conservation and Management) Rules, 2017, the conservation and management of wetlands may be improved since the state government has got more power for taking decisions to identify and notify the wetlands within their jurisdictions and also control prohibited activities.

4.7 Conclusion

The North Bihar region and its water bodies as a whole entity drive the lifeline of the state and support millions of human population as far as food, livelihoods, and daily activities are concerned. The state and its environmental resources, particularly wetlands, have been ignored by the authoritative bodies and also been exploited endlessly by the people dependent on them for various services. The people still do not realize the ecological and functional roles that wetlands play at a larger level. The flood mitigation benefits, food, fodder, fuel, and construction materials that Bihar wetlands provide have been unable to drive the people to actually think toward using them judiciously for a sustainable future. The thriving poverty and lack of education are a big factor toward this negligence. The wanton destruction of wetland areas and river floodplains has reduced the environmental resilience of the ecosystem. The high population density, unethical and exploitative cultivation practices, drainage of wetlands for irrigation, overfishing, poaching of migratory birds, anthropogenically induced organic as well as chemical pollution (trace metals and pesticides), and eutrophication induced by weed infestation are some of the looming challenges faced by the wetlands of Bihar. The problem has been aggravated due to limited attention being received on this front of protection and restoration of wetlands of Bihar. There has been a severe decline in the avifaunal diversity in the past few decades which indicate the unfeasible circumstances that persist. Not only avifauna, but even planktonic, fish and gastropod communities have altered owing to environmental pollution. The livelihoods of local people depending on the resources of the wetlands have been severely affected leading to further poverty and people switching employment. Mass constructions, encroachments, sedimentation, and administrative negligence have caused important wetlands like Baraila jheel of Vaishali, Kabartal of Begusarai, and Gogabeel wetland of Katihar to suffer shrinkage of the wetland area. The type of research studies done on the wetlands of Bihar has been limited to limnological studies and documentation of biodiversity. The institutional limitations, lack of inventories, and laboratories for full-fledged research can be a primary reason for limited knowledge databases on Bihar wetlands. Therefore, more robust physiochemical, geological, and biogeochemical research studies need to be given priority for proper assessment of hydrological and nutrient dynamics in the wetlands. This way, better regulatory frameworks, policies, and action plans can be strategized for conservation and ensuring the sustainable future of wetlands and the people associated with them.

Acknowledgments

Dr. R.K. Ranjan would like to thank the Principal Chief Conservator of Forests, Government of Bihar and Divisional Forest Officers (Begusarai and Vaishali), Bihar for granting permission to conduct sampling in Kabartal and Baraila jheel wetlands. Alvia Aslam would like to thank UGC, New Delhi for providing JRF for PhD research work. We are grateful to all anonymous reviewers for their thoughtful suggestions, which greatly improved this book chapter. The study is partially funded by SERB, Department of Science and Technology, Government of India (Grant no.: SR/FTP/ES-01/2014).

References

Abbasi, S.A., Abbasi, N., and Rajendra, S. (1998). *Heavy Metals in the Environment. Mittal Publications.*

Adhishwar, A. and Choudhary, S. (2014). Article-43 diversity of macrophytic species of Gogabil lake wetland in Katihar, Bihar, India. *Ecology, Environment and Conservation* **19**: 1165–1172.

Ambastha, K., Hussain, S.A., and Badola, R. (2007). Social and economic considerations in conserving wetlands of Indo-genetic plains: a case study of Kabartal wetland, India. *Environmentalist* **27**: 261–273.

Anon. (2004). *Management Action Plan for Kanwar Lake 2004-2005 to 2008-2009*. Department of Environment & Forest, Government of Bihar.

Banerjee, M. (1999). A report on the impact of the Farakka barrage on the human fabric. World Commission on Dams – Thematic Review. South Asia Network on Dams. Rivers and People. New Delhi, India.

Bassi, N., Kumar, M.D., Sharma, A., and Pardha-Saradhi, P. (2014). Status of wetlands in India: a review of extent, ecosystem benefits, threats and management strategies. *Journal of Hydrology: Regional Studies* **2**: 1–19.

Bihar State Profile. (2016). Ministry of Micro, Small and Medium Enterprises, Delhi. Government of India. http://dcmsme.gov.in/dips/state_wise_profile_16-17/Bihar%20-%20State%20Profile.pdf (accessed 23 October 2020).

Bird Life International (2001). *Threatened BIRDS of Asia: The Bird Life International Red Data Book. Part A.*, 1e, vol. **1** (eds. N.J. Collar, A.V. Andreev, S. Chan, et al.). Cambridge, UK: Bird Life International.

Bluffstone, R., Boscolo, M. & R. Molina. (2001). How does community forestry affect rural households? A labor allocation model of the Bolivian Andes. Working Paper, University of Redlands, USA.

Boyd, J. and Banzhaf, S. (2007). What are ecosystem services? The need for standardized environmental accounting units. *Ecological Economy* **63** (2–3): 616–626.

Bystrom, O., Andersson, H., and Gren, I. (2000). Economic criteria for using wetlands as nitrogen sinks under uncertainty. *Ecological Economy* **35** (1): 35–45.

Cavendish, W. (2000). Empirical regularities in the poverty-environment relationship of rural households: evidence from Zimbabwe. *World Development* **28** (11): 1979–2003. https://doi.org/10.1016/S0305-750X(00)00066-8.

Choudhary, D.N. and Ghosh, T.K. (2004). Sighting of Greater Adjutant Storksin the wetlands of north Bihar. *Newsletter for Birdwatchers* **44**: 62–63.

Choudhary, D.N. and Mishra, A. (2006). Sighting of some threatened birdspecies in Vikramshila Gangetic Dolphin Sanctuary (VGDS) Bhagalpur, Bihar. *Newsletter for Birdwatchers* **46**: 68–70.

Choudhary, S.K., Smith, B.D., Dey, S. et al. (2006). Conservation and biomonitoring in the Vikramshila Gangetic Dolphin Sanctuary, Bihar, India. *Oryx* **40**: 1–9.

Chowdhury, S.R., Nayak, A.K., Brahmanand, P.S. et al. (2018). *Delineation of Waterlogged Areas Using Spatial Techniques for Suitable Crop Management in Eastern India*. Bhubaneshwar, Odissa: ICAR.

Cowardin, L.M., Carter, V., Golet, F.C., and LaRoe, E.T. (1979). *Classification of Wetlands and Deepwater Habitats of the United States*. Washington, DC: U.S. Fish and Wildlife Service. FWS/OBS-79/31.

Dandekar, P., Bhattacharya, S., and Thakkar, H. (2011). *Wetland (Conservation and Management) Rules 2010, Welcome, But a Lost Opportunity: This Cannot Help Protect the Wetlands (Sir Note)*. New Delhi: *South Asia Network on Dams, Rivers & People*.

Daniel, J.C. (2002). *The Book of Indian Reptiles and Amphibians*. Mumbai, India: *Bombay Natural History Society and Oxford University Press*.

Das, I. (1998). *Turtles and Tortoises of India*, 186. Mumbai, India: *Oxford University Press*.

De, U.S. and Prakas Rao, G.S. (2004). Urban climatic trends, 2004. *Journal of Indian Geophysical Union* **8**: 199–203.

Devi, T.M. and Sharma, B.M. (2007). Studies on the distribution of macrophytes of Yenapatlake, Bishupur, Manipur. *Indian Journal of Environment and Ecological Planning* **14** (1-2): 311–318.

Dey, S., Dey, S., Choudhary, S., and Kelkar, N. (2014). An annotated bird checklist of the Vikramshila Gangetic Dolphin Sanctuary, Bhagalpur, Bihar, India, with an assessment of threats to bird conservation. *Forktail* **30**: 34–40.

Dudgeon, D. (2000). Large-scale hydrological changes in tropical Asia: prospects for riverine biodiversity. *Bioscience* **50**: 793–806.

Flint, E. (2002). Historical reconstruction of changes in land use and land cover of vegetation in the Gangetic Plain 1880-1980: methodology and case studies. In: *Land Use – Historical Perspectives: Focus on Indo-Gangetic Plains* (eds. Y.P. Abrol, S. Sangwan and M.K. Tiwari), 189–248. New Delhi: *Allied Publishers Private Limited*.

Ganga Flood Control Commission (GFCC). (2012). Annual Report 2010–2011. Ganga Flood Control Commission, Ministry of WaterResources, Government of India, Patna.

Ghosh, A.K., Bose, N.A., Singh, K.R.P., and Sinha, R.K. (2004).Study of spatio-temporal changes in the wetlands of north Bihar through remote sensing. *13th International Soil*

Conservation Organisation Conference, Brisbane, Australia. http://citeseerx.ist.psu.edu/viewdoc/download?doi=10.1.1.566.5537&rep=rep1&type=pdf. (accessed 10 October 2020).

Gore, A. (2006). *An Inconvenient Truth*. London: *Bloomsbury Publishing Plc.*

del Hoyo, J. and Collar, N.J. (2014). *HBW & Bird Life International Illustrated Checklist of the Birds of the World, 1*. Barcelona: *Lynx Edicions.*

IPCC (2000). *Intergovernmental Panel on Climate Change, Special Report on Land Use, Land-Use Change, and Forestry: Summary for Policymakers*. Geneva, Switzerland: *Intergovernmental Panel on Climate Change.*

Irfan, S. 2015. A study of rural wetlands of Bihar state: effect of nutrient dynamics in physiochemical properties of water, sediment and the aquatic flora of the wetland. In (Mishra, G.C. (etd.) *Strategic Technologies of Complex Environmental Issues-A Sustainable Approach*. Krishi Sanskriti, Excellent Publishing House, New Delhi. ISBN: 978-93-83083-85-5

Islam, M.Z. and Rahmani, A.R. (eds.) (2004). *Important Bird Areas in India: Priority Sites for Conservation*. Bombay and Cambridge, UK: Indian Bird ConservationNetwork, Bombay Natural History Society & Bird Life International.

Islam, M.Z. and Rahmani, A.R. (2008). *Potential and Existing Ramsar Sites in India*. New York: *Oxford University Press.*

IUCN (1996). IUCN red list of threatened animals. In: *World Conservation Monitoring Centre* (eds. J. Baillie, B. Groombridge, U. Gärdenfors and A. Stattersfield). Washington, DC: Gland, Switzerland, Conservation International.

Jain, C.K., Singhal, D.C., and Sharma, M.K. (2007a). Estimating nutrient loadings using chemical mass balance approach. *Environmental Monitoring and Assessment* **134** (1-3): 385–396.

Jain, S.K., Agarwal, P.K., and Singh, V.P. (2007b). *Hydrology and Water Resources of India*. The Netherlands: *Springer.*

Jha, V. (2012). Aquatic biodiversity as a basis for development in the flood plains of North Bihar. In: *Identifying Resource Complex Regions and Regional Development Strategy for Bihar* (eds. R.B.P. Singh et al.), 69–84. Patna: Centre for Geosheelitic Studies, Department of Geography, Patna University.

Jha, B.C. and Chandra, K. (1997). *Kusheshwar Sthan Chaur (North Bihar): Status and Prospects of Fisheries Department*. Barrackpore, Calcutta, India: CIFRI.

Jha, V., Verma, A.B., Jha, P. Jha, M. & Kumar, R.2014. Wetlands in North Bihar provide a basis to its sustainable development. *Journal of Aquatic Biology and Fisheries*, (2):843–851. http://keralamarinelife.in/Journals/Vol2-2/132.pdf (accessed 10 February 2020).

Kant, S. and Anand, V.K. (1978). Interrelationship of phytoplankton and physical factors in Mansar Lake, Jammu (J & K). *Indian Journal of Ecology* **5** (2): 134–140.

Kant, S. and Kachroo, P. (1975). Limnological studies in Kashmir lakes, II Diurnal movements of phytoplankton. *Journal Indian Botanical Society* **54**: 9–12.

Kelkar, N. and Krishnaswamy, J. (2010). Keeping rivers alive. *Seminar* **613**: 29–33.

Krishna, G. and Sinha, R. (2014). Algal spectrum of a wetland and its correlation with the physic- chemical parameters. *International Journal of Environmental Sciences* **3** (3): 27–30.

Kumar, M. (2013). Resource inventory analysis of Kabartal wetland. *International Journal of Research in Humanities and Social Sciences* **1** (8): 13–26.

Kumar, S. (2019). Primary production of the diverse freshwater ecosystems: an overview. *Review of research* **8**: 7.

Kumar, R. and Ambastha, K. (2016). Wetlands of the Ganga-Brahmaputra Basin. In: *The Wetland Book* (eds. C. Finlayson, G. Milton, R. Prentice and N. Davidson). Dordrecht: Springer.

Kumar, R. and Ambastha, K. (2016). Wetlands of the Ganga-Brahmaputra basin. In: *The Wetland Book* (eds. C. Finlayson, G. Milton, R. Prentice and N. Davidson). Dordrecht: Springer.

Kumar, B., and Choudhary, S. (2010). Avifauna of Jagatpur wetland near Bhagalpur (Bihar, India). T.M. Bhagalpur University, Bihar, India.

Kumar, N. and Singh, N.P. (2013). Studies on the ichthyofauna of Kararia lake of Motihari, East Champaran, Bihar, India. *Research Journal of Animal Vetenary and Fishery Science* **1** (9): 8–12.

Kumar, M., Prusty, A.K., and Jeniffer, P.N. (2013). Ecosystem (provisioning) services and resource utilization pattern of an oxbowlake in Muzaffarpur urban Agglomeration, Bihar. *The Ecoscan* **3**: 289–295.

Lal, R. (2008). Carbon sequestration. *Philosophical Transactions of the Royal Society of London B* **363** (1): 815–830.

Manju, G., Chowdary, V.M., and Srivastava, Y.K. (2005). Mapping and characterization of inland wetlands using remote sensing and GIS. *Journal of the Indian Society of Remote Sensing* **33**: 51. https://doi.org/10.1007/BF02989990.

McCarthy, T.S., Tooth, S., Jacobs, Z. et al. (2011). The origin and developmentof the Nyl River floodplain wetland, Limpopo Province, South Africa: trunk–tributary river interactions in a dryland setting. *South African Geographical Journal* **93** (2): 172–190.

Millennium Ecosystem Assessment (MEA). (2005). Ecosystems and human well-being: wetlands and water synthesis. World Resources Institute, Washington, DC. https://www.millenniumassessment.org/documents/document.358.aspx.pdf (accessed 14 October 2020).

Ministry of Environment and Forests (MoEF) (2006). *National Environmental Policy*. New Delhi: MoEF, Government of India.

Ministry of Environment and Forests (MoEF) (2007). *Conservation of Wetlands in India: A Profile (Approach and Guidelines)*. New Delhi: MoEF, Government of India.

Ministry of Environment, Forest and Climate Change (MoEF&CC) (2017). *Wetlands (Conservation and Management) Rules 2017*. New Delhi: MoEF&CC, Government of India.

Mishra, A. and Mandal, J.N. (2010). Discovery of a breeding ground of the GreaterAdjutant *Leptoptilos dubius* and their conservation in the floodplains ofBihar, India. *Journal of the Bombay Natural History Society* **106**: 190–197.

Mitsch, W.J., Nahlik, A.M., Wolski, P. et al. (2010). Tropical wetlands: seasonal hydrologic pulsing, carbon sequestration, and methane emissions. *Wetlands Ecological Management* **18**: 573–586.

Munawar, M. (1974). Liminological studies of fresh water pond of Hyderabad. *Hydrobiologia* **45**: 1–32.

Nandan, S.N. and Aher, N.H. (2005). Algal community used for assessment of water quality of Haranbaree dam and Mosam river of Maharashtra. *Journal of Environmental Biology* **26**: 223–227.

National Wetland Atlas, SAC/EPSA/ABHG/NWIA/ATLAS/34. (2011). *Space Applications Centre (ISRO)*. Ahmedabad, India, 310.

National Wetland Atlas: Bihar, SAC/RESA/AFEG/NWIA/ATLAS/23. (2010). *Space Applications Centre (ISRO)*. Ahmedabad, India, 222.

Palmer, C.M. (1969). Composite rating of algae tolerating organic pollution. *Journal of Phycology* **5**: 78–82.

Pandey, R., and Prakash, B. (2019). Strategies for Wetland Conservation: Study of Kanwar Wetland, Begusarai. 10.21275/ART20194346.

Prabhakar, A.K. and Roy, S.P. (2009). Ethnomedicinal uses of some shell fishes by people of Kosi river basin of north Bihar, India. *Ethno-Medicine* **3** (1): 1–4.

Prasad, S.N., Ramachandra, T.V., Ahilya, N. et al. (2002). Conservation of wetlands of India – a review. *Tropical Ecology* **43** (1): 173–186.

Prasad, S., Kumar, S., Gorai, T., and Kumar, A. (2020). To determine the structure and function of Bhagar Oxbow Lake in Dumraon, Buxar, South Bihar, India. *Current Journal of Applied Science and Technology 39* (28): 145–153. https://doi.org/10.9734/cjast/2020/v39i2830948.

Rai, L.C. (1978). Ecological studies of algal communities of the Ganges River at VaranasiIndian. *Journal of Ecology* **5** (1): 1–6.

Ramsar. (2020). The list of wetlands of international importance. The secretariat of the convention on wetlands (Ramsar, Iran, 1971). Gland, Switzerland. https://www.ramsar.org/sites/default/files/documents/library/sitelist.pdf (accessed 23 October 2020).

Ranjan, R.K. and Kumari, P. (2018). Impact of land use and land cover changes on nutrients concentration in and around Kabar tal wetland, Begusarai (Bihar), India. In: *Geosptaial Applications for Natural Resource Management* (ed. C.K. Singh), 243–250. *CRC Press*.

Ranjan, R.K., Sinha, A.K., Gupta, D. et al. (2016). Sedimentary geochemistry of Kabar Tal Wetland, Begusarai, Bihar, India. *Journal of Applied Geochemistry* **18**: 414–429.

Ranjan, R., Srivastava, S.K., and Ramanathan, A.L. (2017). An assessment of the hydrogeochemistry of two wetlands located in Bihar State in the subtropical climatic zone of India. *Environmental Earth Sciences* **76** (16) https://doi.org/10.1007/s12665-016-6330-x.

Roy, S.P., Ramanand, R., Prabhakar, A.K., and Singh, J.P. (2008). Sustainable development of Kawar lake, Begusarai (North-Bihar, India). In: *Proceedings of Taal 2007: The 12th World Lake Conference* (eds. M. Sengupta and R. Dalwani), 1578–1585.

Shardendu, S., Sayantan, D., Sharma, D., and Irfan, S. (2012). Luxury uptake and removal of phosphorus from water column by representative aquatic plants and its implication for wetland management. *International Scholarly Research Network, Soil Science*, Article ID 516947.

Sharma, S.C., Purohit, S.C., and Jeetendra, K. (2007). Aquatic macrophytes of Bikaner District (Rajasthan). *Journal of Indian Botanical Society* **86**: 137–140.

Shekhar, T.R., Kiran, B.R., Puttaiah, E.T. et al. (2008). Phytoplankton as index of water quality with reference to industrial pollution. *Journal of Environmental Biology* **29** (2): 233–236. PMID: 18831381.

Singh, A.K. and Jayakumar, S. (2016). Water quality assessment of Kanwar Lake, Begusarai, Bihar, India. *Imperial Journal of Interdisciplinary Research* **2** (4): 793–803.

Singh, M. and Sinha, R. (2020). Distribution, diversity, and geomorphic evolution of floodplain wetlands and wetland complexes in the Ganga plains of north Bihar, India. *Geomorphology* **351**.

Singh, M., Bir Singh, I., and Muller, G. (2007). Sediment characteristics andtransportation dynamics of the Ganga River. *Geomorphology* **86**: 144–175.

Singh, A.K., Sathya, M., Verma, S., and Jayakumar, S. (2020). Spatiotemporal variation of water quality index in Kanwar wetland, Begusarai, India. *Sustainable Water Resources Management* **6** (44): 1–8.

Sinha, S.K. (2007). Readers' forum: more ornithological surveys needed. *Sanctuary Asia* **27**: 84.

Sinha, C.P. (2011). Climate change and its impacts on the wetlands of North Bihar, India. *Lakes & Reservoirs: Research & Management* **16**: 109–111. https://doi.org/10.1111/j.1440-1770.2011.00457.x.

Sinha, R.K., Behera, S.K., and Choudhary, B.C. (2010). *Conservation Action Plan for the Gangetic dolphins*, 44. National Ganga River Basin Authority, Ministry of Environment and Forests, Govt of India.

State Action Plan on Climate Change (SAPCC). (2015). Environment & forest department. Government of Bihar. http://moef.gov.in/wp-content/uploads/2017/08/Bihar-State-Action-Plan-on-Climate-Change-2.pdf (accessed February 2019).

State of Environment Report. (2007). Bihar State Pollution Control Board, Patna, Department of Environment and Forest, Government of Bihar.

The Wetland Report. (2012). The wetland report for Ganga River Basin. *Environment Management Plan (GRBMP)*, Indian Institute of Technology. http://cganga.org/wp-content/uploads/sites/3/2018/11/031_ENB_Wetands_Draft.pdf (accessed 23 October 2020).

Tiwary, N., Kumar, P., Prasad, R. et al. (2009). Prospects and constraints of crab culture in wetlands of Bihar, India. *Our Nature* 7 (1): 230–232.

Tripathi, B.D. and Shukla, S.C. (1991). Biological treatmentof waste water by selected aquatic plants. *Environmental Pollution* **6**: 69–78.

Turner, R.K., van der Bergh, J.C.J.M., Soderqvist, T. et al. (2000). Ecological-economic analysis of wetlands: scientific integration for management and policy. *Ecological Economy* **35** (1): 7–23.

United Nations Educational, Scientific and Cultural Organization (UNESCO) (2007). *Case Studies on Climate Change and World Heritage*. France: *UNESCO World Heritage Centre*.

Vass, K.K., Das, M.K., Tyagi, R.K. et al. (2011). Strategies for sustainable fisheries in the Indian part of the Ganga-Brahmaputra river basins. *International Journal of Ecology and Environmental Sciences* **37** (4): 157–218.

Verhoeven, J.T.A., Arheimer, B., Yin, C., and Hefting, M.M. (2006). Regional and global concerns over wetlands and water quality. *Trends of Ecological Evolution* **21** (2): 96–103.

Verma, M. (2001). *Economic Valuation of Bhoj Wetlands for Sustainable Use*. [EERC Working Paper Series: WB-9]. Bhopal: *Indian Institute of Forest Management*.

Vyas, L.N. (1968). Studies in phytoplankton ecology of Pichhola Lake. *Udaypur Proceedings of Symposium, Recent Advancement in Tropical. Ecology*, 334–347.

Ward, J.V., Tocker, K., Arscott, D.B., and Claret, C. (2002). Reverine landscape diversity. *Freshwater Biology* **47**: 517–539.

WII. (2017). Management plan for Baraila Wetland, Bihar. Technical Report. Wildlife Institute of India, Dehra Dun, and Department of Environment and Forests, Government of Bihar.

Yahya, H.S.A. (1995). Prospective bird sanctuaries in North Bihar-A survey of wetlands and their conservation strategies, pp. 24–31. In: J.R. Howes (ed.), *Conservation and Sustainable Use of FloodplainWetlands. Asian Wetland Bureau, Kuala Lumpur*, p. 123 (Proceedings of the Workshop on the Conservation and Sustainable Use of Floodplain Wetlands, December 1993, Calcutta-*AWB Publication* No: 113).

Yim, K.Q. and Ni, J.R. (1998). Review of wetland studies. *Acta Ecologica Sinica* **18** (5): 539–546.

Zargar, S. and Ghosh, T.K. (2006). Influence of cooling water discharges from Kaiga nuclearpower plant on selected indices applied to plankton population of Kadra reservoir. *Journal of Environmental Biology* **27**: 191–198.

Zhao, S., Peng, C., Jiang, H. et al. (2006). Land use change in Asia and the ecological consequences. *Ecological Research* **21** (6): 890–896.

ZSI (2011). *Zoological survey of India. Faunal Diversity of Chaurs of North Bihar (Zooplankton, Molluscs, Pisces and Aves).* Kolkata: *The Director, Zoological, Survey of India.*

5

Recognizing Economic Values of Wetland Ecosystem Services

A Study of Emerging Role of Monetary Evaluation of Chandubi Ecosystem and Biodiversity

Deepak Kumar[1] *and Moharana Choudhury*[2]

[1] *United Nations Development Programme (UNDP), New Delhi, India*
[2] *Voice of Environment (VoE), Guwahati, Assam, India*

5.1 Introduction

Chandubi Wetland is facing a serious ecosystem marginalization where ecosystem services have not been priced and reflected in decision making and which proves a complete market failure. Agriculture yield from transformed/converted/encroached lake does not reflect values lost due to flood protection, fisheries, biodiversity, etc. People who degrade are not the same whose livelihoods are affected leading to continued deterioration of wetland. Wetland governance has been ineffective to address sectoral policies providing incentives leading to wetland depletion. Chandubi Wetland has potentially abundant biodiversity of flora and fauna occupying some important medicinal herbs, forests, fishes, and others which we have aforementioned in the ecology of Chandubi Wetland. It is one of the potential ecotourism destinations and it gets the easily available logistic benefit of being near to the state capital Guwahati, Assam. An economic valuation is a powerful tool since it provides means of measuring and quantifying tradeoffs among multiple uses of wetlands (Barbier et al. 1997).

It has been a challenge for us to evaluate a monetary figure for Chandubi Wetland. Valuation of ecosystem services nowadays is a pertinent tool not only to signify the importance of wetlands in human well-being but it incorporates stakeholder engagement prioritizing nature's externalities when we exclude a significant role of the ecosystem in the process of valuation. Chandubi Wetland is one of the beautiful forms of nature's creation which had a seismic origin. So far, we have not many records of valuation of wetland ecosystems in Assam. One of such rare papers likewise "Valuation of Ecosystem Services and Benefits of Son Beel Wetland in Assam, India: A Case Study of Natural Solutions to Climate Change and Water (Kumar et al. 2020)" is only available as a case study for us in the State of Assam, India. We conducted a comprehensive valuation study approach including GIS study, social, economic and ecological study in Chandubi Wetland during January 2019 to February 2020. This particular paper signifies the role of Chandubi Wetland

Wetlands Conservation: Current Challenges and Future Strategies, First Edition.
Edited by Sanjeev Sharma and Pardeep Singh.

in offering multiple ecosystem services that ensure climate security, water security, vulnerability reduction against water-related disaster risks, etc.

The benefits that individuals, society, and the economy have access to from nature are ecosystem services. We will see how we use wetlands, water supply and cleaning, flood and storm protection, storage of carbon and climate management, provision of food and materials, scientific understanding, leisure, and tourism (Millennium Ecosystem Assessment 2005; TEEB 2010a; TEEB 2010b, TEEB 2010c; TEEB 2011). Wetlands are a natural infrastructure and a network of natural ecosystem resources (Krchnak et al. 2011). And where wetlands have disrupted successful restoration interventions, they either recover slowly (over decades or centuries) or switch to alternative conditions, which vary from the original (pre-sturbation) state (Moreno-Mateos et al. 2012). In any event, wetland degradation or destruction results in economic erosion of ecosystem resources and benefits, restoration of wetlands will recover some of these benefits. Chandubi's water management and ecosystem services will help recognize opportunities: (i) to better leverage and preserve the numerous benefits ecosystem water and wetland services provide; (ii) to establish methods that are more cost-effective than traditional technological solutions provide, and (iii) to reduce costs related to deepening biodiversity and habitats.

The Ramsar Convention's "wise use" approach has been internationally accepted as a central feature of wetland management. The approach recognizes the need to integrate ties between wetlands and humans, limiting wetland depletion and degradation, and thus stresses that sustainable human use of those habitats is consistent with conservation (Finlayson et al. 2011). This approach provides a new dimension for wetland usage because certain natural disturbances and interference are important in preserving resilience in wetland ecosystems.

The Ramsar Convention on Wetlands defines wise use as "maintenance of their ecological character achieved through implementation of ecosystem approaches, within the context of sustainable development". Ecological character is the "combination of ecosystem components, processes, and services/benefits that classify a wetland at a given point of time." Ecosystem services/benefits framework has been considered into the definition of ecological character as one of the facets of linking wetland ecosystem functioning and their human use for well-being (Finlayson et al. 2011). Wise use of wetlands needs participation by stakeholders and accountability in the negotiation of the trade-in ecosystem resources in connection with the diversity of wetland use to determine fair conservation outcomes (Finlayson et al. 2011). Economic assessment increases the ability to make rational decisions on the use and management of wetlands by serving as a tool for social input to inform society of the effects of consumer choices and behavior (Zavestoski 2004).

The economic interests of wetlands can be understood and measured by applying the current natural resource evaluation system. The economic interests of ecosystem services can be measured with a bio-physical or preferential approach. Biophysical methods include estimating the inner properties and the prime productivity of wetland habitats, including energy analysis, as well as the material flows (Costanza 1980; Odum 1996). Subjective preferences applied by individuals as a basis for assessment include preferences-based approaches.

The economic value consists of production value (benefits derived from ecosystem service provision within a given ecosystem state) and insurance value (ecosystems' ability to sustain performance values through resilience and reorganization capabilities) (Holling 1973; Walker et al. 2004). Complete Economic Benefit (TEV) is the amount of the benefit of all resources to the wetland ecosystems from suppliers across spatial and temporal scales to beneficiaries. In the TEV context, we defined two categories to derive from ecosystem services provided by Chandubi; the value of use involves a direct/indirect use of ecosystem services, while non-use value is other than use-value; the psychological satisfaction that a well-managed wetland is a significant asset for future generations.

Chandubi Wetland's economic value was calculated primarily based on knowledge obtained directly from market transactions or transactions in related markets. The direct market valuation approach accesses real market data to extract ecosystem services' economic value. The first form of market price based on values based on quantity and prices sold in a perfect market can be split into three broader categories. Secondly, the cost-based approach is based on cost estimates if ecosystem resources have been recreated by alternative means. Thirdly, the production function-based approach derives values from our awareness and information about the contribution of ecosystem services to economic activity.

The valuation method based on revealed preferences derives values based on "preferences revealed" through the purchase of goods and services at different income and price circumstances. Travel cost and hedonic price methods are two major methods based on revealed preferences. Revealed preferences observe real market behavior likewise market price, production function approach including travel cost and hedonic pricing.

Established methods of preference derive a willingness to pay or agree in hypothetical and built-in state decisions. The preferences mentioned are generally for nonuse values of wetland ecosystems with two generalized methods in general: (i) contingent valuation and (ii) modeling of choices.

Though we have limitations of valuing ecosystem services, some critics have been expressing concerns over the method of valuations. The complexity of ecosystem services coupled with the nature of various ecosystem services renders their individual classification impossible (Costanza and Folke 1997).

While valuating the monetary value of Chandubi Wetland, we studied some of the valuation studies in India and we considered them as case studies. Early economic valuation techniques were based on the use of the Travel Cost Method to assess the consumer surplus for Keoladeo National Park (Chopra 1998). James and Murty (1998) used Contingent Valuation Method for nonuser benefits from cleaning Ganga, while valuating Pallikarnai Marshlands in Chennai was based on a contingent valuation method to estimate a payback for ` 2096 per annum for improving the ecological status of wetland (Venkatachalam and Jayanthi 2016). (Anoop et al. 2008) the imputed use value of ` 1924 million to Asthamudi estuary using a mix of direct market and value transfer-based method for fisheries, husk retting, inland navigation, recreation, and carbon sequestration. Guha and Ghosh 2009 used an estimation of the annual recreational value of Indian people traveling by Indian Sundarbans, valued at $377 000, in the field of zonal travel. The cost of saving mangroves' lives by chance was ` 11.7 million a lifetime (Das and Vincet 2009). In an Indian Sunderbans analysis of habitat destruction and loss of biodiversity, losses of Rs. 6.2 billion were estimated to amount to 4.8 percent annually of regional GDP in 2009 (World Bank 2014).

Economic valuation in India is in a state of brainchild and it's been emerging spontaneously when wetland ecosystem services are being recognized globally as an important resilient infrastructure of climate change and disaster risk reduction. Governments are considering wetlands as an indirect and direct source of potable drinking water. Wetlands and their particular and distinct importance wetlands are habitats worth maintaining, enhancing, and sustaining. The loss of wetlands means an imbalance in the ecological balance. They protect the environment from the extremes of climate, disasters like cyclones, flash floods, etc. Being the world's highly productive ecosystems, they are rightly called the cradles of biodiversity. The fast-growing population, emissions, huge factories, deforestation, and massive unplanned construction projects are factors responsible for India's wetland dwindling. Thus owing to immense significance, the wetlands need immediate attention for their conservation and Ramsar Convention is one such step in this direction. In the Indian context, the wetland loss acute and chronic acute failure means the filling up of wet areas with soil and the gradual removal of forest cover followed by erosion and sedimentation of the wetlands over many years is the chronic loss (Choudhury et al. 2020).

5.2 Methodology of Ecosystem Valuation

Dollar ($)-based Ecosystem Valuation adopts three significant approaches estimating values of ecosystem services and benefits in dollars. Each approach is based on three different paradigms. Firstly, market rates are dependent on the willingness to pay disclosed. The second is introduced on the basis of circumstantial evidence that is imputed readiness to pay. The last survey is focused on an explicit willingness to pay.

5.2.1 Market Prices – Revealed Willingness to Pay

There are some ecosystems where the valuation of natural capital in terms of ecosystem services and benefits can be estimated based on market prices. Some of the products like fish, farm, and wood are directly traded markets. So, their prices can be estimated directly by understanding consumer and producer surplus dynamics as with any other market goods. Other ecosystem services such as clean water are used inputs in production and their value can be measured from profit-sharing amid final good production.

There are some ecosystem services comprising aesthetic and recreational values that may not be directly measured based on market prices. However, the prices people are willing to pay in the market for related goods and services can be used to estimate their values. People often pay a much higher value for viewing/siting/spotting significant natural ecosystem and biodiversity. Even the price of land or home is much higher in proximity to such prominent ecological sites.

We normally use different methods to evaluate market prices where there is a revealed willingness to pay. These methods are being predominantly discussed in this particular chapter along with the valuation of Chandubi Wetland.

5.2.1.1 Market Price Method

It is an estimate used when a particular natural capital of ecosystem goods and services is directly supplied in the commercial market for trade. This method estimates ecosystem products and services directly bought and sold in the commercial market.

The market price approach can be used to assess improvements in ecosystem products and services' quantity or quality. This method uses a traditional economic approach to determine economic benefits from sold goods based on the supply and demand chain, taking into account the quantity of goods purchased at different prices and the quantity of goods produced at different prices. Total net economic profit is the amount of user and producer surplus. We measure service value based on market price and quantity data of consumer surplus and product surplus.

Market price application includes data to estimate customer surplus and producer surplus. Estimating consumer surplus must be determined. This includes time-series data on the quantity of goods demanded at various prices, besides data on other factors influencing demand such as income and demographics. Estimating producer surplus requires data on variable production costs and requires revenue from products.

Market price approach represents our revealed willingness to pay for costs and benefits of directly purchased and sold products and services in markets such as timber, fish, farm and fuelwood, etc. This approach is simple to implement as price, quantity, and cost data are reasonably easy to access for developed markets. This method uses observed consumer preference data. This approach uses ordinary, common economic demand and supply chain techniques.

This system, however, has some significant limitations. Market data may be available for a limited number of ecosystem products and services provided by an ecosystem and may not represent a resource's gross productive uses. The real economic value of ecosystem products and services may not be determined by market imperfections or paralysis. Seasonal and spatial variations can cause fluctuations in the valuation of goods and services by the ecosystem. This approach fails to calculate the importance of large-scale changes that are likely to cause sudden demand or supply side effects of either products or services. Normally, the market price approach may not subtract the market value of other associated services used to market ecosystem goods, and, therefore, can overemphasize benefits.

5.2.1.2 Productivity Method

This approach estimates economic values for products and services from the environment that contribute natural resources to commercial goods. It refers to net income/derived value form. It is generally applicable when considering ecosystem products or services along with additional inputs to generate market goods. It simply means that water quality affects crop yield or the cost of treating and purifying municipal water. Wetlands act as purifying natural water. Such habitats naturally enhance surface and groundwater quality. Thus, the economic benefits of improved water quality can be measured by expanding revenue from high crop yield or decreasing the cost of providing clean water. Wetland ecosystem reduces additional heavy machine and management costs, serving as a natural filter.

We can safely presume that alongside such important habitats, there are a variety of settlements where agricultural runoff is routinely deposited to rural agglomeration in wetlands. Some wetlands exist alongside densely populated urban bodies that directly impact the environment municipal waste and city drainage. If we handle such habitats properly, we will mitigate the heavy cost of water quality accessible to our agricultural land and safe drinking water for people.

Application of efficient method involves data collection as to how changes in quantity and quality of natural resources affect – first, production costs for the final good; second, supply and demand for the final good; and third, supply and demand for other production factors. This approach is generally implemented in two particular cases – one where the resource in question substitutes other inputs. We may exemplify such a scenario as improved water quality in a reservoir would minimize additional purification costs through chlorination or other water purification techniques. In this case, an increase in resource quantity or quality would result in reduced production costs. In another case, producers of the final good get benefitted from changes in quality or quantity of the resource. Consumers are not at all affected. Improved quality of water for irrigation may increase high crop yield. If the market price of the crops to the consumer does not change, benefits can be estimated from changes in producer surplus resulting from increased income from the other inputs. The major advantage of this method lies in its simple and straightforward approach to estimation. This method requires limited data and significant data can be accessed easily that puts this method to be relatively less expensive to implement.

This method is limited to valuable resources that can be utilized as an input in the production of market goods. When valuing an ecosystem, not all services and benefits will be correlated in the production of marketed goods. Therefore, the inferred value of a particular ecosystem may not state an absolute value in society in totality. Scientific information is needed to understand relationships between action to improve the quantity or quality of the resources and the actual outcomes of those actions. In most of the cases, such information and relationship are lacking and we need such awareness at the local and ground level. This method becomes complicated and complex when the changes in the natural resource affect the market price of the final good or the prices of any other production inputs.

5.2.1.3 Hedonic Pricing Method

This method is prominently significant in estimating economic values for ecosystems or ecosystem services that directly influence market prices. It proficiently applies to the housing price fluctuation that is largely influenced by environmental attributes in proximity to the house. It can be used to estimate economic benefits or costs linked with the following major environmental attributes – firstly, environmental quality includes air pollution, water pollution, and noise pollution comprising other major regulatory services of wetland; secondly, cultural services where people access benefits through spiritual enrichment, cognitive development, reflection, recreation, and aesthetic experiences.

The basic assumption of this method lies in the price of a marketed good and is confined to its characteristics and the nature of the services it offers.

This method is normally used to estimate the values of environmental services and benefits that directly affect the prices of market goods. Most applications use housing prices to figure out the value of environmental services and benefits. The core assumption is that people value the characteristics of a good or service it offers rather than the good. Therefore, the final price will reflect the value of the ecosystem services and benefits that people avail while purchasing a good. This method is relatively straightforward and acceptable to apply because this method is based on actual market prices and readily accessible data that tend to be a relatively inexpensive method to apply.

The major advantage of this method lies in its strength that it can be used to estimate values based on actual choices. Property records and property markets are reliable sources of a good value indication. Data on property sales and characteristics are easily accessible through various sources and can be correlated to other secondary and tertiary data sources to obtain an extensive analysis of different attributes together. It is a versatile method and it can be useful to consider various possibilities and probabilities of interactions between market goods and ecosystem services and benefits.

There are wide ranges of limitations in applying this method of ecosystem valuation. Firstly, it is limited to housing prices. This method will capture only people's willingness to pay for ecosystem attributes in terms of services and benefits and their direct consequences. Therefore, if people are not aware of linkages or relationship between environmental attributes and benefits to them or their property, the value will not be reflected in the housing prices. The housing market is affected by other factors like taxes, interest rate, etc. The result depends on model specification where a large section of data can be gathered and manipulated.

5.2.1.4 Travel Cost Method

This method is applied to estimate the economic values of an ecosystem or sites that are used for recreation. This method can be applied to estimate the economic benefits or cost from changes in access costs for a recreational site first and second by eliminating an existing recreational site. The third is an inclusion of a new recreational site and, lastly, a change in environmental quality at a recreational site. The fundamental paradigm of the travel cost method is that the "price" to reach a site in which time and travel cost have consumed from the place one is traveling to the site of visit. Thus, a visitor's willingness to pay to travel a site can be estimated based on the number of trips that they expense at different travel costs.

The travel cost method is closely related or more likely to copy the most conventional empirical methods used by economists to estimate economic values based on market prices. The major advantage of this method is to develop an understanding of actual behavior what people do rather than the stated willingness to pay and what people say they do in a hypothetical situation. It is also a relatively low-cost method to apply. It provides a scope of on-site survey study where a large number of visitors are eager to participate. The results are relatively easier in interpreting and explaining.

This method has however some extraordinary and genuine issues and limitations. This method assumes that visitors perceive and respond to changes in travel costs in a similar way they respond to changes in entrance cost. This method assumes the stated willingness to pay by visiting a recreational site by an individual for a single purpose. If the trip has more than one purpose, the value of such a site may be overestimated. It can be difficult to correlate various trips for more than one purpose of a certain recreational site. This method cannot be able to define and measure the opportunity cost of time or time consumed traveling has so far been problematic throughout estimation. If people travel at self, it is obvious that time spent in traveling a cost would be less and travel cost may vary that could lead to overestimation. The availability of a substitute site shall affect values and this method has a limitation estimating the value of a site in case there remains the availability of multiple substitutes. This model has some limited scope when a visitor decides to stay in proximity to a site other than at the actual site which can lead to a low travel cost and this method fails to include them in the process of value estimation. Another critical aspect is that when you are interviewing a visitor at the site, make the sampling biases to the analysis. It is always a difficult approach to measure a recreational quality and relating it with environmental quality. It is quite difficult sampling visitors traveling all across to recreational site from different places. When a recreational site resides in a densely populated regime, people may have a wide range of differing travel costs as they may visit a recreational site with different destinations. The travel cost method is limited up to the scope of certain applications where there are direct participation users. It cannot be used to estimate values to on-site environmental attributes and functions that users of the site do not consider valuable. It cannot be used to value off-site values well supported by the recreational site. Significantly, it cannot be used to estimate nonuse values. Therefore, such recreational sites have intrinsic and unique qualities that on being valued by nonusers will be undervalued.

5.2.2 Circumstantial Evidence – Imputed Willingness to Pay

The importance of such ecosystem services can be calculated by calculating what people are willing to pay or the cost of actions they are willing to take, mitigating adverse effects that might prevail if these services are extinct/missing or replacing the services missing. Wetland function as a disaster risk reduction-based ecosystem. Such an ecosystem provides floodwater resistance. Estimates of willingness to pay for wetland flood mitigation programmes may be used to prevent flood damage in areas close to those covered by wetlands.

5.2.2.1 Damage Cost Avoided, Replacement Cost, and Substitute Cost Methods

These methods are related methods of estimating values of ecosystem services and benefits based on either the cost of preventing loss of services, the cost of replacing ecosystem services, or the cost of offering replacement services. These approaches do not have strict measures of economic values based on people's ability to pay for and support an ecosystem service. Rather than assuming the cost of preventing degradation or ecosystem replacement, or the cost of offering an alternative, provides useful estimates of ecosystem services and benefits. This approach is used appropriately in situations where damage prevention or repair expense or substitute costs have been or may actually be incurred.

This method gives a rough indicator of economic value, subject to data constraints. It is easier to produce costs of benefits than the benefits themselves. These approaches need less data resource. Data or resource limitations may rule out valuation methods that estimate willingness to pay.

However, this method has limitations. These approaches assume that expenditures to repair or to replace ecosystem services are valid measures of the benefits provided. However, costs are normally not an accurate measure of services. These methods do not consider social preferences for ecosystem services, or individual's behavior in the absence of those services. Thus, they should be used as a last resort to value ecosystem services. This method may be inconsistent because certain environmental actions and regulations are dependent solely on benefit-cost comparisons at the national level. The replacement cost method requires information on the degree of substitution between the market good and the natural resource. Certain environmental resources have such direct or indirect substitutes. The goods or services being replaced perhaps represent only a part of the full range of services provided by a natural resource. Thus, the benefits of an action to protect or restore ecological resource would be understated. Just because an ecosystem service is eliminated, it is no guarantee that the public would be willing to pay for the assigned least-cost alternative merely because it would offer the same benefit level as that service. This method is not an economically appropriate estimate of an ecosystem value of services and benefits.

5.2.3 Surveys – Expressed Willingness to Pay

Most ecosystem services are not traded in markets and they are not connected to consumer products. In such situations, surveys may be used to specifically ask people what they're willing to pay based on a hypothetical questionnaire-based scenario. Respondents may be asked to make tradeoffs between various alternatives to estimate their willingness to pay alternatives.

5.2.3.1 Contingent Valuation Method

It is used to estimate economic values grossly for all kinds of ecosystem and environmental services. It can be used to estimate both use and nonuse values. This is the most prominently used method for estimating nonuse values it is also the most controversial and disputable of the nonmarket valuation method. This method is based on questionnaire-based sampling, in a survey, how much they would be willing to pay for ecosystem and environmental services. There are cases where people were asked for the amount of compensation they would be willing to agree for giving up the specific ecosystem services. It is known as "contingent" valuation because people are asked to state their willingness to pay, contingent on a specific hypothetical assumption and description of the ecosystem and environmental services. This method is also referred to as a "stated preference" method because it asks people to directly state their values rather than considering values from actual choices as the "revealed preference" methods do. This method is the only way to consider dollar values to nonuse values of the environment – values that do not include market purchases and may not involve direct participation. These values are sometimes referred to as "passive use" values. It considers every attribute based on procuring, regulatory, cultural, and supporting functions of a wetland ecosystem as well as comprising biological diversity of such particular ecosystem. It is known of the fact that people are willing to pay for passive use/nonuse ecosystem services and benefits. However, these values are considered as zero till we won't have dollar value estimation. Since people do not reveal their willingness to pay for them through their purchases or by their behavior, the only option left is to ask people questions directly. However, this method is of asking people questions directly rather than assuming their actual behavior is highly controversial. It is highly debatable in economics conceptually, empirically, and practically to apply dollar-based estimates of economic values on the basis of how people respond to hypothetical questions about hypothetical market situations. This raises a doubtful state for the contingent valuation method. Policymakers often deny accepting the result of this method. It is such a controversial method that users must be cautious of investing money and time in contingent valuation studies. This method is extensively used for estimating "total economic value" including all types of nonuse or passive-use values. This method can estimate use values, existence values, option values, and bequest values.

5.2.3.2 Contingent Choice Method

This method is similar to the contingent valuation method in which it can be used to estimate the economic values for virtually any ecosystem or environmental services and can be used to estimate nonuse as well as use values. A contingent choice method is also a hypothetical method where people are asked questionnaire-based hypothetical scenario to make choices. However, this method does not directly ask people to say their values in dollars; instead, values are inferred from the questionnaire-based hypothetical choices or tradeoffs that people make. This method asks the respondent to state a preference between one set of environmental services or characteristics, at a given price or cost to the individual, and other sets of environmental services or characteristics at a different price or cost. This method is suited to policy decisions where a set of possible actions might result in different impacts on natural resources or environmental services. With improved management of wetland, we can access the improved nature of multiple environmental services likewise provisioning, regulating, cultural, or supporting.

This method offers various formats for applying for the economic valuation of ecosystem services and benefits. First are the contingent-ranking surveys that call on people to compare and rank alternate programme outcomes with various characteristics, including costs. People might be asked to prioritize and rank different ecological improvement programmes on consideration for a wetland ecosystem, each of which has a different outcome and different cost. Respondents are asked to rank alternatives in the order of preference. Second is the discrete choice where respondents are simultaneously given two or more different alternatives and their characteristics and asked to sort out the most preferred alternative in the choices. The third is the paired rating which is a variation on the discrete choice format. In this format, respondents are asked to compare a pair of alternate situations and rate them in terms of strength of preference. Respondents may be asked to compare two environmental improvement programmes and their outcomes and to mention which programme is preferred the most, moderate or slight compared to other programmes.

Whatever format is opted, the choices that respondents make are statistically analyzed using discrete choice statistical techniques to determine the relative values of the different attributes. If one of the characteristics is monetary price, it is possible to compute the respondent's willingness to pay for the other attributes. The major advantage of this method is that it can be used to value the outcomes of action as a whole as well as the various attributes or effects of the action. This method gives respondents a time frame to think about tradeoffs, which may be easier than directly stating dollar values. Respondents may respond in a better way to state preferences for different improved environmental activities. This is an advantage over the contingent valuation method in that it does not ask the respondent to make a tradeoff directly between environmental quality and money. Respondents here are more comfortable providing qualitative rankings and rating of attributes that include prices rather than dollar valuation.

However, there are limitations as well where respondents may find some tradeoffs difficult to evaluate because they may be unheard of or unfamiliar. Respondents may find a simple choice to prefer which may lead to a biased statistical interpretation. If the number of attributes or level of attributes is increased, the sample size or the number of comparisons each respondent makes must be increased. Contingent ranking needs more sophisticated statistical techniques to measure willingness to pay.

5.3 Ecosystem Services of Wetland

Wetlands are one of the most productive ecosystems in the world. However, immense invisible ecosystem services of the wetlands go unnoticed and non-assessed. Wetlands provide diverse sets of Provisioning, Regulating, and Cultural as well as Supporting Services.

Provisioning services include those benefits and products obtained from wetlands such as freshwater, food, fiber, fuel, genetic resources, biochemical, natural medicines, and pharmaceuticals.

Regulating services include those benefits obtained from the regulation of the ecosystem services likewise water regulation, erosion regulation, water purification, waste regulation, and climate and natural hazards regulation (e.g. floods, storms, and droughts).

Cultural services include nonmaterial benefits people obtain from wetlands through spiritual enrichment, cognitive development, reflection, recreation, and aesthetic experiences likewise cultural diversity, knowledge systems, educational values, social relations, sense of place, cultural heritage, and ecotourism.

Supporting services include those that are necessary for the production of all other ecosystem services. The beneficial impacts of such ecosystem services are indirect and on a long-time basis likewise primary production, water cycling, and nutrient cycling, etc.

5.4 Chandubi Wetland: Introduction, Impact, and Introspection

Chandubi Beel is one of such naturally perennial landscapes that were formed as a post-catastrophic consequence of tectonic submergence of forests during the massive earthquake of 1897 in Assam. It reflects a cultural ecotone of two bordering states, namely Assam and Meghalaya. Chandubi Lake is situated at the foothill of Garo Hills under Rabha Hasong Autonomous Council, Kamrup District, Assam, which is about 60 km distant from Guwahati. This wetland is in proximity to Borduar Reserved Forest to its north and Mayong Hill Reserved Forest to its south. River Kulsi, a southern tributary of River Brahmaputra, links downstream to Chandubi Wetland.

Assam Remote Sensing Application Centre (ARSAC) disclosed in its satellite imaging pictures that the wetland had a catchment area of 448.08 hectares in 1911–1912, which observed a drastic decline and which had led to the wetland's water-spread area up to only

186.52 hectares in 2007. This shrink persists till now. Dying wetlands and depleting biodiversity are a major threat globally when such ecosystems are acting as a natural structure of "Disaster Risk Reduction."

Chandubi Wetland does not fall in between urban and industrial agglomerations which led to minimal biotic interventions and "No Disturbance Zone" between man and wetland. This makes it a "secluded wetland" and it is a favorable habitat for birds' conservation of A1 category.

Pallas's fish eagle (*Haliaeetus leucoryphus*) and lesser adjutant (*Leptoptilos javanicus*) are the threatened species found in this lake, which attracts a large number of waterfowl. In the adjoining tropical moist deciduous forests, white-cheeked hill partridge (*Arborophila atrogularis*), mountain bamboo partridge (*Bambusicola fytchii*), Blyth's kingfisher (*Alcedo hercules*), blue-throated barbet (*Megalaima asiatica*), white-throated bulbul (*Criniger flaveolus*), grey peacock pheasant (*Polyplectron bicalcaratum*) are various biome species of conservation interest.

Chandubi Lake had been an abundantly rich habitat of many extinct fishes. Shrinking *Piscean* species is one of the major alarms to the ecological balance of the wetland and, ultimately, expanding human settlement is posing an undue pressure to life and livelihoods to a large extent. Studies disclosed that Chandubi Lake has around 90% of Piscean fauna and a 20% decline in the precious ornamental species of fish was which have already become extinct. Apart from a habitat of critically endangered fish species Nandhani and ornamental fish Phutkiputhi, this seismic wetland is home to Chital. River Kulsi is one of the habitats of unique Gangetic Dolphins (*Platanista gangetica*) which is linked with Chandubi Wetland and its hydrological ecosystem. Chandubi Wetland offers a rich ecosystem to such unique Dolphin species and fish species which later on migrates to River Kulsi through a long channel of around 2.5 km. Experts accept that Chandubi Lake is supposed to be a home to around 70 different species of fish.

Chandubi Wetland is a waterlogged productive ecological wealth which offers "Ecosystem-Based Disaster Risk Reduction" against consecutive flood every year. This wetland offers a balance to the hydrological cycle. This lake supports life and livelihoods to settlements in proximities like village Rajapara and other rural agglomerations. Being an extremely significant ecosystem of high economic and biodiversity value, native people show serious resentment over governance the way they are looking after this wetland ecosystem and biodiversity. The villagers blame that there has been operating a logger syndicate imposing serious sustained threat over forest resources. Illegal logging of timber trees in a protected reserve forest habitat raises serious concerns over the role of forest administration. Persistent logging has led to biodiversity erosion and it is imposing deleterious effect on the food chain system. Aging people say that there were once numerous numbers of clouded leopard and Bengal Tiger. Scarcity of food happened there and "food diversity" had reached a dismal level. Such top predators on a higher trophic pyramid have reported extinct because Chandubi Wetland is not offering a conducive habitat for wild pigs, swamp deer, barking deer, Himalayan black bear, hoolock gibbon, porcupine, etc., once the availability of Gangetic crocodiles (*Gravialis gangeticus*) was reported till 1960 but now they are completely extinct from this wetland ecosystem.

Chandubi Wetland is not facing urban municipal dumping or corporate invasion till now. Borduar-Myong protected forest regimes surround this tectonic wetland that develops a beautiful scenic wetland site in India. Such a scenic ecotourism site entices a large number of tourists. Winter (April–November) is the best phase to visit this seismic wetland. Chandubi hasn't been developed into a full-fledged ecotourism spot yet; one can move early morning from Guwahati hiring a private cab or arrive early by public transport so that they could return by evening. The tourists should carry an adequate amount of food with themselves. If one can stay for a day or two, there is a well-furnished government guest house but one would have to bring food, emergency light, battery power bank, and other necessary accessories as there may not be electricity or canteen. One can hire a boat to go fishing or watching the scenic beauty of the lake and forest coverage around this natural tectonic structure.

Ecotourism sometimes deteriorates the quality of such a precious and productive ecosystem but giving a nod to people to interact with the natural ecosystem, specifically a wetland, live is a need of the hour. University, schools, nongovernment organizations, etc., should make it necessary to visit wetlands for the purpose of research analyzing the significance of such a productive ecosystem for the social and economic welfare.

While assessing "The Economics of Ecosystem and Biodiversity (TEEB)" of Chandubi Wetland, we find that cultural and ecological externalities operate in an enormous way to invent multiple dynamics of livelihoods for locals and an enormous economic source for the state without disturbing the normal discourse of the ecosystem. This tectonic wetland has been a sacred site for Rabha-dominant inhabitations. The state administration with locals organizes the famous Chandubi Festival in the first week of every year. This festival designates the fringe culture of various tribes constituting in the Borduar-Mayong Forest Range in the Garo Hills neighbouring Meghalaya and Assam. The tourists visit the great native festival all across the north-east states, India and the world. We witnessed overseas tourists during the festival. This festival comprises all about diverse food, fabric, traditional war-tactics, dance, art and craft, traditional herbal medicines and treatments, traditional fishing and hunting tactics, boating, farming, and harvesting, etc., of multiple ethnic tribes.

The simplification of TEEB valuation is our preliminary step of this whole exercise. We have limited baseline data and Chandubi Beel is facing serious anthropogenic detrimental constraint. People are not informed about its high productivity and its wide range of ecosystem services and benefits. We work on a comprehensive evaluation regime to bring larger community's association and make them informed about the economic status of Chandubi Beel adopting a six-step approach:

Step 1: Specify and approve Chandubi Beel's stakeholder concerns.
Step 2: Identify a list of relevant ecosystem services on a priority basis for locals.
Step 3: Identify the baseline information.
Step 4: Assess anticipated improvements in the ecosystem services' availability and delivery.
Step 5: Identify and assess policy options based on analyzing potential changes in the ecosystem services.
Step 6: Assess the social and environmental effects of policy options as a transition in the ecosystem resources affects wise citizens

We worked on demarcating, identifying, and listing out the diversity of ecosystem services being availed of directly or indirectly either by locals and nearby communities, or visitors, research groups, etc.

We use this particular method for the fact that Chandubi Wetland is developing as a beautiful touring destination that links native ethnic communities and their crafts to access a larger market value. The annual Chandubi festival attracts around half a million tourists

Table 5.1 We have identified and listed ecosystem services that were being put in tabular statistics below:

Chandubi Wetland	Ecosystem services assessed	Policy and decision making
Inland	• Food • Fiber and fuelwood • Freshwater • Moderation of diverse water regimes • Erosion resilience • Mitigating pollution and detoxification • Erosion protection • Groundwater recharge • Recreation • Species habitat	• Wetland restoration • Land-use planning and regulation • Integrated Water Resources Management • Property rights • Use of market-based instruments • Financing • Livelihood generation

Source: Choudhury et al. (2021).

Table 5.2 Provisioning ecosystem services and monetary valuation of Chandubi Wetland.

Provisioning services (PS)	Number of used attributes	Minimum values ($/hectare/year)	Maximum values ($/hectare/year)
1) Food	10	10	1813
2) Freshwater supply	02	02	1730
3) Raw materials	06	06	1218
4) Genetic resources	N.A.	N.A.	N.A.
5) Medicinal resources	N.A.	N.A.	N.A.
6) Ornamental resources	N.A.	N.A.	N.A.
Total	18	18	4761

in January alone around India and overseas which explores their ethnic crafts open to market access at a satisfactory price. Chandubi Wetland is surrounded by five major village agglomerations, mainly Raja Para, Joramukhia, Kothalguri, Bargaon, and Sonapara comprising a population of around 2651 people according to census 2011. Most of the inhabitants are engaged in agriculture, fishing, boating, private services, government jobs, and daily wage labor. This wetland supports an extensive range of provisioning, regulating, supporting and cultural ecosystem services. We studied an extensive pattern of ecosystem services and how wise policy and planning could work to create a healthy ecosystem that could be capable of offering a vast range of services and benefits (Table 5.1). We observe 18 number of different used attributes as a part of provisioning services (Table 5.2). We observe 11 number of used attributes comprising regulating services of the Chandubi Wetland. (Table 5.3). There are 2 used attributes during evaluation study as a part of wetland's supporting services (Table 5.4). Chandubi Wetland comprises total 19 number of used attributes in our ecosystem valuation (Table 5.5).

Table 5.3 Regulating ecosystem services and monetary valuation of Chandubi Wetland.

Regulating services (RS)	Number of used attributes	Minimum values ($/hectare/year)	Maximum values ($/hectare/year)
1) Influence on air quality	N.A.	N.A.	N.A.
2) Climate regulation	03	03	210
3) Moderation of extreme events	03	03	1899
4) Regulation of water flows	01	01	2342
5) Waste treatment/water purification	03	03	952
6) Erosion control	N.A.	N.A.	N.A.
7) Nutrient recycling	01	01	918
8) Pollination	N.A.	N.A.	N.A.
9) Biological control	N.A.	N.A.	N.A.
Total	11	11	6321

Source: Choudhury et al. (2021).

Table 5.4 Supporting ecosystem services and monetary evaluation of Chandubi Wetland.

Supporting services (SS)	Number of used attributes	Minimum values ($/hectare/year)	Maximum values ($/hectare/year)
1) Life cycle maintenance	01	01	459
2) Gene pool protection (Conservation)	N.A.	N.A.	N.A.
Total	01	01	459

Source: Choudhury et al. (2021).

Table 5.5 Cultural ecosystem services and monetary evaluation of Chandubi Wetland.

Cultural services (CS)	Number of used attributes	Minimum values ($/hectare/year)	Maximum values ($/hectare/year)
1) Aesthetic information	04	04	7812
2) Opportunities for recreation and tourism	07	07	2877
3) Inspiration for culture art and design	05	05	1985
4) Spiritual experience	03	03	175
5) Cognitive information (Education and research)	N.A.	N.A.	N.A.
Total	19	19	12849

Source: Choudhury et al. (2021).

We can observe scenic and recreational nature of Chandubi Wetland (Photo 5.1). We need an upgradated business model for making it a classic destination for tourism and research. We observe a traditional ticket counter for boating (Photo 5.2).

Total number of used attributes = PS + RS + SC + CS = 18 + 11 + 01 + 19 = 49

Total minimum values = PS + RS + SC + CS = 18 + 11 + 01 + 19 = 49 ($/hectare/year)
Total maximum values = PS + RS + SC + CS
= 4761 + 6321 + 459 + 12849
= 24390 ($/hectare/year)

And, we conducted this ecosystem evaluation exercise during January 2019 to February 2020. During this period, the average dollar value was around ` 71 (in Indian currency)

Total minimum values = 49 * 71 = ` 3479/hectare/year
Total maximum values = 24390 * 71 = ` 17 31 690/hectare/year

We estimated the monetary value of Chandubi Wetland ranges from a minimum of $49 hectare/year to a maximum of $24 390/hectare/year.

Photo 5.1 Scenic view of Chandubi Wetland, Assam, India.

Photo 5.2 Boating ticket collection center at Chandubi Wetland.

5.5 Scaling up Wetland Conservation, Wise Use, and Restoration for Achieving Sustainable Development Goals

Wetlands are pertinent for our well-being, inclusive economic growth, and climate mitigation and adaptation. They are the biggest sources of freshwater for our consumption, agriculture, and maintaining our groundwater table by naturally recharging and filtering it. They act as a natural water sink. They are the biggest terrestrial ecosystem for carbon sequestration and acting as a natural systematic carbon sink system. They act as an "Ecosystem System-Based Disaster Risk Reduction" structure protecting shores and providing cities and settlements with a safe and climate-resilient prospect. They provide sustainable livelihoods for the community welfare and offer a healthy ecosystem for exploring multiple ecosystem services and benefits in parallel to abundant biodiversity such ecological systems support. These multiple benefits and services provided by wetlands are significant in achieving Sustainable Development Goals (SDGs).

The Ramsar Convention's fourth strategic plan (2016–2024) identifies 4 overarching goals and 19 specific targets in the wake of Agenda 2030.

Goals and Specific Targets of 4th Ramsar Strategic Plan:

1) Addressing Drivers of Wetlands' Loss and Degradation
   - 1.1 Featuring wetland benefits in national/local policy strategy and plans focusing on key sectors like water, energy, mining, agriculture, tourism, urban development, infrastructure, industry, forestry aquaculture, and fisheries at the national and local levels (Target 1).
   - 1.2 Ensuring the appropriate amount of water a wetland ecosystem needs for providing benefits and services (Target 2).
   - 1.3 Implementing wise-use planning guidelines in public and private sectors (Target 3).
   - 1.4 Prioritizing systematic identification of invasive alien species and thereafter managing a planning framework for controlling and eradicating such species (Target 4).
2) Effectively Conserving and Managing Ramsar Site Network
   - 2.1 Effective planning and integrated management to restore and maintain the ecological character of Ramsar sites (Target 5).
   - 2.2 Expanding wetland areas under Ramsar Designation of Wetlands of International Importance (Target 6).
   - 2.3 Addressing the risk of threats of change in ecological character (Target 7).
3) Wisely Using All Wetlands
   - 3.1 Completing national wetland inventories for promoting the conservation and effective management of all wetlands (Target 8).
   - 3.2 Strengthening wise use of wetlands for integrated management of river basin or coastal zone (Target 9).
   - 3.3 Respecting and using indigenous knowledge resource for the wise use of wetland management plan (Target 10).
   - 3.4 Documenting wetland's functions, services, and benefits (Target 11).
   - 3.5 Restoration of degraded wetlands prioritizing biodiversity conservation, livelihoods, disaster risk reduction, and climate change mitigation and adaptation (Target 12).

3.6 Enhancing sustainability in key sectors – water, energy, mining, agriculture, tourism, urban development, infrastructure, industry, forestry aquaculture, fisheries when they affect wetlands, contributing to biodiversity conservation and human livelihoods (Target 13).

4) Enhancing Implementation
   4.1 Developing guidance and political methodologies at global and regional levels (Target 14).
   4.2 Reinforcing Ramsar Regional Initiatives for the full implementation of the convention (Target 15).
   4.3 Mainstreaming wetland conservation through Communication, Capacity Development, Education, Participation and Awareness (CEPA) (Target 16).
   4.4 Mobilizing resources for implementation (Target 17).
   4.5 Strengthening international cooperation (Target 18).
   4.6 Building capacity to implement the convention and strategic plan (Target 19).

5.6 Wetlands' Role in Achieving SDGs

SDG 1 End Poverty in All Its Forms Everywhere

Building resilience of the poor and vulnerable is one of the targets under this goal and one where the role of wetlands can be clearly illustrated. Wetlands provide a clean potable drinking source of water (even during drought period) for cattle, agriculture and human consumption. In Cameroon, the restoration of the Waza Floodplain, a Ramsar site, helped reinstate the flooding regime. It resulted in an improvement of gross livelihoods, agricultural yield, grazing, and fishing, thus generating economic benefits estimated at $2.3 million per year.

Changes in tidal marshes and mangroves have led to an estimated "Global Financial Depletion" of $7.2 trillion per year and those from the declining coral reef is $11.2 trillion annually. This reflects an enormous depletion of revenue and potential livelihoods of marginalized and vulnerable communities globally. In Malawi, Lake Chilwa Wetland is home to extremely poor and marginalized communities. Yet this wetland provides an abundant productive fish protein source to the surrounding inhabitants. Its gross monetary value is estimated to total $21 million per year. Recently, the Kingdom of Bhutan has identified and quantified the values (including food, genetic resources, DRR, and cultural values) of Bhutan's inland wetlands for a total monetary value of USD 50 million per year. On a per hectare basis, inland wetlands are estimated to provide the highest values @ $14 183/hectare/year.

Target 1.5 of SDG 1 is committed to building the resilience of the poor and those in vulnerable situations and reducing their exposure and vulnerability to climate-related extreme events and other economic, social, and environmental shocks and disasters by 2030. Loss of wetlands, whether it is from climate change, upstream large hydropower establishments, or anthropogenic causes, severely affects the way of life and livelihoods of local surrounding communities. While it is hard to establish direct causality, there are a number of studies linked between migration and water scarcity.

SDG 2 End Hunger, Achieve Food Security and Improve Nutrition and Promote Sustainable Agriculture

Food production relies largely on water from man-made and natural wetlands. Rice, which is one of the staple crops feeding half of the world population, is grown mainly in man-made and natural wetlands. Wetlands store an enormous mass of water to be utilized to irrigate land under cultivation. Wetlands are an important source of food protein for many people around the world. In Cambodia, fish from large Tonle Sap Lake and associated floodplains, which include two Ramsar sites, provide communities with 60–80% of their animal protein. Wetlands are the home of many edible species which provide food security to the local communities.

SDG 5 Achieve Gender Equality and Empower All Women and Girls

Internationally we are aware of how women are predominantly responsible for food collection and agriculture, and for water collection and management. Women's role and their instinctive knowledge resource have been predominantly unrecognized and unnoticed and social and economic norms often impose discriminatory and unequal approach toward bringing women's participation in decision making.

Wetlands' conservation, management, and restoration projects need to be gender-sensitive recognizing the differentiated knowledge, roles, needs, and vulnerabilities of men and women and continuing to empower women in governance and decision making. Conference of Parties/COP 13 considers a draft resolution which confers a process to increase awareness among contracting parties on the linkages between gender equality and wetland management. Women have different access to and control over natural resources and information about how conservation and wise use shapes the way in which wetlands are managed, affecting their rights and customary uses of wetlands products and services.

SDG 6 Ensure Availability and Sustainable Management of Water and Sanitation for All

Wetlands are critical to ensure water availability. Almost all of the freshwater is drawn directly or indirectly from wetlands. Wetlands provide a fundamental natural infrastructure for freshwater making wetlands a ground base to ensure water security for all. Wetlands are natural water filters. Wetlands' vegetation captures nutrients, pollutants, and sediments, and thus, maintains the quality of groundwater and surface water. Though mismanaged wetlands are heavily polluted and if used for irrigation or drinking water, these may prove lethal for human health. So, restoring the functional quality of wetlands is critical for providing some life-supporting intangible ecosystem services. Wetlands around Kampala act as a natural filter and prevent contamination to reach Lake Victoria which is a critical drinking water source for 1.3 million inhabitants of the capital city.

The basin-level approach provides a framework for Payment for Ecosystem Services (PES). Large scale deforestation in the Atlantic Forest deteriorated the quality of water of River Parana which provides drinking water solely to Sao Paolo's inhabitants. The Nature Conservancy started this PES Project led to the major water users such as water supply companies and industries to pay a fee to support farmers and ranchers to plant trees among riparian zones in the river's headwaters. Target 6.6 seeks to protect and restore ecosystems. The Ramsar Convention is a co-custodian of Indicator 6.6.1 that monitors the change in the extent of water-related ecosystems over time. The convention provides data submitted by contracting parties on wetlands critical to monitoring the status and taking decisions on managing water ecosystems.

SDG 8 Promote Sustained, Inclusive and Sustainable Economic Growth, Full and Productive Employment and Decent Work for All

Economic growth models have so far been based on deteriorating, depleting, and exploiting natural resources and we have created massive disaster risks over a period of time. Currently, we have successfully designed concepts of "Green Economy," "Green Growth," and "Natural Capital" recognizing the significance of securing economic growth models that conserve and use sustainably our natural resources. Decoupling economic growth from ecological degradation is one of the important targets and wetlands contribute to this target by providing water to agriculture and industrial production such as nutrient recycling, protection against flooding and water filtration naturally, otherwise it would have needed a massive cost of engineering to install such technologies. One of the targets of creating employment in the tourism sector and wetlands prove 266 million jobs in tourism and travel services alone.

SDG 9 Build Resilient Infrastructure, Promote Inclusive and Sustainable Industrialization and Foster Innovation

Intensifying climate uncertainties and extremities require advanced infrastructural protection measures. Since 1995–2015, flood and storms globally affected about 3 billion people damaging 87 million residences as well as 130 000 public buildings. Ecosystem-Based Disaster Risk Reduction (DRR) could have been an effective and efficient solution-building to mitigate and adapt to these mounting risks. Wetlands provide a cost-effective and efficient natural infrastructure against water-related catastrophes. Despite restoring such natural pivotal ecosystem-based DRR structures is a costly affair, they are less costlier than constructing engineered infrastructures for protecting coastal habitations from flooding and storms. South Africa's "Working For Wetlands" Programme combines water conservation with employment generation. It aims to restore degraded wetlands for smooth and continuous public water supply and generate jobs for the most marginalized people in the community. In Vietnam, Red Cross has paid $1.1 million to restore, protect, and conserve 12 000 hectares of mangroves to protect coastal zones and habitation from typhoons as well as offering subsidiary benefits of culturing molusks, shrimps, and seaweeds to supplement diets and income where it was estimated $7.3 million for dike maintenance earlier. A sustainable drainage system contributes to this target by using the landscape to control the flow and volume of surface water from rainfall, preventing and controlling pollution downstream and promoting groundwater recharge.

SDG 11 Make Cities and Human Settlements Inclusive, Safe, Resilient, and Sustainable

Water-related disasters including flood, drought, cyclone, etc., account for over 90% of overall disasters in the last two decades. Ecosystem-based disaster risk reduction infrastructure such as mangroves, coral reefs, and salt marshes provide low-cost natural immunity to the coastal regime by reducing wave height and potential strength, reducing storm surges, and absorbing some of the excess water. It was observed during Hurricane Sandy in the USA during 2012, these wetlands mitigated its dangerous and devastating impact in 12 provinces saving $625 million in flood damage. These low-cost EcoDRR structures provide not only community resilience against water-related risks but enabling communities to better adapt to climate change, and provide multiple ecosystem services. Wetlands serve as upstream retention basins protecting downstream cities from flood risk.

SDG 13 Take Urgent Action to Combat Climate Change and Its Impacts

"Wetlands in all parts of the world share a significant role in DRR if wetlands are effectively managed and restored wherever necessary." This is how parties to the Ramsar Convention agreed in 2015. Wetland soils contain 35% of the world's organic carbon. Coastal ecosystems, in particular, mangroves, coral reefs, salt marshes, sea-grass beds sequester two to four times more carbon than terrestrial forests and these "Blue Carbon Ecosystem" play a significant role in climate change mitigation. This carbon is stored for a persistent period in wetland soil preventing further degradation, drainage, and depletion of the wetland ecosystem which is critical to prevent further Green House Gas Emissions (GHGEs).

Peatlands contribute 3% of the earth's surface but they hold twice as much carbon as the world's forests. Losing and mismanaged wetlands contribute to intensify climate change while restoring and conserving them shall help in building resilience and climate change mitigation.

SDG 14 Conserve and Sustainably Use the Oceans, Seas, and Marine Resources for Sustainable Development

Around 3 billion people depend on marine resources for their essential primary source of protein supplement in their food. Recognizing the significance of marine resources to half the global population, this goal stresses to reduce pollution, sustainably manage and protect coastal ecosystems, reduce overfishing, and conserve at least 10% of coastal and marine areas by 2020.

SDG 15 Protect, Restore, and Promote Sustainable Use of Terrestrial Ecosystems, Sustainably Manage Forests, Combat Desertification and Halt and Reverse Land Degradation and Halt Biodiversity Loss

Terrestrial ecosystems including forests and wetlands and the biodiversity they hold can be considered as nature's gift to the human. Target 15.1 specifically deals with conservation, restoration, and sustainable use of terrestrial and freshwater ecosystems and their services including wetlands in specific. Wetlands have been estimated to provide 40% of global renewable ecosystem services. Though assessing gross ecosystem services of the wetlands in its monetary value is quite difficult, the Economics of Ecosystem and Biodiversity has assessed and estimated potential natural capital value of ecosystem services and benefits these wetlands provide worldwide is at $36.2 trillion and that of forests at $19.5 trillion.

Agenda 2030 provides a broader roadmap for national and international policy action for governments, civil society, private sector, and other state/non-state actors to achieve SDGs for our present and future generations. Wetlands provide a wide range of natural capital flow in terms of ecosystem services for the life and livelihood of people and community. The commitment of parties to the Ramsar Convention designating Ramsar sites, wetlands of International Importance to conserve and wisely use all their wetlands is critical in achieving the SDGs.

We need to ensure that wetland conservation, wise use, and restoration are an integral part of SDGs' planning and implementation. Contracting parties to the Convention on wetlands should ensure that their efforts to implement Ramsar Strategic Plan are integrated into their SDGs' planning and implementation efforts including wetlands and Ramsar Convention in National SDGs' plan with an additional inventory on how wetland conservation, protection, restoration, and wise use are critical to achieving SDGs so that we can give larger emphasis on conserving and managing wetlands at local levels. Integrating wetland services and benefits in Nationally Determined Contributions for the Paris Agreement on Climate Change is critical for achieving SDGs. Adopting policies and practices for the conservation and wise use of wetlands to prevent further depletion and degradation of wetlands are keys to scale up wetland intervention and minimizing local conflicts at the community level. Identification of services and benefits these Ramsar sites provide for people and environment when they are being designated to help improve understanding of a site's values and subsequent proposed actions related to conserve, wise use, and restore it.

Developing multi-stakeholders' partnerships is a critical means of implementation of wetland conservation, wise use, and restoration. It is a much-needed alignment to involve partnerships between the different sector and level of society Ramsar sites. Seizing opportunities and synergies with other sectors, conventions, and priorities is the need to facilitate collaboration between different focal points at the national convention level and SDGs' implementation level to advance broader landscape approaches to conservation and sustainable development related to Ramsar sites and other World Heritage sites. Increase funding of comprehensive actions in wetlands including Wetlands of International Importance' is a reflection of the significance of such natural ecosystems not only from an abundant biodiversity perspective but also with the prospective and role these ecosystems play in implementing, managing, and achieving sustainable development for current and future generations.

5.7 Conclusion

Wetlands have social, economic, and ecological significance. They offer a diversity of ecosystem services and benefits. Wetlands are proving to be a watershed; the intensity of urban flash flood risk has magnified multiple times now because of dying wetlands. People's ignorance and knowledge deficit have put such a vital ecosystem of disaster risk reduction at stake. Wetland governance lacks an institutional framework at the policy level. It takes capacity-building measures to involve communities to understand the significant role of wetlands in people's lives and livelihoods. Economic valuation of ecosystem services and benefits shall give governments a far-sighted vision to formulate sustained policies to protect and conserve such significant ecosystems and their biodiversity. Wetlands offer water security as well as a formidable structure of an ecosystem-based disaster risk reduction. Chandubi Wetland has an immense potential to be designated as a "Ramsar Site" of wetlands of international importance. We tried to produce a broader picture of the monetary evaluation of this wetland within our limited resources so that people could understand the used and nonused values of wetlands and their potential role in driving sustainable development goals.

Acknowledgments

The authors are thankful to **Rabha Hasong Autonomous Council** for their support during the field study. The authors are also grateful to **Voice of Environment** (Youth Environmental Organization) for the necessary support for data collection, field study survey time at **Chandubi Wetland** area.

References

Anoop, P., Suryaprakash, S., Umesh, K.B., and Amjath Babu, T.S. (2008). Economic valuation of use benefits of ashtamudi estuary in South India. In: *Proceedings of Taal 2007: The 12th World Lake Conference*. Jaipur, India, 28 October–2 November 2007 (eds. M. Sengupta and R. Dalwani), 1822–1826. New Delhi: Ministry of Environment and Forests, Government of India.

Barbier, E.B., Acreman, M.C., and Knowler, D. (1997). *Economic Valuation of Wetlands: A Guide for Policy Makers and Planners*, 78–85. Gland: Ramsar Convention Bureau.

Chopra, K. (1998). *The Valuation of Biodiversity Within Protectedareas: Alternative Approaches and Case Study*. New Delhi: Institute of Economic Growth.

Choudhury, M, Sharma, A, Chakravorty, A., and Dutta, J. (2020). Wetlands are in Peril – a case study of son beel Wetland of Assam. Natural Resources and their Ecosystem Services, 102.

Choudhury, M., Sharma, A., Singh, P., and Kumar, D. (2021). Impact of climate change on wetlands, concerning Son Beel, the largest wetland of North East, India. In: *Global Climate Change* (eds. S. Singh, P. Singh, S. Rangabhashiyam and K.K. Srivastava), 393–414. Elsevier.

Costanza, R. (1980). Embodied energy and economic valuation. *Science* 210: 1219–1224.

Costanza, R. and Folke, C. (1997). Valuing ecosystem services with efficiency, fairness and sustainability as goals. In: *Nature's Services: Societal Dependence on Natural Ecosystems* (ed. G. Daily), 49–70. Washington, DC: Island Press.

Das, S. and Vincent, R.V. (2009). Mangroves protected villages and reduced death toll during Indian super cyclone. *Proceedings of the National Academy of Sciences* 106 (18): 7357–7360.

Finlayson, C.M., Davidson, N., Pritchard, D. et al. (2011). The ramsar convention and ecosystem – based approaches to the wise use and sustainable development of Wetlands. *Journal of international Wildlife Law and Policy* 14 (3–4): 176–198.

Guha, I. and Ghosh, S. (2009). A Glimpse of the Tiger: How Much are Indians Willing to Pay for It? [SANDEE Working Paper No. 39–09]. Kathmandu, Nepal: South Asian Network for Development and Environmental Economics (SANDEE).

Holling, C.S. (1973). Resilience and stability of ecological systems. *Annual Review of Ecology and Systematics* 4: 1–23.

James, A.J. and Murty, M.N. (1998). Measuring non–user benefits from cleaning ganges. [Working Paper E/202/99]. New Delhi: Institute of Economic Growth.

Krchnak, K.M., Smith, D.M., and Deutz, A. (2011). Putting Nature in the Nexus: Investing in Natural Infrastructure to Advance Water-Energy-Food Security, Bonn2011 Conference: The Water, Energy and Food Security Nexus – Solutions for the Green Economy. Background Papers for the Stakeholder Engagement Process.

Kumar, D, Choudhury, M, and Rathore, A. (2020). Valuation of Ecosystem Services and Benefits of Son Beel Wetland in Assam, India: A Case Study of Natural Solutions to Climate Change and Water. *Climate Change Conference*, Hamburg University, Germany.

Millennium Ecosystem Assessment (MA) (2005). *Biodiversity Synthesis Report*. Washington, DC: World Resources Institute.

Moreno-Mateos, D., Power, M.E., Comin, F.A., and Yockteng, R. (2012). Structural and functional loss in restored wetland ecosystems. *PLoS Biology* 10 (1): e1001247. https://doi.org/10.1371/journal.pbio.1001247.

Odum, H.T. (1996). *Environmental Accounting: Energy and Environmental Decision Making*, 370. New York: Wiley.

TEEB (2010a). *The Economics of Ecosystems and Biodiversity: Ecological and Economic Foundations* (ed. P. Kumar). London and Washington: Earthscan.

TEEB (2010b). *The Economics of Ecosystems and Biodiversity for Local and Regional Policy Makers* (eds. H. Wittmer and H. Gundimeda), 209.

TEEB (2010c). *A Quick Guide to The Economics of Ecosystems and Biodiversity for Local and Regional Policy Makers* (eds. H. Witmer and H. Gundimeda), 8.

TEEB (2011). *The Economics of Ecosystems and Biodiversity in National and International Policy Making* (ed. P. Ten Brink). London: Earthscan.

Venkatachalam, L. and Jayanthi, M. (2016). Estimating the economic value of ecosystem services of pallikaranai marsh in chennai city: a contingent valuation approach. [MIDS Working Paper No. 220]. Chennai, India: Madras Institute of Development Studies (MIDS)

Walker, B.H., Holling, C.S., Carpenter, S.R., and Kinzig, A.P. (2004). Resilience, adaptability and transformability. *Ecology and Society* 9 (2): 5.

World Bank. (2014). Building Resilience for Sustainable Development of the Sunderbans – Strategy Report (Report No.88061–iN). Sustainable Development Department, Environment & Water Resources Management Unit, Washington DC: The World Bank.

Zavestoski, S. (2004). Constructing and maintaining ecological identities: the strategies of deep ecologists. In: *Identity and the Natural Environment: The Psychological Significance of Nature* (eds. S. Clayton and S. Opotow), 297–316. Cambridge, MA: The MiT Press.

6

Ecosystem Services of Lagoon Wetlands System in India

Sadaf Nazneen[1], Gauhar Mahmood[1], Zoha Jafar[1], and Sughosh Madhav[2]

[1] *Department of Civil Engineering, Faculty of Engineering, Jamia Millia Islamia, New Delhi, India*
[2] *School of Environmental Sciences, Jawaharlal Nehru University, New Delhi, India*

6.1 Introduction

Lagoons occupy 13% of the earth's surface and are marine habitats with high biological productivity due to their unique ecological position (Velasco et al. 2018). These coastal wetlands are transitional in nature due to their positioning between the land and the sea. They appear due to the formation of barrier islands which cuts them from the sea; however, they maintain a restricted connection with the sea through one or several inlets. They are often drained by few rivers and hence a range of salinity exits in the water body. The presence of fresh, brackish, and saline water gives rise to rich biodiversity in these ecosystems. Coastal lagoon ecosystems also support a wide range of human activities and form an economic base for the local community, including sectors such as fisheries and aquaculture, as well as leisure and tourism (Newton et al. 2014). The lagoons deliver not only goods but provide various important services like flood management, climate regulation, carbon storage, nutrient cycling, habitat provisioning, to name a few (Newton et al. 2018). The lagoons hold significant ecological importance; hence, overall management including ecologists, environmental scientists, managers, and economists is required for their conservation and management (Newton et al. 2014; Pandisamy et al. 2020). Lagoons are shallow water bodies separated from the seas or oceans by barrier islands or reef. Coastal lagoons occupy 15% of the world's shoreline. They tend to accumulate sediments from the inflowing rivers, run off from the catchment area, and also the sediments carried through the inlets from the tides. Submerged macrophytes, thick algal mats, and marsh plants account for significant primary production and facilitate the accumulation of sediments. The phytoplankton community, benthic macroalgae as well as macrophytes, contribute to the productivity of lagoons.

Rich and diverse fishery resources are the benchmark of lagoons (Duarte et al. 2013; Srichandan et al. 2015b). Both finfish and shellfish thrive in lagoons forming the food and livelihood basis of a large number of local communities. In many lagoons, there are traditional community-based management systems in place which ensure that the

Wetlands Conservation: Current Challenges and Future Strategies, First Edition.
Edited by Sanjeev Sharma and Pardeep Singh.

resources are sustainably used. They are home to a variety of native and migratory birds which feed on various algae and planktons. Lagoons are also crucial for tourism and recreational activities and hold religious significance in many cultures. Coastal regions are substantially diverse and complex, predominantly those integrating estuaries or coastal lagoons as they are transition areas between freshwater and marine systems and between aquatic and terrestrial systems. Moreover, coastal areas are characterized by an intense human presence and activity, being subject to powerful and growing pressures and impacts. Thus, it is important to manage these productive ecosystems for long-term ecosystem services provision (Sousa et al. 2016).

The present study talks about the ecosystem services provided by the two biggest lagoons of India, Chilika lagoon, the largest lagoon of Asia, and the first designated Ramsar site of India. The other lagoon we discuss here is Pulicat lagoon, the second largest lagoon of India. Both these lagoons lie on the east coast of India having narrow connections with the BoB. Chilika lagoon lies in the state of Odisha whereas Pulicat lagoon spans in two states of Andhra Pradesh and Tamil Nadu.

6.2 Chilika Lagoon

Chilika covers an area of 1100 km^2 and lies in the three districts of Odisha namely Khurda, Puri and Ganjam. The lagoon is 65 km long and 20 km wide connected to the BoB through narrow inlets at Satpada on the southeastern side. On the southern end, Chilika is connected to Rushikulya estuary through a canal. The lagoon has a drainage basin of 4300 km^2. The lagoon catchment enjoys a tropical climate. The lagoon is roughly divided into four sectors based on salinity and ecological characters. Chilika lagoon remains a vital source of livelihood for 0.2 million people living in 141 villages around the lagoon (Nazneen et al. 2019b). The Chilika drainage basin covers an area of 4300 km^2. In addition to the 1100 km^2 of the lagoon area, 2325 km^2 is the agricultural land and around 525 km^2 consists of the forests (Figure 6.1).

Within the lagoon, a wide variety of habitats such as marshes, mudflats, freshwater and marine water with varying depths, and mangroves patches and significant area of seagrasses exists supporting a wide range of biodiversity (Barik et al. 2017). The ecosystem of the lagoon is predominantly determined by the salinity regime. During a large monsoon inflow of freshwater from Mahanadi, tributaries lower down the salinity values in the northern and central sectors. The southern sector remains relatively undisturbed because of low water renewal; hence, brackish water conditions prevail even during the monsoon (Khandelwal et al. 2008). Chilika lagoon is drained by distributaries of Mahanadi River – Daya, Bhargavi, and Nuna. Therefore, the northern sector of the lagoon has freshwater.

6.3 Ecosystem Services Provided by Chilika Lagoon

Coastal lagoons are complex ecosystems capable of delivering many ecosystem services that provide livelihoods, food security, general well-being, and welfare to humans. The natural conservation of coastal lagoons is important for their ecological importance and

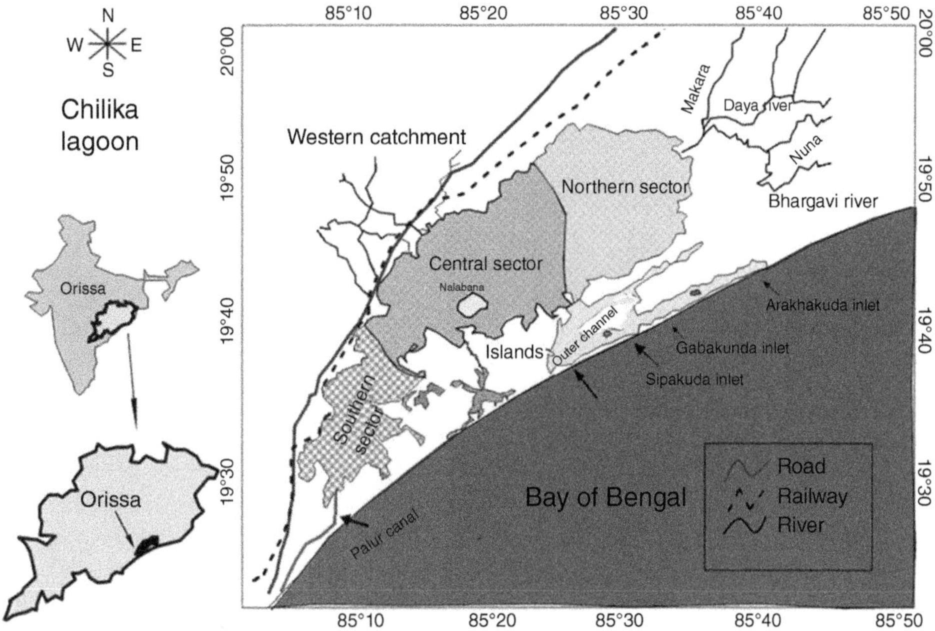

Figure 6.1 Physiographic map of Chilika.

the valuable ecosystem services they provide for human well-being. These are shallow semi-enclosed systems that support important habitats such as salt-marshes, seagrass meadows, and mangroves, which further support rich biodiversity (Newton et al. 2018). The Millennium Ecosystem Assessment categorized four major ecosystem services, i.e. provisioning services, regulatory services, cultural services, and supporting services. Coastal lagoons provide food provisioning (fish and shellfish), climate regulation, nutrient cycling, flood regulation, hydrological balance, recreation, and tourism. Chilika lagoon provides all the four broadly classified ecosystem services, i.e. provisioning of food in the form of fish and shellfish, regulating services such as carbon sequestration and nutrient cycling, coastal protection, cultural services like recreation and tourism, and supporting services include nutrient cycling and habitat provision. The significant provisioning services include commercial fisheries, aquatic vegetation for economic use, and means for inland navigation. Chilika is hydrologically dynamic due to several influences and this adds to the ability to regulate hydrological regime, an important regulating service. The religious and touristic values of Chilika add to its cultural services. The lagoon has rich floral diversity which supports different kinds of fish and other aquatic organisms. Various bacteria, protists, and diatoms further add to its rich microbial community which plays a significant role in nutrient cycling and biodiversity richness. The lagoon exchanges nutrients like carbon, nitrogen, phosphorus, and silica with the BoB along with carbon storage in its sediments which form the basis of supporting services provided by this lagoon.

6.3.1 Provisioning Services

6.3.1.1 Commercial Fisheries

Chilika harbors 336 fish, 29 prawn and 3 crab species out of which many are of commercial value (Suresh et al. 2018). The annual average harvest of 11 961.37 MT (average for 2001–2011) supports livelihoods of 0.14 million fishers living in 152 villages spread around Chilika. The commercial shellfish of Chilika is popular internationally. The fishery is a good source of revenue for the local fishermen sustaining their livelihood. The lagoon's fishery resources have been used sustainably by the formation of several cooperatives which help regulate the resources equitably and in a sustainable way. For the sustainable fishery harvest, traditional fishing methods, along with using ecological zones and different fish-capturing methods are being used currently. Chilika has rich macrophyte beds which form habitats for many fish species. Macrophytes also add to the detrital pool, which forms food for the lower organisms in the food chain (Jaikumar et al. 2011; Bonthu et al. 2016). Nearly 75% of the world shrimp/prawn culture occurs in Asia. Tiger prawns are naturally found in Chilika. In early 1980s the global trend of shrimp aquaculture caught up in Chilika. As the government saw prospects' foreign earnings and investor-oriented profits, shrimp aquaculture expanded rapidly to cover much of Chilika, a large part of the lagoon area according to various reports (Nayak and Berkes 2011). The lagoon has shown evidence for steady fish-landing records through a sustainable fishing strategy, using ecological zones, different contraptions, and traditional methods (Ghosh et al. 2006). Mohapatra et al. (2007) have described the enhancement of fish-landing post-hydrological intervention through the opening of a new mouth of Chilika into the BoB.

6.3.1.2 Other Flora and Fauna of Chilika Lagoon

Chilika Lagoon supports a rich diversity of macrophytic vegetation, several species of which are harvested for use by the local communities and used for several purposes like making of mats, thatching of roofs to preserving the fish catch. As per the assessment of 2007, over 58 000 MT of vegetation is harvested for various uses. Chilika is also rich in various types of microphytic vegetation. The various micro- and macrophytes growing in Chilika form the food source for birds and fishes and support rich biodiversity. Chilika is a habitat for 399 phytoplankton, 14 algae, 729 plants, 37 zooplankton, 61 protozoa, 6 Porifera, 7 Coelenterata, 29 Platyhelminthes, 36 Nematoda, 31 annelids, 136 Mollusca, 62 Crustacea, 5 echinoderms, 1 protochordate, 314 fish, 7 amphibia, 30 reptilia, 224 birds, and 19 mammals. The lagoon is also a habitat to some rare and endangered species like Irrawaddy dolphins, limbless skinks (Kumar and Pattnaik 2012).

Chilika also harbors one million birds during different seasons, many of them migrating from other countries. Chilika is the largest wintering ground for birds in India. The lagoon system hosts over 211 species of birds in the peak migratory season, with 97 being intercontinental migrants from the Caspian Sea, Baikal, Aral Sea, remote parts of Russia, Kirghiz Steppe of Mongolia, Central and South-East Asia, Ladakh, and the Himalayas (Ghosh et al. 2006). The important bird species of the lagoon are *Anas clypeata*, *Aythya ferina*, *Anas querquedula*, *Anas penelope*, *Anas fuligula*, *Anser strepera*, *Limnodromus semipalmatus*, and *Eurynorhynchus pygmeus*. Some rare and threatened species include the spoon-billed sandpiper, spot-billed pelican, Asian dowitcher, Dalmatian pelican and Pallas's fishing eagle.

Chilika has shown bioprospecting potential as novel bacterial species *Streptomyces chilikensis* and *Halobacillus marinus* have been found (Ray et al. 2013; Panda et al. 2018). Chilika harbors almost 20% of the seagrass population found in India. Out of the 16 species of seagrasses, 5 are found in Chilika. These seagrass meadows sequestering 10.1–16.8 tCO_2-equivalent per hectare of carbon annually (Ganguly et al. 2018). The dense growth of marsh species of *Phragmites karka* biomass provides nutrients to the adjoining areas and adds to the nutrient pool. Chilika also supports a large variety of microbiota, some of which have been found associated with rhizosphere of *Phragmites karka* and seagrass *Halophila uninervis* (Behera et al. 2017). The bacterial community found in the lagoon plays a major role in detoxification and nutrient cycling.

6.3.1.3 Navigation

Chilika is used as a mode of navigation through boats, especially for island routes within the lagoon and villages on the periphery (Kumar and Pattnaik 2012; Kumar and Saluja 2019). This is a source of revenue generation for the government as well as the private boat operators who ferry the passengers from one sector of the lagoon to the other. During the period of 2003–2006, 35 670 passengers used this mode of transport on an average annually, generating a revenue of 0.72 million. Chilika forms an important mode of navigation for several villages situated on its periphery. The boats are also instrumental in fishing and carrying the fish catch from remote areas of the lagoon to the fish markets in Balugaon, Barkul, and Satpada.

6.3.2 Regulating Services

Chilika experiences different hydrological regime under the influence of BoB and the rivers draining into it. Chilika plays a significant role in flood regulation in its catchment basin. Since Chilika owing to its large size has a huge storage capacity, it stores significant amounts of water during the rainy months, thus reducing flood in the catchment basin (Kumar et al. 2020). The salinity variation observed in different sectors of Chilika supports rich biodiversity, which, in turn, provide several ecosystem services. Chilika has a variety of macrophytes including seagrasses and macroalgae. This kind of floral diversity supports rich fisheries by providing food and shelter to the fish population. Chilika has seen an increase in the area of seagrass beds which is a useful indicator of the ecological health of the lagoon. Seagrasses are angiosperms which perform multiple functions like carbon storage, enhance oxygen levels in the water, and provide a substrate for epiphytic algae to forming food source for various fishes, birds, and crab species. They are also habitats for several fish and small aquatic organisms. Rich and healthy seagrass meadows are preferred habitats for dugongs. Other macrophytes like Nala grass and Phragmites karka provide fiber and raw material for building roofs and storage. Chilika plays an important role in microclimate regulation in its catchment basin. The lakes and lagoons with large surface areas regulate climate by encouraging evaporation, cloud formation and breeze regulation.

6.3.3 Cultural Services

Chilika lagoon is an important tourist destination owing to its rich biodiversity and scenic beauty. Satpada is home to Irrawaddy dolphins and forms one of the major tourist spots along with Rambha, Barkul, Balugaon, and Manglajodi (Kumar and Pattnaik 2012; Kumar et al. 2020). There are numerous small islands within the lagoon that form a major tourist attraction. Nalabana Island is the place where the largest congregation of birds occurs; this number swells up during winters. Towards the southern end of the lagoon, Eastern Ghats present a picturesque sight with Khalikote hill range as a backdrop. The other scenic place is the sea mouth where the lagoon opens into the BoB. Chilika also holds a religious significance for the local population. Kalijai temple situated in the southern sector of the lagoon is visited by a large number of people throughout the year. The local people also celebrate Bali yatra as Chilika was once a trade route from India to Indonesia. These cultural, religious, and tourist activities provide a livelihood to the local fishermen through boat services and other related activities. Small fishing towns, Satpada, Barkul, and Rambha, situated on the banks of Chilika, form the base for tourists benefiting tourist boats, eateries, and hoteliers. This generates revenue and supports a vibrant economy (Kumar et al. 2020). With rich biodiversity and scenic beauty, Chilika is a popular tourist destination on the Indian east coast, accounting for 8–10% of the total tourist arrivals in the state (Kumar and Pattnaik 2012). The majority of the tourists visit Satpada, Rambha, Barkul, Balugaon, and Manglajodi to watch Irrawaddy dolphins, various migratory waterbirds, picturesque islands, and eastern ghat Khalikote hills in the background.

6.3.4 Supporting Services

Coastal ecosystems, including lagoons, provide a range of supporting services like carbon storage, nutrient cycling, shore protection, and primary production to name a few (Duarte et al. 2013; Temmerman et al. 2013). Likewise, Chilika provides several vital supporting services like nutrient cycling, habitat provision, biodiversity sustenance, and protection. Chilika receives a large number of nutrients through river discharge and sediments reaching along with the river water. The sediment becomes the sink for carbon, nitrogen, and other chemicals (Nazneen and Raju 2017; Nazneen et al. 2019a). These essential nutrients become the basis of the food chain in the lagoon. A large part of these nutrients also reaches the sea through the flushing of sediments into the BoB (Nazneen et al. 2019b). Lagoons lie at the interface of land and sea and receive a large amount of terrestrial organic matter along with the primary production within the system; thus, they become a rich source of food for various species and a thriving ecosystem. The extra carbon is stored in the sediments for long term until disturbed; therefore, Chilika serves as a sink for organic matter and nutrients, efficiently recycling the inputs received through various internal processes resulting in the regulation of nutrients, thus enhancing overall productivity (Ganguly et al. 2015; Nazneen and Raju 2017; Amir et al. 2019).

A very important supporting service is of habitat provision. Many species of finfish and shellfish need both saline water and freshwater to complete their lifecycle. The inlets provide a passage for the juvenile fish to enter and exit the lagoon to complete their lifecycles. The rich macrophytic diversity further helps in providing food and habitat to various kinds of fish. Chilika supports different freshwater, brackish, and saline macrophytes which form food source and habitat for various faunal species.

6.4 Threats and Management of Chilika Lagoon

Chilika is a vast lagoon and the largest in Asia, also the first designated Ramsar site of India in 1981. However, Chilika faced a stage of degradation from the 1970s up to the 1990s due to which it was put under Monteux record of threatened wetlands in 1993 (Mohapatra et al. 2007; Mohanty et al. 2009). Chilika, like other lagoons, is a shallow water body that was facing a further decrease in its depth due to sedimentation. The lagoon has a very restricted connection with the sea, and during heavy rains, a large amount of sediments reaches the lagoon; however, the flushing is very low. This led to the accumulation of sediments as well as the choking of mouth connecting the BoB. The other major issue faced by Chilika was a decline in its salinity due to lesser seawater intrusion through the choked inlet. A large amount of sediment input from the watershed eventually choked the lagoon's mouth and inhibited the hydrological exchange between the lagoon and the ocean. This limited interaction with the ocean leads to a lowering of salinity and less exchange of nutrients, further decreasing the overall productivity of lagoon biota, including fish. The lagoon gradually started turning into a freshwater ecosystem which resulted in the proliferation of invasive freshwater aquatic weeds (Panigrahi et al. 2007; Mohanty et al. 2009). The development gradually altered the ecology of the lagoon, which is very much dependent on the seawater inflow and a salinity gradient that allows various habitats within the lagoon. Abundant freshwater plants' growth, death, and decay added a large amount of organic matter and the lagoon turned marshy in the extreme north. The lowering salinity did not support seawater fishes that venture into the lagoon for their lifecycle completion. All this increased the pressure on the communities which depend on this lagoon for their livelihood through the fish catch. All this severely affected the ecology and biological diversity of the lagoon.

Further, the fishery resources in the lagoon experienced a declining trend due to extensive fish catch by the use of mechanized boats, prawn aquaculture in large areas of the lagoon by nontraditional fishers, and harvesting of juvenile fish through a very fine mesh (Pattanaik 2007; Mishra and Griffin 2010; Nayak and Berkes 2014). During the early 90s, large areas of the lagoon were leased for prawn culture, which altered the natural fish population by area encroachment and was economically not suitable for small-time fishermen who were not into prawn culture business. A threat to their livelihood leads to many conflicts between the state government and Chilika fishermen.

The formation of the Chilika Development Authority (CDA) in 1998 was the turning point in the lagoon's revival. The CDA opened a new mouth in 2000 which enhanced the lagoon's connection with the BoB. This new mouth-opening revived the salinity increasing fish production, increased seagrass beds, decreased the spread of freshwater weeds, and enhanced recruitment of brackish species from the ocean (Sahu et al. 2014; Kim et al. 2016).

6.5 Pulicat Lagoon

Pulicat lagoon is the second largest brackish water lagoon after Chilika situated on the east coast formed out of backwaters of the BoB. Pulicat lies north of Chennai at 13°25′–13°55′N, 80°03′–80°19′E (Farooqui and Vaz 2000; Nagarjuna et al. 2010; Kumar and Natesan 2015). The lagoon came into being about 6650 years ago, during the Holocene geological period. The lagoon has an approximate length of 60 km whereas the width widely varies from 0.2 to 17.5 km. The lagoon is drained by three major rivers, namely Swarnamukhi, Arani, and Kalangi, besides many minor inflows (Basuri et al. 2020). The Buckingham Canal, a navigation Channel, is part of the lagoon on its western side (Rajyalakshmi and Basha 2016). The lagoon is the second largest lagoon in India, covering an area of 450 km^2 situated 60 km north of Chennai city. The lagoon is separated from the BoB by a large spindle-shaped island known as Sriharikota. Pulicat is a shallow lagoon with an average water depth of 1 m. Pulicat lagoon is spread between the two states of Andhra Pradesh and Tamil Nadu. Nearly two-thirds of the lagoon area is in Andhra Pradesh, and one-third lies in Tamil Nadu. The lagoon supports more than ten thousand traditional fishermen families living in nearly 52 villages around the lagoon. The lagoon owing to its rich flora and fauna is a good breeding ground for a variety of fish. Pulicat lagoon's biodiversity is rich and flourishing with researchers recording 168 species of fish and around 12 species of prawns in the lagoon's brackish waters (Thirunavukkarasu et al. 2011). The lagoon is also a designated bird sanctuary, flocked by flamingos, pelicans, and painted storks during the winter months (Figure 6.2).

The lagoon owing to its rich ecology has influenced and sustained the economy of the regional communities for thousands of years, which reflects and has shown its expression in the food, commerce, trade, customs, and language (Benedict 2019).

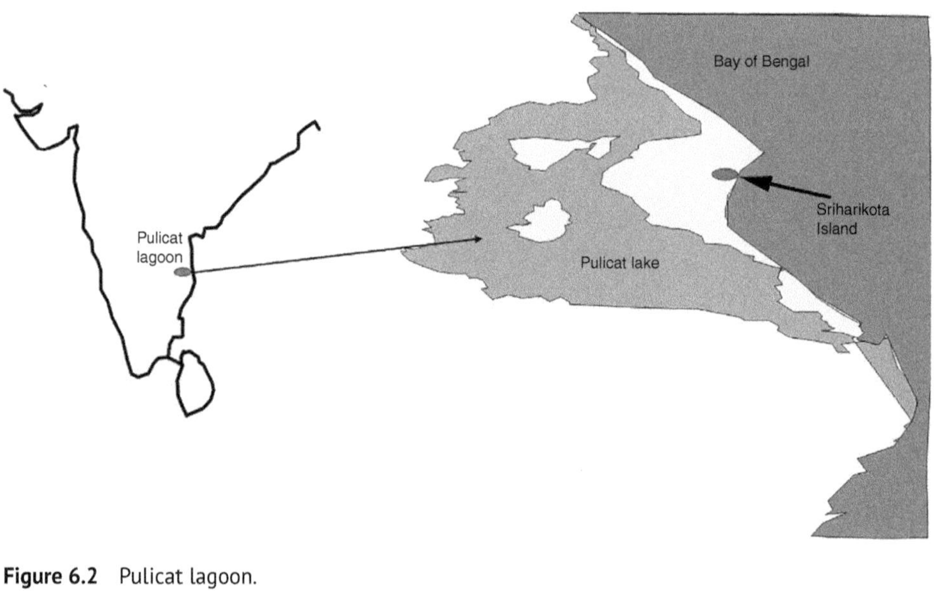

Figure 6.2 Pulicat lagoon.

The lagoon receives a significant amount of industrial waste as many industries are located on the islands and adjoining areas of the lagoon. The wetland also receives large industrial waste from Buckingham Canal and Ennore Creek. The anthropogenic pressure is affecting the ecology of the lagoon.

6.6 Ecosystem Services Provided by Pulicat Lagoon

The ecosystem services in Pulicat lagoon have been less studied. Yet this lagoon provides all the four major categories of ecosystem services mentioned in the context of Chilika lagoon, i.e. provisioning, regulating, cultural, and supporting services. Like Chilika lagoon, Pulicat lagoon also harbors a vast variety of flora and fauna including native and migratory birds. The lagoon is also a designated bird sanctuary. Pulicat witnesses a large number of flamingos that feed on algae and other phytoplankton and zooplankton. Other important migratory birds in this area are painted storks, pelicans, open-billed storks, grey herons, cormorants, white ibises, spoonbills, egrets, reef herons, spot-billed ducks, pintails, and sandpipers (Ramesh and Ramachandran 2005). The birds feed upon the phytoplankton and zooplankton, and, in turn, enrich the lagoon through their excretions, which gives rise to thick algal mats which form food for several faunal species. Flamingoes are the most prominent and frequent visitors to the lagoon, nearly 15 000 of them visiting here every year feeding on phytoplankton and crustaceans. The lagoon boasts of a rich fishery. Prawns, crabs, and mullets are harvested in large quantities. The lagoon is also used for salt mining. Calcium carbonate has also been obtained from the lagoon. The lagoon has rich microbiota. Nearly 41 diatom species, 88 zooplankton species, and 59 benthos species have been recorded in different studies. The lagoon also has a rich cultural and maritime history. Pulicat is known for trade by the Dutch people. Many foreigners settled in the banks of the lagoon and did trade in silk.

6.6.1 Provisioning Services

Pulicat lagoon harbors rich fishery resources along with several macrophytes provisioning food and resource material for other uses. Pulicat lagoon with flora and fauna representing various salinity ranges, attracting several species of birds and serving as a breeding ground for various marine fauna, is a thriving fishery resource sustaining the livelihood of several thousand fisherfolks. The lagoon harbors mullets, prawns, crabs, clams, oysters, catfish, and several other fish species of commercial value which along with food form the mainstay for an income of the local communities. The lagoons are home to various kinds of planktons and algae which form a food source for many local and migratory birds.

6.6.1.1 Fisheries in Pulicat

Lagoons being at the interface of land and sea function as an essential nursery space for several marine organisms including fish, prawns, and crabs and they play an essential role in the export of organic matter to the sea they are connected with (Santhanam and Amal Raj 2019). Pulicat lagoon is rich in flora and fauna and is a good breeding ground for commercial fish. The lagoon is an important habitat for over 160 different fish species, belonging to 26 families, more than 100 varieties of terrestrial and aquatic birds, and small mammals and reptiles. Pulicat is home to prawns, crabs, mullets, oysters, clams, and catfishes of economic value. More than 10 000 traditional fisherfolk families live in 52 villages and fish in Pulicat lagoon waters. Pulicat mud crab has a delicious taste and is world-famous (Francis and Aram 2018). There are many traditional fishing systems followed in Pulicat for ages to maintain a sustainable fishery resource. One such system is the Padu system where few villages manage the fishing rights by giving an equal opportunity to the community to harvest fishery. This ensures a sustainable and equitable share in earnings (Coulthard 2008; Francis and Aram 2018). In Padu system, villages are assigned a particular area in the lagoon for limited fishing rights. This system ensures all villages get an equal opportunity of fishing and the resource is distributed equitably. Padu is an ancient fishing practice followed in Sri Lanka and coastal regions of Tamil Nadu alongside Pulicat. This system also has its downside as it gives entitlements to eligible members of a particular fishing community for undertaking fishing activities in the designated areas. However, this traditional practice is disliked and not respected by other village communities who take it as the monopoly of certain communities. In Pulicat, the fishing grounds fall within a radius of 5 km from the mouth of the lagoon with a salinity level well maintained with sufficient water levels even during low tides.

6.6.2 Aquatic Flora and Fauna of Pulicat

Pulicat lagoon owing to its transitional position between the land and the sea and having both freshwater and saline influence harbors various species of floral and faunal diversity ranging from zooplankton and phytoplankton to macrophytes like seagrasses and macroalgae.

6.6.3 Regulatory Services Provided by Pulicat Lagoon

Hydrological cycle maintenance and flood regulation are some of the main regulating services provided by the lagoons. The lagoon also holds the local climate and provides suitable plants and animals to thrive in and around the lagoon. The lagoon provides a habitat provision for many fishes and other sea organisms. The lagoon is rich in seagrass beds which help in carbon sequestration, and food and habitat provision. The lagoon also has fringe mangroves which help in shoreline protection. Nutrient cycling is another important regulatory service provided by Pulicat lagoon.

6.6.4 Historical and Cultural Importance of Pulicat Lagoon

The lagoon also boasts of a rich historical heritage. Many Arabs settled on the banks of the lagoon. The Pulicat shores still have eminences of Arab infrastructure and literature. The first reference of Pulicat is found in Chola Dynasty, and the Vijayanagara Empire also talks about this important trade route. Lying on the Coromandel Coast between Andhra Pradesh and Tamil Nadu, Pulicat was one of the notable centers on the eastern coast of South India for European trading companies for spices, textiles, etc. (Munuswamy 2011; Benedict 2019). In the fifteenth and early sixteenth century, Pulicat rose to great commercial importance due to the stabilization of the Vijayanagara Empire. Pulicat became the most important port in Southeast India and a well-developed trading centre with a population exceeding 50 000. The Coromandel Coast was a major producer of textiles and yarns for export to Malacca and Burma (Munuswamy 2011). The canal and the three rivers – Arani, Kalangi, and Swarnamukhi – flowing into the lagoon are part of the maritime history due to the popularity of Coromandel cotton and its textile products, which are made in the hinterlands connected by this water system. Arabs, East Asians, and Europeans have been trading this fine cotton for gold since the sixth century (Benedict 2018).

The lagoon is also a bird sanctuary and hosts a sizable number of the bird population. Around 80–100 species visit the lagoon every winter, some from Ladakh, Tibet, and China. It is noteworthy that several near-threatened species like the spot-billed pelican, painted stork, and the white ibis breed in the vicinity of Pulicat lagoon (Ramesh and Ramachandran 2005; Basha et al. 2012). About 15 000 greater flamingoes spend the winter on the lagoon. The lagoon abounds with plenty of zooplankton and phytoplankton during the monsoon season because of large amounts of nutrient-rich silt and water reaching the wetland. A large amount of food attracts numerous wetland birds, especially greater flamingoes along with pelicans, little cormorants, open-billed storks, kingfishers, herons, painted storks, spoonbills, ducks, black drongoes, blue jays, common teals, coots, curlews, dabchicks, large egrets, little egrets, garganeys, little stints, painted storks, pond herons, herring gulls, white ibises, and sandpipers (Ramesh and Ramachandran 2005; Nagarjuna et al. 2010; Saraswathi and Pandian 2016). Rich avian population resulted in two bird sanctuaries established in Pulicat, one in Andhra Pradesh with an area of 172 km^2 and another one in Tamil Nadu covering an area of 60 km^2 (Saraswathi and Pandian 2016).

6.6.5 Supporting Services Provided by Pulicat Lagoon

The east coast of India is naturally endowed with large wetlands of ecological significance. The five massive wetlands lying on the eastern coast starting from Point Calimere (Kodiakarai) and Pulicat in Tamil Nadu, the Krishna-Godavari basin in Andhra, Chilika in Odessa, and Sundarbans in West Bengal provide the necessary moisture for monsoon winds to precipitate; therefore, these wetlands need to remain wet for rain clouds to emerge and develop (Sahoo and Bhaskaran 2015). Besides this, Pulicat also forms home to seagrasses which are angiosperms in coastal habitats, stores large amounts of carbon dioxide, and plays a role in nitrogen biogeochemistry (Purvaja et al. 2008). The thick beds also form a habitat for many aquatic faunas. The lagoon also has fringe mangroves which help in shoreline protection. The lagoon provides the migratory route of sea fishes to complete their lifecycles. The Buckingham Canal stretches for 796 km along the south-east coast of India and its water levels are stabilized by Pulicat lagoon (Benedict 2019).

6.6.6 Threats and Management of Pulicat Lagoon

Pulicat, the second largest lagoon in India, has held a significant position on the east coast since the sixth century as a trade route and safe haven for migratory birds and thriving fishery (Kannan and Pandiyan 2010; Kumar and Natesan 2015). Pulicat lagoon forms an important flyaway for migratory birds linking Point Calimere in Tamil Nadu to Chilika lagoon in Odisha on the eastern coast of India (Ramesh and Ramachandran 2005). In spite of its rich ecological, social, economic, and cultural history, the lagoon has not received the attention it deserves. The lagoon has not yet been designated a Ramsar site, which could put it on the international focus. It lacks national and international assistance for the protection and judicious use of its resources. In the recent decades, partly due to natural pressure and majorly because of population growth, land-use changes, pollution from the industries in the vicinity, the lagoon is under tremendous pressure and has not received the care and protection needed for its ecological stability and long-term sustenance (Reddy et al. 2012). Studies by several scholars reveal a major decline in its biological status as a consequence of several adverse changes over the past decades, especially since the 1970s under the influence of both natural as well as anthropogenic influences.

Pulicat lagoon faces industrial, anthropogenic, and developmental threats which affect not only the livelihood activities of fisherfolk but also the very survival of the lagoon. Like other lagoons in India, Pulicat receives a large silt load, especially from rivers Swarnamukhi, Arani, and Kalangi. The flushing of sediments and silt does not take place from narrow connections, leading to the closure of connection with the BoB. This leads to the lowering of salinity and the development of mudflats with little water. This is gradually turning large portions of the lagoon marshy and not suitable for the fish population. The natural process of littoral drift forms sandbars, which further aggravates the situation by closing of the lagoon–sea connection.

The preservation of biodiversity including avian and aquatic life would be possible with the help of widening of the lagoon–sea connectivity for unrestricted flow of seawater which is vital for the ecology of the lagoon (Nirmala et al. 2016). Another major threat that any lagoon faces is sedimentation which reduces its depth which affects the ecology in a significant way. Further reduction in tidal exchange due to the change in dynamics of the inlets and anthropogenic impacts in the form of various kinds of pollution and unsustainable fish catch add to the deterioration of the lagoon. These problems are very much being faced by Pulicat lagoon. At present, the lagoon is connected with the sea with only one inlet, which is hampering its exchange with the BoB. The lagoon area is shrinking and the depth has decreased from 6 to 1 m due to sedimentation and lesser flow of seawater, putting the livelihood of the people at stake (Francis and Aram 2018). Significant losses in the water-spread area, depth, and water quality are occurring in this critical marine ecosystem due to poor management, climate change, coupled with a lack of policy priority to address current and future threats. The authorities should rally to put this lagoon under Ramsar Convention so that it receives international support and attention. This step will provide it with much-needed attention. Declining water quality, drainage patterns, eutrophication, and catchment disturbances such as development, loss of natural vegetation, and poor agricultural practices are changing their fundamental ecology (Kumar and Natesan 2015; Nirmala et al. 2016). The conservation of lagoons also depends on managing its hydrological influences. The other most important aspect depends largely on the assessment of their natural characteristics, especially biodiversity, which is one of the main criteria used when formulating wetland protection policies. In order to assess the conservation status of the lagoons, it is necessary to incorporate the socioeconomic status of the fishing community. Pulicat lagoon faces ecological problems like the closure of sea mouth, siltation, shrinkage of the lagoon area, pollution, overfishing, shortage of food (fishes, crustaceans, and planktons) required for migratory birds and fishermen, and degradation and destruction of natural habitats in the environment (Nanda Kumar et al. 2008; Nanda Kumar et al. 2010). The major environmental threats to the lake are pollution from sewage, pesticides, agricultural chemicals, and industrial effluents. It is speculated that the Arani and Kalangi rivers draining into the lake bring in fertilizers and pesticides with the runoff from the farm field in the drainage basin (Nirmala et al. 2016). The domestic sewage forms a more diffuse input. Effluents and wastes from numerous fish-processing units are also significant sources of pollution. The oil spill from the mechanized boats is always a potential hazard. Because of the environmental and ecological threats in Pulicat lagoon, the social and economic status of the fishing community is very much affected.

The coastal activities are largely governed by the Coastal Regulation Zone (CRZ) notification of the Environmental Protection Act of 1986. As per the CRZ of Tamil Nadu, Pulicat lagoon is categorized under CRZ I which implies it is an ecologically sensitive ecosystem. In spite of Pulicat being an ecologically sensitive zone, it is not being managed and protected the way it should have been. The lagoon is facing multiple issues which can threaten its sustainability. An accelerated decline in the water quality of the lagoon is being observed since the 1970s. Aquaculture, the use of fine fishing gear by non-fishing communities, limestone mining, and agriculture in the Venadu and Irakkam islands are among the many reasons causing the plight of the lagoon. The Buckingham Canal (BC) is extremely polluted because it brings in urban runoff carrying household sewage, industrial effluent, and hazardous wastes. Siltation from rivers, namely Swarnamukhi and Kalangi and sandbar development at the mouth of the lagoon are the major threats to the ecological balance of this fragile ecosystem (Farooqui and Vaz 2000; Sanjeeva Raj 2006; Nanda Kumar et al. 2008). Because of siltation and lesser flushing, the silt gets deposited in the northern part which has led to the drying up of this region and has turned the lagoon into extensive mudflats. As a consequence of silt deposition and drying up of water, large parts of the lagoon become unproductive for fisheries and are available for shorter periods for the birds to inhabit the lagoon. Bio-wealth of the lagoon is on a decline and deterioration as a result of various kinds of stressors from restricted seawater exchange, pollution from various sources, unsustainable fishing methods, and land reclamation of agriculture and urbanization. An apparent understanding of the lagoon ecosystem is necessary for efficient management. For multidimensional ecosystems like lagoons with several influences, an integrated approach to determine suitable management solutions is required to maintain the biodiversity of coastal lagoons. An interdisciplinary scientific approach is also necessary to develop the required tools for the conservation and management of biodiversity (Bassi et al. 2014; Kumar and Natesan 2015).

6.7 Conclusion

Coastal lagoons are fragile ecosystems owing to their restricted connection with the sea and shallow depth. Lagoons are parts of the coastal zone under the direct influence of anthropogenic activities, such as increased nutrient discharges leading to eutrophication and other adverse impacts. Another important impact is sedimentation and the narrowing of their connection with the adjacent sea. It is therefore pertinent to address these issues by catchment basin management, widening of the inlet, regular dredging, and abatement of pollution for maintaining a healthy ecosystem. Chilika was the first designated Ramsar site in India, and the formation of CDA had a positive impact on the management of this lagoon. Ramsar Convention and national policies have played a significant role in the protection and management of Chilika. A similar approach needs to be taken in the case of Pulicat lagoon. With the changing climate and increasing human population near the coasts, it is significantly important that such rich ecosystems be managed in a sustainable way and that their resource base is not damaged or destroyed. These ecosystems need to be monitored and managed holistically to obtain their ecosystem services for a longer period of time in a sustainable manner. Their important functions of nutrient cycling, carbon storage, and climate regulation need to be maintained optimally. They should be brought under a strongly integrated policy ambit for their protection and preservation.

Acknowledgments

The first author acknowledges the financial support from the University Grants Commission (UGC), Government of India under **Dr. D.S. Kothari Postdoctoral Fellowship (Grant No. 4-2/2006(BSR)/ES/18-19/0042)** to carry out this work.

References

Amir, M., Paul, D., and Samal, R.N. (2019). Sources of organic matter in Chilika lagoon, India inferred from stable C and N isotopic compositions of particulates and sediments. *Journal of Marine Systems* 194: 81–90.

Barik, S.K., Muduli, P.R., Mohanty, B. et al. (2017). Spatio-temporal variability and the impact of Phailin on water quality of Chilika lagoon. *Continental Shelf Research* 136: 39–56.

Bassi, N., Kumar, M.D., Sharma, A., and Pardha-Saradhia, P. (2014). Status of wetlands in India: a review of extent, ecosystem benefits, threats and management strategies. *Journal of Hydrology: Regional Studies* 2: 1–19.

Basuri, C.K., Pazhaniyappan, E., Munnoroo, K. et al. (2020). Composition and distribution of planktonic ciliates with indications to water quality in a shallow hypersaline lagoon (Pulicat Lake, India). *Environmental Science and Pollution Research* 27: 18303–18316. https://doi.org/10.1007/s11356-020-08177-6.

Behera, P., Mahapatra, S., Mohapatra, M. et al. (2017). Salinity and macrophyte drive the biogeography of the sedimentary bacterial communities in a brackish water tropical coastal lagoon. *The Science of the Total Environment* 595: 472–485.

Benedict, X. (2018). Cultural landscape of Pulicat Lagoon. In: *Cultural Landscape of India, Understanding and Managing Heritage Values*, 131–134. Aryan Books International, New Delhi Pollution Research.

Benedict, X. (2019). Nature–culture linkages of pulicat lagoon: a cultural landscape protecting the coromandel coast. *Journal of World Heritage Studies*·Special Issue 2019·Disasters and Resilience: 78–84.

Bonthu, S., Ganguly, D., Ramachandran, P. et al. (2016). Both riverine detritus and dissolved nutrients drive lagoon fisheries. *Estuarine, Coastal and Shelf Science* 183: 360–369.

Coulthard, S. (2008). Adapting to environmental change in artisanal fisheries – insights from a South Indian Lagoon. *Global Environmental Change* 18: 479–489.

Duarte, C.M., Losada, I.J., Hendriks, I.E. et al. (2013). The role of coastal plant communities for climate change mitigation and adaptation. *Nature Climate Change* 3: 961–968.

Farooqui, A. and Vaz, G.G. (2000). Holocene sea-level and climatic fluctuations: pulicat lagoon – a case study. *Current Science* 79: 1484–1488.

Francis, S.J.M. and Aram, I.A. (2018). Management for sustaining the fishery resources in Pulicat estuary, India. *Journal of Media and Communication Studies* 10 (8): 87–94.

Ganguly, D., Patra, S., Muduli, P.R. et al. (2015). Influence of nutrient input on the trophic state of a tropical brackish water lagoon. *Journal of Earth System Science* 124 (5): 1005–1017.

Ganguly, D., Singh, G., Purvaja, R. et al. (2018). Valuing the carbon sequestration regulation service by seagrass ecosystems of Palk Bay and Chilika, India. *Ocean and Coastal Management* 159: 26–33.

Ghosh, A.K., Pattnaik Ajit, K., and Ballatore Thomas, J. (2006). Chilika Lagoon: restoring ecological balance and livelihoods through re-salinization. *Lakes & Reservoirs: Research and Management* 11: 239–255.

Jaikumar, M., Chellaiyan, D., Kanagu, L. et al. (2011). Distribution and succession of aquatic macrophytes in Chilika Lake-India. *Journal of Ecology and the Natural Environment* 3 (16): 499–508. https://doi.org/10.5897/JENE11.057.

Kannan, V. and Pandiyan, J. (2010). Pulicat – threatened lake of the year 2010. *Current Science* 99: 1496–1497.

Khandelwal, A., Mohanti, M., García-Rodríguez, F., and Scharf, B.W. (2008). Vegetation history and sea level variations during the last 13,500 years inferred from a pollen record at Chilika Lake, Orissa, India. *Vegetation History and Archaeobotany* 17: 335–344.

Kim, J.Y., Bhatta, K., Rastogi, G. et al. (2016). Application of multivariate analysis to determine spatial and temporal changes in water quality after new channel construction in the Chilika Lagoon. *Ecological Engineering* 90: 314–319.

Kumar, E.J. and Natesan, U. (2015). Status and critical knowledge gaps for integrated lagoon management – Pulicat lagoon. *Indian Journal of Coastal Conservation* https://doi.org/10.1007/s11852-015-0401-0.

Kumar, R. and Pattnaik, A.K. (2012). *Chilika – An Integrated Management Planning Framework for Conservation and Wise Use*. New Delhi: Wetlands International South Asia and Chilika Development Authority.

Kumar, R. and Saluja, R. (2019). Coastal Wetlands in India and climate change: status, trends and policy response options. In: *Climate Change and the Vulnerable Indian Coast* (eds. R. Ramesh and J.R. Bhatt), 239. Ministry of Environment, Forest and Climate Change.

Kumar, R., Pattnaik, A.K., and Finlayson, C.M. (2020). Ecosystem services: implications for managing Chilika. In: *Ecology, Conservation, and Restoration of Chilika Lagoon, India. Wetlands: Ecology, Conservation and Management*, vol. 6 (eds. C. Finlayson, G. Rastogi, D. Mishra and A. Pattnaik). Cham: Springer https://doi.org/10.1007/978-3-030-33424-6_4.

Mishra, S.R. and Griffin, A.L. (2010). Encroachment: a threat to resource sustainability in Chilika Lake, India. *Applied Geography* 30: 448–459.

Mohanty, R.K., Mohapatra, A., and Mohanty, S.K. (2009). Assessment of the impacts of a New artificial lake mouth on the hydrobiology and fisheries of Chilika Lagoon, India. *Lakes & Reservoirs: Research and Management* 14 (3): 231–245. https://doi.org/10.1111/j.1440-1770.2009.00406.x.

Mohapatra, A., Mohanty, R.K., Mohanty, S.K. et al. (2007). Fisheries enhancement and biodiversity assessment of fish, prawn and mud crab in Chilika lagoon through hydrological intervention. *Wetlands Ecology and Management* 15 (3): 229–251.

Munuswamy, A. (2011). Pulicat through the ages. Thesis submitted to the Department of Indian History, University of Madras.

Nagarjuna, A., Kumar, N.V.N., Kalarani, V., and Reddy, D.C. (2010). Aquatic and avian biodiversity of pulicat brackish water lake and ecological degradation. *World Journal of Fish and Marine Sciences* 2 (2): 118–123.

Nanda Kumar, N.V., Nagarjuna, A., Reddy, D.C. et al. (2008). Satellite remote sensing and field studies on a sea mouth in the northern part of Pulicat Lagoon. *Current Science* 95 (10): 1405–1406.

Nanda Kumar, N.V., Nagarjuna, A., and Reddy, D.C. (2010). Ecology of Pulicat lake and conservation strategies. *The Bioscan: An International Quarterly Journal of Lifesciences* 2: 461–478.

Nayak, P.K. and Berkes, F. (2011). Commonisation and decommonisation: understanding the processes of change in the Chilika lagoon, India. *Conservation and Society* 9 (2): 132.

Nayak, P.K. and Berkes, F. (2014). Linking global drivers with local and regional change: a social-ecological system approach in Chilika Lagoon, Bay of Bengal. *Regional Environmental Change* 14: 2067–2078. https://doi.org/10.1007/s10113-012-0369-3.

Nazneen, S. and Raju, N.J. (2017). Distribution and sources of carbon, nitrogen, phosphorus and biogenic silica in the sediments of Chilika lagoon. *Journal of Earth System Science*: 126–213. https://doi.org/10.1007/s12040016-0785-8.

Nazneen, S., Singh, S., and Raju, N.J. (2019a). Heavy metal fractionation in core sediments and potential biological risk assessment from Chilika lagoon, Odisha state. India. *Quaternary International* https://doi.org/10.1016/j.quaint.2018.05.011.

Nazneen, S., Raju, N.J., Madhav, S., and Ahamad, A. (2019b). Spatial and temporal dynamics of dissolved nutrients and factors affecting water quality of Chilika lagoon. *Arabian Journal of Geosciences* 12: 243. https://doi.org/10.1007/s12517-019-4417-x.

Newton, A., Icely, J.D., Cristina, S. et al. (2014). An overview of ecological status, vulnerability and future perspectives of European large shallow, semi-enclosed coastal systems, lagoons and transitional waters. *Estuarine, Coastal and Shelf Science* 140: 95–122. https://doi.org/10.1016/j.ecss.2013.05.023.

Newton, A., Brito, A.C., Icely, J. et al. (2018). Assessing, quantifying and valuing the ecosystem services of coastal lagoons. *Journal for Nature Conservation* https://doi.org/10.1016/j.jnc.2018.02.009.

Nirmala, K., Ramesh, R., Ambujam, M.K. et al. (2016). Geochemistry of surface sediments of a tropical brackish water lake in South Asia. *Environment and Earth Science* 75: 247. https://doi.org/10.1007/s12665-015-4964-8.

Panda, A.N., Mishra, S.R., Ray, L. et al. (2018). Taxonomic description and genome sequence of *Halobacillus marinus* sp. nov., a novel strain isolated from Chilika Lake, India. *Journal of Microbiology* 56 (4): 223–230.

Pandisamy, R., Kathiresan, K., Mohan, P.M. et al. (2020). Ensuring the adaptive potential of Coastal wetlands of India- a need of the hour for sustainable management. *Wetlands Ecology and Management* https://doi.org/10.1007/s11273-020-09742-z.

Panigrahi, S., Acharya, B.C., Panigrahy, R. et al. (2007). Anthropogenic impact on water quality of Chilika Lagoon RAMSAR site: a statistical approach. *Wetlands Ecology and Management* 15 (2): 113–126. https://doi.org/10.1007/s11273-006-9017-3.

Pattanaik, S. (2007). Conservation of environment and protection of marginalized fishing communities of Lake Chilika in Orissa, India. *Journal of Human Ecology* 22: 291–302.

Purvaja, R., Ramesh, R., Shalini, A., and Rixen, T. (2008). Biogeochemistry of Nitrogen in seagrass and oceanic systems. *Memoir Geological Society of India* 73: 435–460.

Rajyalakshmi, E. and Basha, S.K.M. (2016). Wetland Flora of Pulicat Lake- Southern Coastal Wetland of Spsr Nellore District of Andhra Pradesh, India. *International Journal of Scientific and Engineering Research* 7 (2): 1333–1346.

Ramesh, D.A. and Ramachandran, S. (2005). Factors influencing flamingo (*Phoenicopterus roseuis*) distribution in the Pulicat Lagoon ecosystem, India. *Wetlands Ecology and Management* 13: 69–72.

Ray, L., Suar, M., Pattnaik, A.K., and Raina, V. (2013). *Streptomyces chilikensis* sp. nov., a halophilic streptomycete isolated from brackish water sediment. *International Journal of Systematic and Evolutionary Microbiology* 63: 2757–2764.

Reddy, S.R.B.C., Jayaraju, N., and Reddy, K.R.R. (2012). Anthropogenic impact on the Pulicat Lagoon monitoring with foraminifera, east coast of India. *Marine Sciences* 2: 66–76. https://doi.org/10.5923/j.ms.20120205.05.

Sahoo, B. and Bhaskaran, P.K. (2015). Assessment on historical cyclone tracks in the Bay of Bengal, east coast of India. *International Journal of Climatology* 36: 95–109.

Sahu, B.K., Pati, P., and Panigrahy, R.C. (2014). Environmental conditions of Chilika Lake during pre and post hydrological intervention: an overview. *Journal of Coastal Conservation* 18: 285–297. https://doi.org/10.1007/s11852-014-0318-z.

Sanjeeva Raj, P.J. (2006). Macro fauna of Pulicat lagoon, NBA Bulletin No.6. National Biodiversity Authority, Chennai, p. 67.

Santhanam, H. and Amal Raj, S. (2019). Spatial and temporal analyses of salinity changes in Pulicat lagoon, atransitional ecosystem, during 1996–2015. *Water Science* 33 (1): 93–104. https://doi.org/10.1080/11104929.2019.1661944.

Saraswathi, R. and Pandian, P.K. (2016). Pulicat lake: a fragile ecosystem under threat. *Slovak Journal of Civil Engineering* 24 (3): 8–18. https://doi.org/10.1515/sjce-2016-0012.

SKM, B., Rajya Laksmi, E., Ratneswara Rao, B. et al. (2012). Biodiversity and conservation of Pulicat lake – Andhra Pradesh. *International Journal of Geology, Earth & Environmental Sciences* 2: 129–135.

Sousa LP, Sousa AI, Alves FL, Lillebo AI (2016). Ecosystem services provided by a complex coastal region: challenges of classification and mapping. DOI: https://doi.org/10.1038/srep22782

Srichandan, S., Kim, J.Y., Kumar, A. et al. (2015b). Interannual and cyclone-driven variability in phytoplankton communities of a tropical coastal lagoon. *Marine Pollution Bulletin* https://doi.org/10.1016/j.marpolbul.2015.11.030.

Suresh, V.R., Mohanty, S.K., Manna, R.K. et al. (2018). Fish and Shell Fish diversity and its sustainable management in Chilika Lake, ICAR- Central Inland Fisheries Research Institute, Barrackpore, and Kolkata and Chilika Development Authority, Bhubaneswar, 376p.

Temmerman, S., Meire, P., Bouma, T.J. et al. (2013). Ecosystem-based coastal defence in the face of global change. *Nature* 504: 79–83.

Thirunavukkarasu, N., Gokulakrishnan, S., Premjothi, P.V.R., and Inbaraj, R.M. (2011). Need of coastal resource management in Pulicat Lake–challenges ahead. *Indian Journal of Science and Technology* 4 (3): 322–326.

Velasco, A.M., Pérez-Ruzafa, A., Martínez-Paz, J.M., and Marcos, C. (2018). Ecosystem services and main environmental risks in a coastal lagoon (Mar Menor, Murcia, SE Spain): the public perception. *Journal for Nature Conservation* 43: 180–189.

7

Sustainable Practices for Conservation of Wetland Ecosystem

Krishna Rawat and Amit Kumar Yadav

School of Environment and Sustainable Development, Central University of Gujarat, Gandhinagar, Gujarat, India

7.1 Introduction

The ecosystem may be understood as the living and nonliving components' interaction zone. This interaction may be in the form of energy exchange or in the form of a nutrient flow cycle. Here living and nonliving forms are defined as biotic and abiotic components of the ecosystem. It is broadly categorized as the terrestrial and aquatic ecosystem. In between these two forms, the wetland ecosystem may be considered as a distinct transitional zone, where one form seems to merge another form with an extremely rich diversity of different living forms. They are also rich sources of different mineral forms. Millions of people survive on these wetlands for fulfilling their basic needs that can be in either terms of food, medicines, fiber supply, fuel, etc. As per Ramsar Convention, wetlands may be defined as the "area of marsh, fen, peatland, or water, whether natural or artificial, permanent, or temporary, with water that is static or flowing, fresh, brackish, or salt, including areas of marine water, the depth of which at low tide does not exceed six meters." On the basis of the ecological, hydrological, and geological characteristics, they can be classified as estuaries (that may consist of tidal marshes, deltas, and mangrove swamps), marine (coastal wetlands), riverine, palustrine (marshes, swamps, and bogs). They are of both aesthetic and economic importance. According to EPA or Wetland Reserve Program, wetlands may be described as those lands that are saturated or inundated by groundwater or surface water at a frequency and duration sufficient to support and it further supports the vegetation that is adapted to such a condition of saturation under normal circumstances (Wetland Conservation 2020). While in some other definitions, wetlands are regarded as a general term to those areas of land that are permanently or seasonally waterlogged, low-lying landmass periodically inundated by saline or freshwater, they can be either submerged in the form (Mac et al. 2010; Phukan and Saikia 2014).

Cowardin (1979) firstly developed a wetland classification system that was widely used. This classification was based on the ecological, geological, and hydrological characteristics of an area. According to Ramsar Convention (report Ramsar 2020a), 2391 sites covering 253 879 235 hectares of the area have been identified all over the world as wetlands of

Wetlands Conservation: Current Challenges and Future Strategies, First Edition.
Edited by Sanjeev Sharma and Pardeep Singh.

international importance on account of their international significance in terms of ecology, botany, zoology, limnology, or hydrology (ramsar.org.com).

India has currently 37 Ramsar sites, with the area occupying 1 067 939 hectares. Figure 7.1 shows the current wetland status in India. Sites identified as wetland sites under Ramsar Convention are spread to almost the whole length and breadth of the country (Table 7.1).

7.2 Role of Wetlands in the Ecosystem

Comparing diverse ecospheres, wetlands are one of the most productive ecosystems, providing various services to mankind (Ghermandi et al. 2008; Brink et al. 2012). In general, these sites are very sensitive; their diversity range depends on their geographical location and other physicochemical parameters (Turner et al. 2000; Bassi et al. 2014). Owing to its functionality, wetlands are also termed as a natural sponge of nature for both dry and wet periods of water streamflow. Wetlands provide multiuse water services such as water for irrigation, recreational uses, fishing, and other domestic uses (Sarkar and Ponniah 2000; Das et al. 2012; Kumar et al. 2012). Despite their important role, they are now in an endangered and vulnerable state. They are the critical natural environment that provides habitat to various plants and animals of different life forms that are found nowhere else. They are also recognized as a significant part of the tourism industry (MEA 2005; Ramsar Convention on Wetlands and WTO 2012).

There is no substitution for the services given by wetlands; they are an important vital bond between land and water. They perform various important roles in nature such as streamflow regulation (by absorbing extra water during a wet period and release of water in dry period), trapping of sediments (retaining sediments of water runoff, these sediments are trapped by the vegetation of wetland by controlling the velocity of water flow), supporting water supply in the water-deficient area, and assimilating phosphate and nitrate (oxygen-deficient conditions of wetland promote the formation of phosphate and nitrate). These wetlands are rich in diverse life forms, so they can act as natural gene banks of diverse life forms. Wetland vegetation slows down the water flow hence controlling erosion. They are habitat for diverse living beings, hence helping in maintaining the diversity of living forms. Besides these, they can be research, recreation, and tourism center for knowledge and economic welfare of corresponding regions. The oxygen-deficient nature of these sites makes them a natural source of methane in the atmosphere. They play an important role in carbon sequestration. Coastal wetlands also play an important role in carbon sequestration and in India, they extend up to about 43 000 km^2 (Kathiresan and Thakur 2008). There are many more tangible, intangible, and ecological contributions given by the wetland to various living beings. Wetlands are a hotspot of numerous small creatures like amphibians, reptiles, mammals, fishes, insects, various plant forms, microbes, etc. These support diverse nutrient-rich ecosystem to support the life forms. These act as a shelter for birds and for migratory birds these act as temporary primary habitat (Agarwal 2011; Lalchandani 2012). Wetlands also serve as a sink for various pollutants. Various studies have been done to study wetlands as a sink for point and nonpoint sources of pollution and they are considered as cost-effective strategies to combat pollution (Bystrom et al. 2000; Groffman and Crawford 2003; Verhoeven et al. 2006; Kaur et al. 2012).

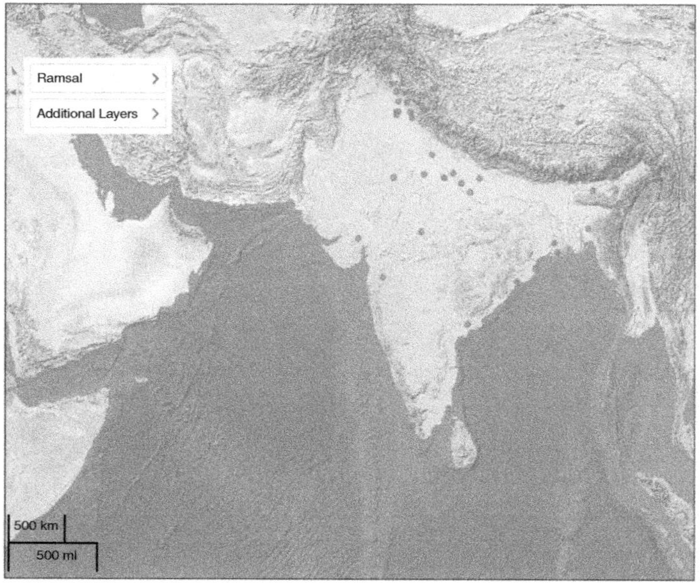

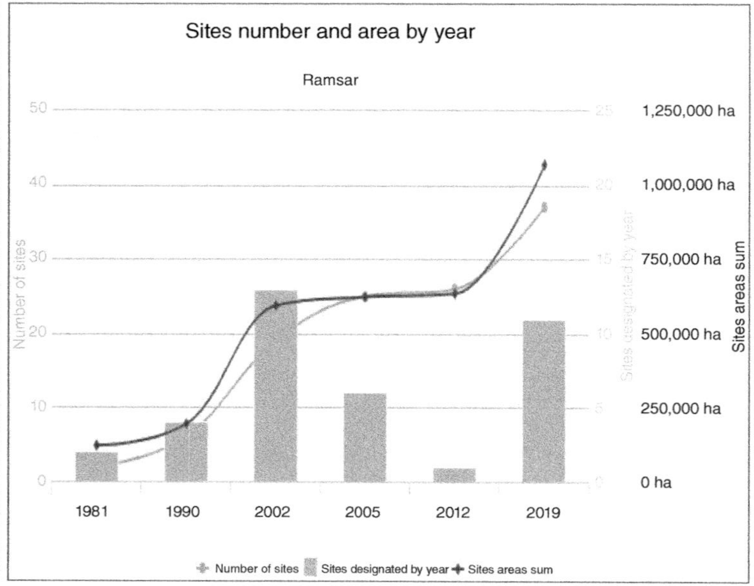

Figure 7.1 Current Wetland status in India. *Source:* https://rsis.ramsar.org/ris-search/?f%5B0%5D=regionCountry_en_ss%3AAsia&f%5B1%5D=regionCountry_en_ss%3AIndia&pagetab=2

Table 7.1 Ramsar Convention sites in India.

Site	Region	Area	Coordinates
Ashtamudi Wetland	Kerala	6140 ha	08°57′N 076°34′E
Beas Conservation Reserve	Punjab	6429 ha	31°23′N 075°11′E
Bhitarkanika Mangroves	Odisha	65 000 ha	20°39′N 086°54′E
Bhoj Wetland	Madhya Pradesh	3201 ha	23°13′N 077°19′E
Chandertal Wetland	Himachal Pradesh	49 ha	32°28′N 077°36′E
Chilika Lake	Odisha	116 500 ha	19°42′N 085°21′E
Deepor Beel	Assam	4000 ha	26°07′N 091°39′E
East Calcutta Wetlands	West Bengal	12 500 ha	22°27′N 088°27′E
Harike Lake	Punjab	4100 ha	31°13′N 075°12′E
Hokera Wetland	Jammu and Kashmir	1375 ha	34°04′N 074°42′E
Kanjli	Punjab	183 ha	31°25′N 075°22′E
Keoladeo National Park	Rajasthan	2873 ha	27°13′N 077°31′E
Keshopur-Miani Community Reserve	Punjab	344 ha	32°05′N 075°23′E
Kolleru Lake	Andhra Pradesh	90 100 ha	16°37′N 081°12′E
Loktak Lake	Manipur	26 600 ha	24°25′N 093°49′E
Nalsarovar	Gujarat	12 000 ha	22°46′N 072°02′E
Nandur Madhameshwar	Maharashtra	1437 ha	20°01′N 074°06′E
Nangal Wildlife Sanctuary	Punjab	116 ha	31°23′N 076°22′E
Nawabganj Bird Sanctuary	Uttar Pradesh	225 ha	26°36′N 080°39′E
Parvati Arga Bird Sanctuary	Uttar Pradesh	722 ha	26°56′N 082°09′E
Point Calimere Wildlife and Bird Sanctuary	Tamil Nadu	38 500 ha	10°19′N 079°37′E
Pong Dam Lake	Himachal Pradesh	15 662 ha	32°01′N 076°04′E
Renuka Wetland	Himachal Pradesh	20 ha	31°37′N 077°27′E
Ropar	Punjab	1365 ha	31°01′N 076°30′E
Rudrasagar Lake	Tripura	240 ha	23°28′N 091°16′E
Saman Bird Sanctuary	Uttar Pradesh	526 ha	27°00′N 079°10′E
Samaspur Bird Sanctuary	Uttar Pradesh	799 ha	25°59′N 081°23′E
Sambhar Lake	Rajasthan	24 000 ha	27°00′N 075°00′E
Sandi Bird Sanctuary	Uttar Pradesh	309 ha	27°18′N 079°58′E
Sarsai Nawar Jheel	Uttar Pradesh	161 ha	26°58′N 079°15′E
Sasthamkotta Lake	Kerala	373 ha	09°01′N 076°37′E
Sundarban Wetland	West Bengal	423 000 ha	21°46′N 088°42′E
Surinsar-Mansar Lakes	Jammu and Kashmir	350 ha	32°45′N 075°12′E
Tsomoriri	Jammu and Kashmir	12 000 ha	32°54′N 078°18′E
Upper Ganga River	Uttar Pradesh	26 590 ha	28°33′N 078°12′E
Vembanad-Kol Wetland	Kerala	151 250 ha	09°49′N 076°45′E
Wular Lake	Jammu and Kashmir	18 900 ha	34°16′N 074°33′E

Source: Ramsar 2020b. https//www.ramsar.org/sites/default/files/documents/library/sitelist.pdf?utm_content=buffer2ce19&utm_medium=social&utm_source=twitter.com&utm_campaign=buffer

Wetlands serve to be a vital link between water and land. Compared to other ecosystems, their services are of a high order. A large part of valuable products is obtained from these spheres. Above all, they have scientific, recreational, cultural, as well as historical richness. Therefore, they play an important role in cultural and human development all around the world. These ecosystems ensure water security, natural hazards, safety, as well as climate change adaptation.

7.3 Challenges to Conserve Wetlands

Wetlands are very productive system simultaneously, they are very sensitive and adaptive. Wetlands have great potential in order to sustain life. In spite of it, there is a rapid decline globally in its occupied area. Among all other wetland forms, freshwater wetlands have been severely exploited (Molur et al. 2011). McAllister et al. 2001 stated that a major area of wetlands is lost in agriculture, dam construction, and other activities annually. Wetlands are threatened by both, i.e. anthropogenic activities as well as natural hazards. Another reason for wetland degradation is the lack of awareness of the value owed by wetlands to the locals, which causes unidentified social and natural cost in a broad way. Climate change and global warming are the other threats to these valuable ecological zones (Blankespoor et al. 2012; Kumar 2013). Dependence upon wetland services in various spheres of life has led to the deterioration of these ecosystem forms. Wetlands are economically and socially valuable. Therefore, we need to come up with some strategies that may help us wisely use these forms and conserve these for the future too. Wetland degradation may be in the processes such as the conversion of wetland area into other landforms' usage, agricultural land expansion, enhanced siltation, the impact of climate change, carrying out developmental activities in these areas, industries, etc. All these activities and ignorance of wetland have led to complex and diverse wetland losses all around the world. Due to wetland deterioration, we have lost many of the migrant avian species as these wetlands served as the primary habitat to many avian species.

Many research studies have been published addressing wetlands' importance, pollution, and degradation such as Kumar et al. 2020 had published a study on Son Beel wetland, Assam, stating its valuable services and enriched biodiversity. Ahidur 2016 also presented the impact of human encroachment in the area of Tamaranga (beel) wetland in Assam that has made area prone to drought and flood-like condition. Dutta and Sengupta 2015 has presented a study on the Chatra Beel wetland of West Bengal, which is falling under threat due to the increased population of the area. Phukan and Saikia 2014 have studied the wetlands of the Golaghat district of Assam to analyze the impact of human interference on the wetland. He stated that municipal waste that was dumped in the area of the wetland was a major threat that was causing wetland degradation. Davidson (2014) has stated that since 1700, around 87% of wetlands had degraded and this degradation mainly occurred in the twentieth and early twenty-first centuries. Dugan (1993) has estimated a 50% global wetland loss. Ramsar (2015) also conferred a 35% reduction in global wetlands between 1970 and 2015. Various other data and authors have conferred the same degradation worldwide.

The main underlying factor of this loss can be considered as a lack of proper information and management strategies for the conservation of these valuable ecosystems. These losses may have an acute or chronic impact, i.e. either they can be observed in a short or long time period. Acute losses can be seen in the form of hydrological depletion, agricultural conversion, or in the form of deforestation, etc., while chronic impact can be brought up by drivers such as global climate change, depletion of the water table, etc. Different anthropogenic activities such as urbanization, hydrologic activities, pollution, agricultural practices, deforestation, salinization, hybridization, aquaculture, and non-native predators have led to the degradation of the wetlands.

Other challenges for wetland conservation included no proper legal framework available for wetland conservation. Wetland conservation rule came into existence in 2010, that too not encircled natives of the area for conservation strategies. Wetland conservation plans were not strictly implemented; individual duties of each department were not clear. Conflicts of imposed laws were seen that hindered the effective implementation of the wetland conservation law.

7.4 Wetland Management and Sustainable Development

Wetlands are one of the most valuable diversified regions that serve humanity in many ways. They are unique ecological features that provide variable products and other services to humanity (Prasad et al. 2002). Our future, in many ways, directly or indirectly depends on their future. By their conservation, we can conserve our future too. With the increasing population, our dependence on these wetlands has increased manifold in terms of medicines, food, etc., and many intangible services. This makes dependence on one's future with the common future of wetland. So, we need to address the challenges of wetlands that are rapidly declining. This is a global issue that different areas have a different level of degradation.

We need to conserve and manage these beneficial ecological sites through an integrated management plan to maintain the ecological character of these sites. One way to achieve this is by enveloping more area under restricted zone or marked region under wetland designation, where guidelines of Ramsar Convention can be implemented. We must bring in light the threat pose to the ecological nature of these zones. All wetlands must be wisely used up. We must put in the effort to restore the degraded wetland by utilizing traditional knowledge and practices. We need to develop scientific guidelines and sustainable policies in order to implement them for sustainable use and conservation of wetland sectors. One needs to develop international cooperation to restore and conserve these vital ecological zones.

In the light of the above discussion for wetland management, the following suggestions can be taken into consideration for better conservation strategy:

- Common boundaries of jurisdiction should be implemented to conserve wetlands at the state and central levels. To avoid confusion of law implementation, clear jurisdiction power must be imposed to guideline the conservation plans.
- The regulatory agency must ensure the proper enforcement of law regulating and managing natural resources. Violators must be penalized for strict enforcement of the law.
- Wetlands are natural resources for humanity; hence, their security must be endured as common property custody.
- Integrated management of wetland with other natural resources must be ensured.
- The proper regulatory framework needs to be developed for wetland conservation that is to be imposed at the state and national levels.

- Activities such as prevention of industrial sewage and other waste dumping in wetlands, preventing the conversion of wetland area into agricultural lands, land-use transformation prevention, etc., can be promoted to enhance the wetland properties.
- Research facilities must be facilitated to monitor the physical, chemical, and biological characteristics of the ecosystem.
- Encroachment of wetland regions, industrial activities, deforestation activities, topographical amendments, etc., must be prohibited.
- Invasion of alien species to these vital ecosystems must be checked out to endure their natural composition reservation.
- Awareness programs must be carried out for locals for better conservation of the wetland ecosystem.
- Various firms such as NGOs, agencies of government, etc., must step forward for the actual implementation of laws for wetland conservation.
- One needs to formulate interdisciplinary research such as natural sciences, social sciences, ecological engineering, etc., to examine and find possible solutions to wetland degradation.

7.5 Future Strategies for Wetland Conservation

Wetlands that are a vital dynamic ecosystem and serve many of the significant ecological functions are declining due to different factors such as encroachment of land, developmental activities, the introduction of alien species, climate change, deforestation, drainage of water, etc. These factors are vital challenges to sustainable wetland conservation that needs to be drawn from a combination of science and traditional knowledge. It will be successful only when every individual understands their role in wetland conservation. In the context of the sustainable future of wetland, sustainable consumption and production must be promoted. Ramsar Convention has brought in focus to the world information about deteriorating wetlands as an important environmental issue. This condition is much severe in the case of India, where 2–3% of wetlands are lost every year, thus a matter of global concern to protect our precious wetland. We need to see them as our common future and conserve them for our sustainable urban future. At all levels, sound policies, attitude, and efforts are required for the sustainable conservation of wetlands. Majorly threatening factors must be checked in response to wetland deterioration. Management and wetland legislation must be strengthened for improving wetland conservation. Wetland education must be promoted. Lastly, international cooperation must be headed for better conservation strategies.

7.6 Development of the Legal Framework

In order to conserve wetlands, India developed legal guidelines in accordance with the Ramsar Convention for the protection of wetland. Under these guidelines, wetland rules were developed that were to be implemented with the coordination of state and center. It was after the year 2010 when independent wetland rules for conservation and management came into existence. The protected wetland regions were prohibited from any kind of anthropogenic interventions. The shielded area was further categorized into prohibited and restricted zones. This separation may lead to the wise use of the restricted ecosystem.

7.7 Technology Intervention with Baseline Data for Wetland Conservation

Spatial data merged with grassroots data for the development of more effective strategies are needed for wetland conservation. Advance technology must be used up for better and sustainable use of conserved wetland sites in order to meet present requirements without deteriorating the wetland ecosystem, and conserve it for the future. Various researches have been taken up in this regard where satellite data has been used up to identify the wetland hotspots, and proper management strategies have been planned up in these regions.

7.8 Development of National Action Plans

Wetland, although a very important ecosystem, has not been taken seriously due to its complex nature. We need to develop the protocol to protect these dynamic ecosystems and enthusiastic participation of an individual to enact the conservation law. The national wetlands action plan is the guidance for the prudent use of these wetlands guiding and promoting the strategies for the wise use and conservation of this dynamic world. It basically gives us the framework of strategies and actions required to act upon to all sections of society including every individual for the wise use and management of these valuable resources. So, in order to sustain these benefits in future too, national plans such as "national action plan for the conservation of migratory birds and their habitats (2018–2023)" need to be acted to safeguard and boost up migratory birds. It will also help us restore the biodiversity wetland hotspots many more plans are needed to be acted upon and brought to the limelight.

7.9 Promotion of Research for Conservation Setup

In India, wetland management research generally deals with limnological or ecological or environmental economics, while their socioeconomic prospect is less explored. Thus, their invasion must be done for efficient wetland management. So, in order to boost up, wetland management research must be promoted in this area. It must promote sustainable management of wetlands. Local communities must be a participatory section in research implementation to facilitate sustainable wetland conservation.

7.10 Conclusion

Wetland can be considered as the lifeline for any nation owing to the services it provides. Still, it is facing a decline in area and quality. We are fortunate to have the immense reserved area under wetland sites that continue to provide their valuable products and other services to us, but their degradation is of major concern. For our sustainable future, we need to consume these products sustainably in order to secure our survival in the coming years. In this direction, Ramsar Convention has come out to give us a blueprint for the wise use of these sites. Though India is a signatory party of this convention, yet no significant contribution is made for the wise use and conservation of these dynamic sites. For humans, wise use and conservation of these wetland sites are of utmost importance. They are the heart of sustainable development, yet decision makers pay the least consult for their ecological values, which, on other hand, pay in form of wetland degradation. We need to understand their role and make policies for their sustainable use and conservation for our sustainable common future. We must take steps to restore the ecological characters of lost wetland sites and also check out the drivers of wetland loss.

For planning the wetland management process, we must focus on the objectives of site management. We must identify the factors that affect the features of wetlands and take relevant steps to make them positive for wetland conservation. If there are conflicts between locals and the government for its management, usage, etc., those need to be resolved with prime concern. Monitoring requirements of wetlands must be undertaken and effective management must be ensured together with resources' exploration. A proper channel of communication must be ensured between locals, policymakers, etc. In this way, we need to take urgent steps globally to raise awareness regarding wetlands' benefits and safeguard them with their inclusion in our development without their loss.

References

Agarwal, M. (2011). Migratory birds in India: migratory birds dwindling. *New Global Indian*. http://newglobalindian.com/nature.

Ahidur, R. (2016). Impact of human activities on Wetland: a case study from Bongaigaon district, Assam, India. *International Journal of Current Microbiology and Applied Sciences*: 392–396. https://doi.org/10.20546/ijcmas.2016.503.046.

Bassi, N., Kumar, D., Sharma, M. et al. (2014). Status of wetlands in India: a review of extent, ecosystem benefits, threats and management strategies. *Journal of Hydrology: Regional Studies* 2: 1–19. https://doi.org/10.1016/j.ejrh.2014.07.001.

Blankespoor, B., Dasgupta, S., and Laplante, B. (2012). *Sea-Level Rise and Coastal Wetlands: Impacts and Costs*. The World Bank.

Brink, T., Badura, P., Farmer, T. et al. (2012). *The Economics of Ecosystem and Biodiversity for Water and Wetlands: A Briefing Note*. London: Institute for European Environmental Policy.

Byström, O., Andersson, H., and Gren, M. (2000). Economic criteria for using wetlands as nitrogen sinks under uncertainty. *Ecological Economics 35* (1): 35–45.

Cowardin, L.M. (1979). *Classification of Wetlands and Deepwater Habitats of the United States*. Fish and Wildlife Service, US Department of the Interior.

Das, A., Ramkrushna, G.I., Choudhury, B.U. et al. (2012). Natural resource conservation through indigenous farming systems: Wisdom alive in North East India. *Indian Journal of Traditional Knowledge* 11 (3): 505–513.

Davidson, N.C. (2014). How much wetland has the world lost? Long-term and recent trends in the global wetland area. *Marine and Freshwater Research 65*: 936–941. https://doi.org/10.1071/MF14173.

Dugan, P. (1993). *Wetlands in Danger – A World Conservation Atlas*. New York, NY, USA: Oxford University Press.

Dutta, S. and Sengupta, A. (2015). Wetland restoration, a need for sustenance: a case study Chatra Beel of English bazar, district Malda, W.B, India. *International Journal of Applied Research* 1 (8): 810–814.

Ghermandi, A., Bergh, V.D.J.C.J.M., Brander, L.M. et al. (2008). The Economic Value of Wetland Conservation and Creation: A Meta-Analysis. [Working Paper 79]. Fondazione Eni Enrico Mattei, Milan, Italy.

Groffman, P.M. and Crawford, M.K. (2003). Denitrification potential in urban riparian zones. *Journal of Environmental Quality 32* (3): 1144–1149.

Kathiresan, K. and Thakur, S. (2008). *Mangroves for the Future: National Strategy and Action Plan, India*. New Delhi: Ministry of Environment and Forests.

Kaur, R., Dhir, G., Kumar, P. et al. (2012). Constructed wetland technology for treating municipal wastewaters. *ICAR News* 18 (1): 8–9.

Kumar, M.D. (2013). Climate in India: key features of the variables. *Climate Change Impacts on Water Resources Systems*: 104–111.

Kumar, M.D., Panda, R., Niranjan, V., and Bassi, N. (2012). Technology choices and institutions for improving economic and livelihood benefits from multiple uses tanks in western Orissa. In: *Water Management, Food Security and Sustainable Agriculture in Developing Economies* (eds. M.D. Kumar, M.V.K. Sivamohan and N. Bassi), 138–163. London: Routledge/Earthscan.

Kumar, D., Choudhury, M., and Rathore, A. (2020). Valuation of ecosystem services benefits of Son Beel Wetland in Assam, India: a case study of natural solutions to climate change water. *Climate Change Conference*, Hamburg University, Germany.

Lalchandani, N. (2012). Green zones packed as avian guests flocked. *The Times of India* (December 4).

Mac, M.J., Opler, P.A., Puckett Haecker C.E., et al. (2010). Status and trends of the nation's biological resources, 2 vol., US Department of the Interior, US Geological Survey, Reston, Va., 1998, http://www.nwrc.usgs.gov/sandt/index.html.

McAllister, D.E., Craig, J.F., Davidson, N., et al. (2001). Biodiversity impacts of large dams. International Union for Conservation of Nature and United Nations Environmental Programme, Gland and Nairobi. Background paper, 1.

Millennium Ecosystem Assessment, M.E.A (2005). *Ecosystems and Human Well-being*, vol. 5. Washington, DC: Island Press.

Molur, S., Smith, K.G., Daniel, B.A., and Darwall, W.R.T. (2011). *The Status and Distribution of Freshwater Biodiversity in the Western Ghats*. India. Cambridge, UK and Gland, Switzerland: IUCN, and Coimbatore, India: Zoo Outreach Organisation.

Phukan, P. and Saikia, R. (2014). Wetland degradation and its conservation: a case study of some selected wetlands of Golaghat district, Assam, India. *Research Journal of Recent Sciences* 3: 446–454.

Prasad, S.N., Ramachandra, T.V., Ahalya, N. et al. (2002). Conservation of wetlands of India-a review. *Tropical Ecology* 43 (1): 173–186.

Ramsar. (2020a). https://www.ramsar.org/wetland/india (accessed 2 February 2020).

Ramsar. (2020b). https://www.ramsar.org/sites/default/files/documents/library/sitelist.pdf (accessed 19 February 2020).

Ramsar Convention Bureau (2015). *Wetlands for Our Future: Act Now to Prevent, Stop, and Reserve Wetland Loss*. Gland, Switzerland: Ramsar Convention Bureau.
Ramsar, U.N.W.T.O. (2012). Destination wetlands: supporting sustainable tourism. *Secretariat of the Ramsar Convention on Wetlands*, Gland, Switzerland, & World Tourism Organization (UNWTO), Madrid, Spain.
Sarkar, U.K. and Ponniah, A.G. (2000). Evaluation of North East Indian fishes for their potential as cultivable, sport and ornamental fishes along with their conservation and endemic status. Fish Biodiversity of North East India: NBFGR. *NATP Publications* 2 (229): 11–30.
Turner, R.K., van der Bergh, J.C.J.M., Soderqvist, T. et al. (2000). Ecological economic analysis of wetlands: scientific integration for management and policy. *Ecological Economics* 35 (1): 7–23.
Verhoeven, J.T., Arheimer, B., Yin, C., and Hefting, M.M. (2006). Regional and global concerns over wetlands and water quality. *Trends in Ecology & Evolution* 21 (2): 96–103.
Wetland Conservation. (2020). Wikipedia. https://en.wikipedia.org/wiki/Wetland conservation (accessed 21 March 2020).

8

Assessing the Benefits, Threats and Conservation of Reservoir-Based Wetlands in the Eastern Himalayan River Basin

Kausik Ghosh[1] *and Poulomi Chakraborty*[2]

[1] *Department of Geography, Vidyasagar University, Midnapore, West Bengal, India*
[2] *Indian Institute of Technology Kharagpur, Kharagpur, West Bengal, India*

8.1 Introduction

Wetlands are considered as distinct "transitional (ecotonal) systems between upland terrestrial and deep open water systems characterized by aquatic vegetation (hydrophytes) with specific ecological characteristics, functions and values" (Mitsch and Gosselink 1986, 2007; Patten et al. 1990). In addition to that, Ramsar Convention (RCS 2006) defines "wetlands are areas of marsh, fen, peatland or water, whether natural or artificial, permanent or temporary, with water that is static or flowing, fresh, brackish or salt, including areas of marine water and the depth of which at low tide does not exceed six meters." Therefore, wetlands can be both natural and human-made as well. However, the river impoundments have formed human-made reservoir wetlands and, thus, may not fulfil the traditional definition of wetlands (Gopal and Sah 1995). Nevertheless, the Ramsar classification system (RCS 2006) mentioned that wetlands can form out of water-storage areas, reservoirs, barrages, dams, or impoundments (generally over 8 ha) and fall under human-made wetlands. Here, we define such water bodies impounded by human-made engineering structures on the river as reservoir-based wetlands (RBWs). Many studies identified RBWs which contributed to essential ecosystem services (ES) along with the natural wetlands across the globe (Panigrahy et al. 2012; Yuan et al. 2014). However, reservoirs across the world have created several issues more than the ES it delivered. The biodiversity hotspot of the Himalayas has been mostly affected by the encroachment of dams across the mountain river basin. Reservoir storage due to dam regulation has intervened between the upstream and downstream environmental flow links.

Consequently, many countries have withdrawn dams and destroyed the reservoirs, ignoring the essential benefits it serves to the surroundings. Thus, "human destroys, then creates, and then again destroys" an ecosystem by the construction, operation, and then the demolition of such reservoirs. In order to link the impact of dam constructions to functions on the riparian ecology and potential ES provided by the RBW, the present study has argued

Wetlands Conservation: Current Challenges and Future Strategies, First Edition.
Edited by Sanjeev Sharma and Pardeep Singh.

that humans should rethink before construction of new dams and proposed conservation policy measures for the existing RBWs focusing on the Eastern Himalayan river basin. The objectives of the present study are (i) to assess the ES provided by the RBW and their potential challenges across the globe with particular reference to the Eastern Himalayas utilizing the secondary data sources; (ii) additionally, to identify and establish the significance of all the RBWs in the Tista basin with regards to ES and challenges; and (iii) finally, to propose the management and conservation strategies of the existing RBW to maintain the ecosystem.

8.1.1 RBWs' Significance and Ignorance

In the year 2050, global demand for food will increase by 70% with an additional 40% increase in the world population (Lehnar et al. 2011). Reservoirs of the large dams are currently provisioning 12–16% of global food production (WCD 2000). For the additional food production, 11% of irrigation water is also likely to be supplied by the reservoirs. Significantly, there is a substantial impact on the global water cycle due to the reservoirs, which are considered anthropogenic aquatic systems (Barros et al. 2011). Again, hydropower supplied 19% of electricity across 150 countries (WCD 2000). RBWs are ignored mainly with regards to their extensive ES as reservoirs are the outcome of human regulation resulting in an adverse impact on the ecology. Around 16.7 million smaller reservoirs were covered in a large area (Lehnar et al. 2011) which remained underemphasized and unexamined (Downing et al. 2006). On the other hand, small reservoirs significantly removed nitrogen around 19.7 Tg N $year^{-1}$ from surface water at a regional and global scale (Harrison et al. 2009). The Grand Reservoir and Dam Database (GRanD) contains information regarding 6862 dams and their associated reservoirs with a total storage capacity of approximately 8070 km^3 for the reservoir size between 0.01 and 0.1 ha existing globally (Lehnar et al. 2011). These reservoirs have increased the total Earth's terrestrial surface water area by more than 305 000 km^2 contributing around 7.3% (Downing et al. 2006). In India, Gopal and Sah (1995) recognized the significance of human-made wetland functions in their study on wetland inventory and classification. According to the directory of the Ministry of Environment and Forests (MOEF), approximately 65 254 human-made wetlands are covering 2.59 million ha (m ha) area (MOEF 1990) in India.

Despite the extensive ES of RBW, such human-made construction vastly altered the river basin ecology. The flow of almost all the major and minor rivers around the world somehow has been regulated by large dams and barrages (Rao 1975) or else the impoundment of small catchments has generated numerous small-scale reservoirs (Fernando 1984; Sharma 1985). Human intervention structures have affected moderate to severely about 48% of the global river volume either by dam regulation or due to fragmentation, or by both (Grill et al. 2015). The same impact would increase by 93% if all the proposed dams were constructed soon. Take Brazil as an example, where reservoirs were observed in almost all the hydrographic basins which were visibly more than 700 large dams and associated reservoirs (Agostinho et al. 2008). But it does not stop here; as the reservoir starts undergoing fast and intense modifications, with the resultant inconvenience caused out of this, the same constructions are again re-modified at the cost of environmental degradation. Likewise, several alterations usually track the construction of any dam; in fact, this is possibly the reason for the destruction of both upstream and downstream ecosystems which find their replacement from the newly generated system, then again distorted with the further dam removal process. Hence, constructions, destructions, and again reconstructions of new reservoirs are disagreeable; rather, the reappraisal of the existing projects needs to be warranted. The RBWs across the globe have encountered many challenges like anthropogenic stress, water-level variations in the reservoir, and climate change but most important is the dam removal processes, which is considered as a dam management strategy. However, dam removal is not economically practical and the ecological consequences are best understood by viewing it as a disturbance (Stanley and Doyle 2003). Thus, dam removal disrupts and reconfigures the physical environment by eliminating the existing reservoir-wetland ecosystem. The ES provided by the reservoirs in contrast to the natural wetland is ignored primarily due to two reasons. First, the reservoirs are predominantly regulated and controlled by the dam regulation policies; second, delineations of the reservoir area as wetland are sometimes very critical due to variabilities in regulated water level than the seasonal variations. The development and destruction of RBWs and associated functions, goods and services, and threats and possibilities were rarely highlighted in detail in the past studies specifically, the possible impacts of anthropogenic activities, and climate change in general. As a result, this lack of focus on such human-made wetlands has limited the information on RBWs in the Himalayan region, whereas information about the Eastern Himalayan RBWs is very scattered.

8.1.2 RBWs in India

A large population in the Indian subcontinent is directly or indirectly involved with many wetlands. While the dam constructions converted the natural water bodies into human-controlled wetlands, as a result, a new regulated wetland ecosystem formed (Gopal and Sah 1995). In India, many small- and large-scale dams are operating and proposed new dams, barrage, and flow modification schemes for water supply, irrigation, hydropower, and fisheries (Smakhtin and Anputhas 2006; Grumbine and Pandit 2013). To reach the expected electricity demand by 2020, the Government of India (GoI) announced an aspiring plan in 2003 to generate 50 000 MW with the construction of new hydropower infrastructure (Ministry of Power, Hydropower policy, GoI 2008). This would ensure more than double of India's present hydropower capacity in < 15 years (Dharmadhikary 2008). Over the next several decades, the GoI (MoP 2008) aims to construct almost 292 dams alone in Indian Himalayas, projected to contribute ~6% of the national energy needs by 2030 (Grumbine and Pandit 2013). Constructions of these dams will form many large to small reservoirs.

A national-level assessment and wetland inventory was carried out using remote sensing data (RESOURCESAT-1 LISS-III data of 2006–2007) in India for further monitoring by Panigrahy et al. 2012 (Table 8.1). They have found that the reservoirs of dams/barrages covered around 2481 987 ha areas that are more than 16% of the total wetland area in India. More than 43% of seasonal wetland area reduction observed between post-monsoon and pre-monsoon seasons, reflecting the influence of Indian monsoon on RBW hydrology. Both floating and emergent aquatic vegetation on the reservoir surface increased during pre-monsoon conditions (311%) compared with the post-monsoon period in reservoir/barrage, which is four

Table 8.1 Reservoir-based wetland statistics in India and their seasonal change.

Reservoir-Wetland characteristics	Post-monsoon (ha)[a]	Pre-monsoon (ha)	Change in area (ha)	Change in %
Open Area	2 260 574	1 268 237	992 337	−43.90
Aquatic Vegetation	119 711	492 237	372 526	311.19

[a] ha is hectare.
Source: Adapted from Panigrahy et al. (2012). © John Wiley & Sons.

times compared to the post-monsoon. It indicates two significant factors: (i) water-holding capacity increased due to reservoir operation resulting in stagnant water during pre-monsoon seasons; (ii) high-flow condition with variable discharge in monsoon retarded the development of aquatic vegetation immediately during the post-monsoon seasons.

8.1.3 The RBWs in the Eastern Himalayas

The Eastern Himalayan region is a distinct biogeographic area which is a subset of the "Himalayan biodiversity hotspot" (Mittermeier et al. 1998, 2004). It includes the Ganga, Brahmaputra, and Irrawaddy river basin, which is also a subset of Indo-Burma Biodiversity Hotspot (Allen et al. 2010). The predominant wetland types in the Eastern Himalayas are lakes, freshwater swamps, and marshes or bills (Singh 2001) which can be witnessed predominantly across the Brahmaputra river basin. Reservoir-associated wetlands in the Eastern Himalayas are the recent development (Gopal et al. 2010). Additionally, floodplain wetlands, beels, jheels, and pats along with reservoirs in the river basins have constituted significant habitats for fish and fisheries. Therefore, fish is one of the important consumed foods and livelihood opportunities for many in the region. Reservoirs as wetlands of this region are also supplying food (fish and agriculture water), fodder, fuels, medicinal plant, and many valuable ES. Greater diversity of flora and fauna has been witnessed in these wetlands. While many threatened species, considered as the most critical wildlife habitats recognized internationally, belong to the freshwater swamps of the region, more than 13% of waterbirds and around 4% of fish species across the world were recorded in the Eastern Himalayas.

In a report by Dharmadhikary (2008), it was mentioned that the Indian Himalayas needs more than $60 billion funds to produce 50 000 MW of hydroelectricity for 162 dam-building projects till 2018. Only 32% (26 376 MW) of hydropower capacity has been developed in the Indian Himalayas till now and rest 78% remain unexploited out of the total estimated hydropower potential which is 118 210 MW. A total of 318 hydro projects with 93 615 MW capacity were proposed and among which 74 projects are already constructed, and 37 projects are under construction in the Himalayan region. A substantial number of these projects are planned to implement in the Indian state of Sikkim on the Tista river basin. After the establishment of the comprehensive hydropower policy in 1998, the Government of Sikkim gave environmental clearance and approval to 26 hydro projects to private players between 1999 and 2002 and the Sikkim Power Development Corporation Ltd. (SPDCL) was formed to facilitate the joint ventures on the Tista basin (Ahlers et al. 2015).

8.2 The RBWs in the Tista Basin

The Tista basin is located in the North-Eastern hill state of India in Sikkim and it flows ~151 km in Sikkim Himalayas. Further, it flows through another Indian state, West Bengal for ~142 km and then it flows ~121 km to join the mighty Brahmaputra in Bangladesh. Thus, Tista is a transboundary river and covers around 12 000 km^2 of basin area between India and Bangladesh. The Tista basin covers more than 95% of the area of Sikkim and this Indian state is one of the vital biodiversity hotspots, enriched with dense forest cover. However, utilizing the drop-down elevation of about 7000 m in the higher Himalayas to 200 m drop at the mountain front, more than 13 cascading run-of-the-river (RoR) dams with small reservoirs are constructed in the Tista basin to generate hydroelectricity (Figure 8.1). Hence, it becomes one of the emerging hotspots of RoR dams in the Himalayan region. Out of the proposed 29 dams (the Tista basin in Sikkim and West Bengal) with more than 25 MW capacities, 13 dams are already constructed and operational, around ten dams

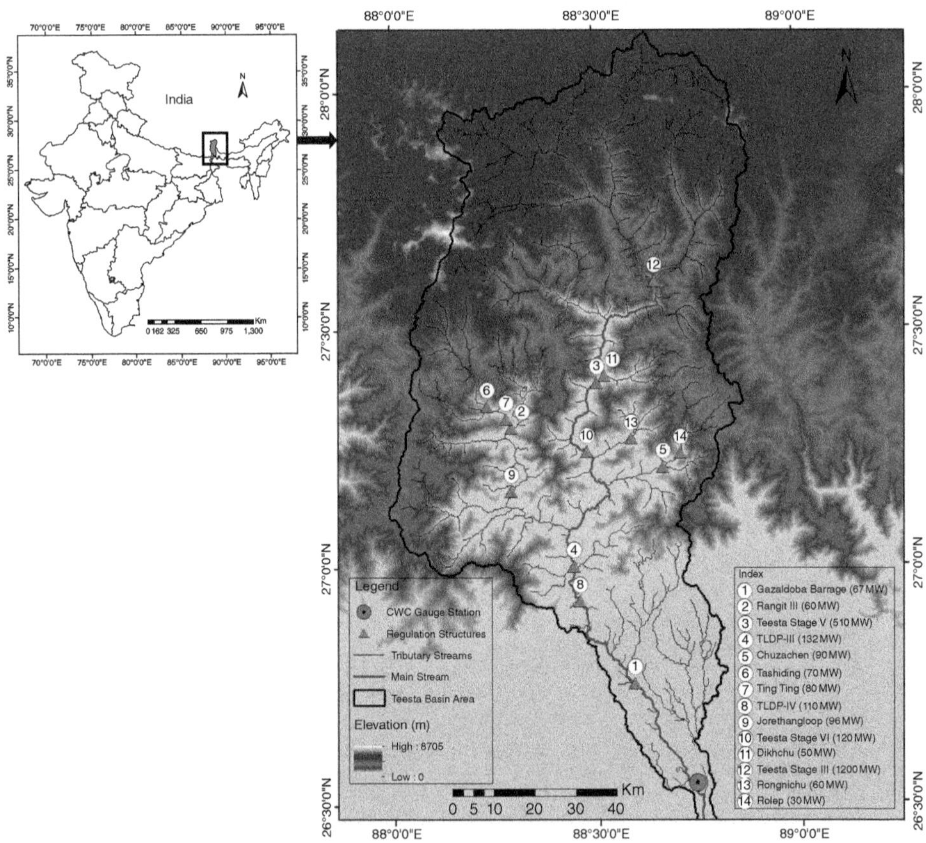

Figure 8.1 Location of the Tista basin in North-East India. The Tista basin spreads across the upstream hill to the downstream plain till the Central Water Commission's (CWC's) gauge station located at Jalpaiguri; the basin shows the distribution of the location and hydropower generation capacities of several dams/reservoirs.

are under proposed construction, and many dams are scrapped. The basin has experienced massive river flow due to monsoonal rainfall (May to October) with 80–90% of annual total rainfall but observed dry conditions during the non-monsoon (November to April) lean season. During monsoon months, reservoirs regulate less due to heavy discharge and high energy flow conditions, relative to the lean seasons' low flow condition to fulfil the rising electricity and water demands. Therefore, the seasonal variation of regulation and construction of reservoirs in the Tista basin has impacted the ES of the river.

8.3 Benefits of Reservoirs as Wetland

8.3.1 Ecosystem Services Provided by the RBWs

The RBWs are providing multiple ES like any other wetlands. The generation of Hydroelectricity as an essential economic service makes it distinct from other wetland services. According to Zarfl et al. (2015), the worldwide construction of 3700 new dams may increase global hydropower production by 73% which is an increase in the exploitation of the technically feasible hydropower potential from a total of 22 to 39%. Similarly, Maavara et al. (2015) propounded that the land-use alterations due to agricultural and human activities enhanced the river water phosphorus, while reservoirs successfully prevented the additional phosphorus amount by maintaining the ecological balance and nutrient cycle. The reservoirs accumulate 5.3% of global reactive silicon (Si) loading in the watershed, however less than the natural lakes (Maavara and Van Cappellen et al. 2014). Reservoirs act as a sink to partially offset the terrestrial CO_2 and carbon flux into the reservoirs is ~0.3 Pg C yr^{-1} (Macklin et al. 2018). Although, a study by Agostinho et al. (2008) on the impact of dams on fish diversity in the Parana River basin in Brazil attributed that the reservoir operations have influenced the ecological functions positively and negatively. Although it provides essential ES, see the list below which shows ES provided by the RBWs across the globe.

List of ecosystem services provided by RBWs across the world:

Broad Categories	Services
Economic services	
Power	Hydropower (WCD 2000; Ahlers et al. 2015); Electricity (Ministry of Power, Hydropower policy, GoI 2008); Fuel (Dharmadhikary 2008; Grumbine and Pandit 2013)
Livelihood and others	Aquaculture and commercial fishing, Medicines and other biochemical products, generates fuelwood and fodder (Buechler et al. 2016)
Tourism	Ecological and surrounding beauty promotes tourism (Zwieten et al. 2011)
Employment generation	Generates various employment and livelihood opportunities for locals (Bro et al. 2018)

(Continued)

(Continued)

Broad Categories	Services
Provisioning services	
Food	Fish (Agostinho et al. 2008); Fruits and grains (WCD 2000)
Water	Drinking water (Yang et al. 2019); Agricultural use (Smakhtin and Anputhas 2006; Gopal et al. 2010; Grumbine and Pandit 2013)
Regulating services	
Water regulation	Hydrological flows – groundwater recharge and discharge, water-level fluctuation control (Cheng and Chau 2004); Irrigation water supply (WCD 2000; Gopal et al. 2010; ICIMOD 2010); Helps increasing Earth's terrestrial surface water area (Downing et al. 2006)
Climate regulation	Helps shrinking greenhouse gases (Harrison et al. 2009; Macklin et al. 2018); Influences local and regional climatic processes (temperature, precipitation, and others; Alternative source of water – savior at the era of climate change (Palmer et al. 2008)
Erosion regulation	Preservation of soils and sediments provides resistance (Abedini et al. 2012)
Natural hazard regulation	Flood control and draught control (López-Moreno et al. 2009)
Water purification	Maintenance, improvement, and quality control through deduction of unwanted nutrients and other pollutants (Maavara et al. 2015; Wiejaczka et al. 2018); reduces heavy metals contain (Wiejaczka et al. 2018)
Waste treatment	Manages wastewater, recovery, recycle, reuse, revival (Ramasar Handbook 2016)
Ecological services	
Habitat	Habitat of numerous plants and species (Datta 2011; Panigrahy et al. 2012); Additionally provides habitat to several migratory species (Datta 2011)
Biodiversity hotspot	Preservation of numerous plants and species (Leria and Cartonati, 2008; Wantzen et al. 2008; Datta 2011; Seth 2016); non-native species (Gehrke et al. 2002); shelter for many threatened species (Allen et al. 2010; O'Neill, 2019)
Cultural services	
Recreation and leisure	Provides an environment for spare time and recreational activities (Daugherty et al. 2011); fishing (Zwieten et al. 2011).
Aesthetic	Wetlands associated with reservoirs enhance the beauty or aesthetic value of the surroundings (Ghosh and Singh 2012)
Cultural landscape	Cultural landscape (Wijesundara and Dayawansa 2011)
Other supporting services	
Soil formation	Sediment retention and accumulation of organic matter (Gopal et al. 2010; ICIMOD 2010)
Nutrient cycling	Storage, recycling, processing, and acquisition of nutrients (Maavara et al. 2015)

8.4 Assessment of Ecosystem Services in the Tista Basin Provided by the RBWs

In general, the RBW regulations can impact in two ways: first, the dam obstruction can alter the riverine biotic community structures and wipe out the fish species from upstream and downstream of the dams (Bhatt et al. 2017); second, in contrast, the reservoirs can create opportunities for the enhancement of non-native species through stocking (Gehrke et al. 2002) resulting in more diversity. However, there is a high chance of replacement of native species due to the invasion of new species and the issue is still debatable. In the Tista basin, 14 reservoirs of several hydroelectric projects with > 25 MW capacity and a barrage are identified. The Gazaldoba barrage reservoir of the Tista Barrage Project (TBP) at Gazaldoba, West Bengal in the foothill of Himalayas is very significant in terms of its area and capacity to hold the water as well as the extensive ES it provides to the surroundings. Remaining reservoirs formed by the RoR cascading dams across the mountainous basin at several elevations (Figure 8.1). The impacts of these reservoirs varied not only due to the physical characteristics like the size, shape, capacity, and operational principle of the dam but also on the location of the dams in the eco-sensitive region and the number of displaced people and their livelihood opportunities surrounding the dams. The total capacity or total volume of water is significant to understand the impact of reservoirs on the ES of the basin. The missing reservoir's capacity in the Tista basin was calculated following Lehner and Doll (2004) from the GranD database using the following equation (8.1):

$$\mathrm{Wv}(\mathrm{MMC}) = 0.678(\mathrm{A} * \mathrm{h})^{0.9229} \tag{8.1}$$

where Wv = wetland volume/capacity in 10^6 m^3 (MMC); A is the area of the reservoir in km^2; and h is the dam height in m.

Cumulatively, the construction and operation of reservoirs have many positive and negative impacts on the river ecology. Thus, the ES of RBWs is ignored in comparison with the natural wetlands across the globe. Surprisingly, the Tista basin has witnessed several ES provided by the RBWs, among which the Gazaldoba wetland is very significant. The total area covered by the reservoirs calculated around 969.32 ha where all these reservoirs are producing more than 3295 MW of hydroelectricity (see Table 8.2). The upstream of the mountainous basin has been recognized as a biodiversity hot spot (O'Neill 2019), surrounding which many RoR dams are being constructed. Likewise, in the alluvial plain of the Himalayan foothill at Gazaldoba, an important RBW has formed due to the barrage operation and provides multiple ES to its surroundings (Ghosh and Singh 2012). RBW area and capacity indicate the efficiency to control flood and the potential to enrich the biodiversity and species richness. A total of 969.32 ha area with 89.15 Mm^3 capacity of the reservoirs can regulate peak flows of a massive flood. After the 1968's massive flood in the basin, no such flood was recorded in the basin. Many dams were started construction since 2000 (except for the barrage in 1987 and Rangit III in 1995), and subsequently all the 14 reservoirs are become functional. Strikingly, the

Table 8.2 Details of the hydroelectric projects in the Tista basin.

Sl No.	Name of the project	Operated since (Year)	Reservoir area (ha[a])	Reservoir capacity (Mm^{3}[a])	HE (MW[a])	Dam height in meter
1	**Gazaldoba Barrage**	1998	260	7.722	67.5	11
2	**Rangit III**	2000	12.9	1.175	60	47
3	**Tessta Stage V**	2008	67.7	13.25	510	88.6
4	**TLDP III**	2013	183.49	18.36	132	32.5
5	**Jorethang Loop**	2016	14.489	0.665	96	17
6	**Tessta Stage III**	2017	25.41	5.08	1200	60
7	**Teesta Stage VI**	2016	24.8225	3.18	500	26.5
8	**Chuzachen HEP**	2013	13	1.28	110	48
9	**TLDP IV**	2016	308	37	160	45
10	**Tashiding**	2015	1.33	0.051	97	12
11	**Dikhchu**	2016	39.13	0.33	96	36
12	**Rolep**	2011	2.9	0.35235	72	45
13	**Rongnichu**	2017	10.7	0.364014	96	12.6
14	**Ting Ting**	2015	5.45	0.342	99	49
	Total		969.32	89.15	3295.5	

[a] ha is hectare, Mm^3 is million cubic meters, MW is megawatt. Further, see Figure 8.1 for locations of the dams on the map.

Source: Recorded from several databases and estimated by the study.

average peak-flow during monsoon reduced to 4434 cumecs from 6189 cumecs, a 28% reduction recorded between 1998 and 2016 (computed from CWC data, 1990 to 2016). The reservoir at Gazaldoba expected to supply irrigation water across 0.92 million ha of agricultural land in the northern part of West Bengal and more than five cumecs of drinking water to the surrounding cities through the canal. A total of 3295.5 MW of hydroelectricity generation from these dams would have a substantial economic value. Besides, the water level in the reservoir (Gazaldoba) has attracted water-based recreation and tourism in the region, which created additional livelihood opportunities for the locals (Ghosh and Singh 2012).

Furthermore, such wetland forms a new habitat that enriches the flora and fauna diversity. More than 20 000 waterfowl species observed near the Gazaldoba (Seth 2016) along with 80 species of birds and more than 73 species of zooplankton (Datta 2011). Newly built reservoirs in the Tista basin enhanced the heavy metals; surprisingly, downstream of the reservoirs reduced the heavy metals which indicate the river water purification due to reservoir operations (Wiejaczka et al. 2018). However, the alteration of physicochemical properties and concentrations of ions due to reservoir operations have substantially normalized by environmental factors before the river debouches into the downstream plain from the mountain part of the basin.

8.5 Adverse Impact of RBWs

8.5.1 Construction and Function of RBWs Across the World

There were only 5000 large dams worldwide in 1950. The same within 50 years in 2000 increased to 45 000 (Khagram 2004), predominantly built to regulate the river flow for hydropower and irrigation. This era showcased the exponential growth of dam-building exploded worldwide due to technical, financial, and sociopolitical reasons (Pandit and Grumbine 2012). These dams are the reason behind the creation of many human-made wetlands like RBWs, but at the same time, they may affect many other riparian wetland ecosystems. A study by Grill et al. (2015) estimated that 43% of the global river volume is moderately to severely affected due to dam operation while the severe impacts of dam-based river fragmentation influenced 24% of the river worldwide. Another study by Lehner et al. (2011) attributed that the dam regulation will affect globally 46.7% of the large rivers with average flows above $1000\,m^3/s$ (cumecs), whereas small rivers will be affected less than 2%. Moreover, if all the future dams will be constructed, then the number of severely affected categories of rivers will increase from the present 9 to 16% in the future (Grill et al. 2015). According to Grill et al. (2015), there are more than 6374 large existing dams that have modified the river connectivity and resulting fragmentation and about 3377 planned or proposed dams across the globe still in count. The increasing age of reservoir operations in the tropical areas will enhance the carbon emission to contribute to the Greenhouse Gasses (Barros et al. 2011). Total organic carbon will be reduced to the ocean by the existing and proposed future reservoirs by 13–19% for the period considered between 1970 and 2030 (Maavara et al. 2017). The average N_2O emission is estimated very high at 9.6 ± 6.0 mmol $N.m^{-2}\,year^{-1}$ for the reservoirs relative to very low at 0.8 ± 0.5 mmol $N.m^{-2}\,year^{-1}$ for the natural lakes (Lauerwald et al. 2019). Similarly, the construction and operation of dams accelerated the process of riverbank erosion and sedimentation, which altered the geomorphology, hydrology, and ecology of the river (Brandt 2000). As a result, a combination of the newly altered hydrological regime such as loss of vegetation and species influenced the ES (Richter et al. 2003; Nilsson et al. 2005) while the invasion of new species is one of the emerging issues due to RBWs affecting the ecological integrity of the riverine ecosystem (Wu et al. 2004; New and Xie 2008; Willison et al. 2013). Significantly, the RBWs also impact the socioeconomic values and bring vulnerabilities to the surroundings. For example, the risky position of the reservoir, situated in a densely populated region (China) when flooded by the rising water, results in the resettlement of hundreds of thousands of residents (Stone 2008; Fu et al. 2010). Likewise, the recent suspected threat of the Three Georges dam break and massive release of discharge downstream during 2020 flood in China. On the other hand, the extent, exposure and adversity of river regulation and associated ecosystems in Indian Himalayas are still largely remained unknown (Pandit and Grumbine 2012).

8.5.2 Adverse Impact of RBWs in the Eastern Himalayas

Particularly, in the Eastern Himalayas and developing countries like India and China, the share of reservoirs and dams is rising (Pandit 2009; Grumbine and Xu 2011). If all the proposed dams (a total of 292) are going to be constructed across the Himalayas, based on the current global number of dams, the region will hold the tag of having the highest density of dams in the world. The average dam density across the Himalayan states was 1.6 dams/1000 km^2 where the Eastern Himalayan river basin Tista in Sikkim will have the highest density covering 4 dams/1000 km^2 if all the proposed dams will be constructed (see Table 8.3). It is suspected that a high percentage (90%) of dams would be disproportionately concentrated in species-rich subtropical and temperate zones in the Eastern Himalayas in the near future (Pandit and Grumbine 2012; Grumbine and Pandit 2013). Yet, at present, due to limited studies and little certainty about the likelihood of all projects being built, it is difficult to precisely quantify the full extent of ecological changes that may result from the proposed dam building. Although, there is a strong correlation with this continuous process of aggressive dam building, the density of proposed dam building and wetland in the Indian Himalaya. More specifically, dam-building activities would have adverse effects on the existing ecosystem as well as it will alter the riverine ecology of the watershed. About 90% of the Indian Himalayan biodiversity region would be affected by the construction processes and 27% of these dams would affect dense forests. The submergence of the forest area and habitat degradation due to dam building could lead to the extinction of 22 angiosperm and seven vertebrate taxa and tree species richness likely to be reduced by 35%, tree density by 42%, and tree basal cover by 30% in the dense forests (Pandit and Grumbine 2012; Grumbine and Pandit 2013). The diversity of the species and birds helps estimate the linkages between RBWs and ecology in different river basins (Gopal et al. 2010); however, their count may decline due to the imposed changes by the dam and the loss of their habitats (Agostinho et al. 2004; Ceregato and Petrere Jr 2003). The Tipaimukh High Dam hydroelectric project located in the northeast Indian state of Manipur, intended to modify the flow and flooding of the Barak basin, has the potential to impact on the habitats of several mountain river fishes and migratory species in the upstream as well as in the downstream parts of the river due to fragmentation (Allen et al. 2010).

8.6 Assessment of Impact on the Tista Basin

Achary et al. (2011) demonstrated that the Tista basin covers diverse habitat with complex vegetation structure and high bird species in the mid-elevational (2000 m) range of the basin. Unfortunately, most of the RoR dams are constructed and planned to be built in the same elevation zone between 500 and 4500 m (Figure 8.1). If all the proposed 29 RoR dams would be constructed, then nearly 70.5 km, around 59% of the mountainous river stretch, would be diverted through Head Race Tunnels (HRTs) (Table 8.3). Subsequently, small reservoirs of the projects will convert 29 km or about 24% of the existing flowing water lotic ecosystem into multiple static semi-lacustrine ecosystems (Bhatt et al. 2017). At the same time, approximately eight dams were observed in a 120 km mountainous basin in Tista, which means one dam in every 15 km river stretch compared with average one dam in every 32 km length of the river stretch across the Himalayan river basin (Grumbien and Pandit 2013). There are many negative impacts of hydropower projects along the river Tista. Such adverse effect can exert pressure on the microlevel aquatic species, fish-fauna diversity, loss of biodiversity, habitat restructuring, forest cover destruction to affect the social and cultural profile of the native people due to displacement and rehabilitation.

Table 8.3 Distribution and density of dams in the Indian Himalayan states and global river basin and dam density in the major river basins in India.

State/basin	No. of dams	Geographic area in km^2	Dam density (dams/km^2 of area of the state/basin)	Dam density (dams/1000 km^2 of the area of the state/basin)
Sikkim	29	7096	0.00400	4.0000
Himalayan states	292	422 232	0.00161	1.6120
Ganga	89	861 452	0.00010	0.1022
Indus	94	321 290	0.00029	0.2895
Brahmaputra	109	194 413	0.00056	0.5825
Himalayan basins	292	1377 155	0.00032	0.3247
Global basins[a]	194	36 591 693	0.00001[a]	0.0053[a]

[a] Based on 23 major basins distributed across the 7 continents.
Source: Northwest Alliance for Computational Science and Engineering (NACSE 2011) and Pandit and Grumbine (2012). © John Wiley & Sons.

The usual impacts of RoR dams on the Tista basin are the downstream flow variabilities, sediment nutrient modification, loss of breeding ground for fauna, habitat degradation, fragmentation, and obstruction of the aquatic species migration. During the reservoir regulation period between 1998 and 2016, the annual flow reduced to around 28%, the monsoon flow reduced to 22%, and the non-monsoon flow around 70% compared to the pre-regulation period (1990–1997) based on calculation from CWC data (Figure 8.1, gauge station). The dam constructions in the basin dramatically increased the suspended sediment to 46 mt (million tons)/year between 1998 and 2010 relative to the average 11 mt/year recorded between 1990 and 1997. Since after the dam operation period witnessed a 60% reduction in the suspended sediment during the non-monsoon lean seasons between 2011 and 2016 while for the same period but during monsoon seasons, the suspended sediment increased compared with the pre-regulation period. It indicates rigorous regulations during non-monsoon seasons; the reservoirs absorbed sediments while a part of the same deposit released during monsoon peak flow along with the sediment delivered from the hillslope erosion in the basin.

Consequently, sedimentation into the reservoirs and decreasing sediment supply into downstream led to riverbank erosion which is one of the crucial issues in the alluvial plain of the basin. Simultaneously, RBWs suffer due to variations of water level by the regulation operations of the dam. The Tista Low Dam Project (TLDP) IV holds water throughout the day and releases less than ten cumecs of streamflow for 20 hours and rest 4 hours it, releases more than 700 cumecs of water into the downstream specifically during non-monsoon season. Such water level fluctuations in the reservoirs and sudden water discharge into the downstream of the reservoir have a dramatic impact on the aquatic ecosystem of the upstream and downstream both (Rudra 2018). At Gazaldoba, less than 6 m-deep reservoir witnessed up to ~5 m of water-level fluctuations due to barrage water diversion through the canal (Figure 8.2). Water-level variations influenced the livelihoods of the local fishermen and boatmen (Mullic et al. 2013). The river basin attracts many

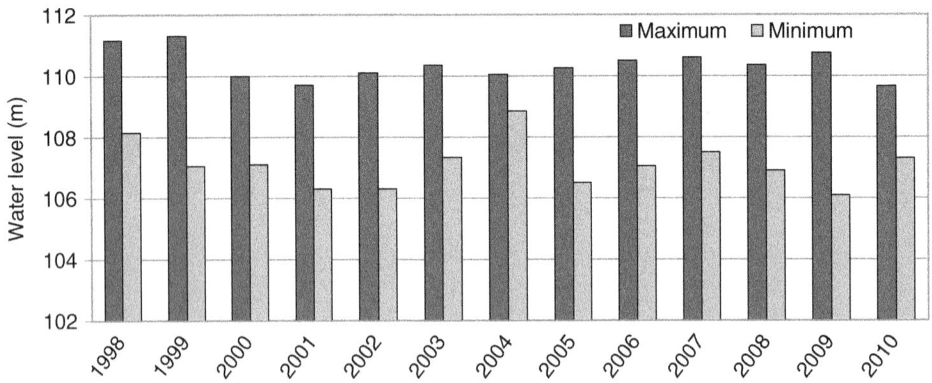

Figure 8.2 Maximum and minimum water-level fluctuations between 1998 and 2010 in the Gazaldoba reservoir, computed from the Central Water Commission (CWC), India data at Gazaldoba Barrage.

migratory bird species; high regulation and sudden water-level variation influenced the breeding of the bird species, resulting in lower nesting of waterfowls (Datta 2011). The hydropower projects in Tista are likely to affect more than 103 fish species that belong to 13 families; meanwhile, 40% species are designated as vulnerable to critically endangered and almost 55% bottom-dwelling fishes are facing difficulties to adapt to the new way of reservoir system of life (Bhatt et al. 2017). The density of the fish species increased in the lower elevation of the basin due to high discharge concentration with large basin area and relatively higher water temperature. However, dam operations across the basin affected these species. Bhatt et al. (2012) estimated that 59% of the Himalayan fish endemics are clustered between the elevation of 700–2000 m with high species' richness, while most of the RoR dams of the basin are located within this elevation zone. The same study suspected that the hydropower generation will reduce the fish species in the basin. Noteworthy to mention, a vulnerable fish species, named Schistura inglisi has been found in River Tista at the mountainous basin of Sikkim and Darjeeling, but now likely to be susceptible to sedimentation (Allen et al. 2010).

8.7 Potential Challenges and Threats to RBW

Singh (2001) demonstrated that the major challenges of the Eastern Himalayan wetlands are the uncontrolled anthropogenic activities like urbanization, enormous population growth, wetland conversion to serve agriculture and aquaculture, massive deforestation, and the major engineering constructions to produce hydroelectricity.

8.7.1 Anthropogenic Activities

Not only have RBWs influenced ways of people's lives, but the human has also affected RBWs in many ways. The tourism industry is one of the attractive economic activities in Sikkim Himalaya, which attracts many tourist populations. Consequently, increasing population pressures along with local population growth and rapid urbanization process enhanced the built-up area from 1.9 to 2.5% between 1990 and 2013 in Sikkim (Kanade and John 2018). More than 30 million populations across the Tista basin in India and Bangladesh are directly and indirectly dependent upon the river water for their lives and livelihoods (The Asia Foundation 2013). There was a dramatic increase in the number of tourists (including domestic and foreign tourists) from 99 323 in 1994 to 14 24 965 in 2017, meaning thereby a total 58.03% increase (Department of Tourism, Government of Sikkim, Gangtok; Tourism and Civil Aviation Department 2017). Surprisingly, many tourists are hardly sensitive enough to wetland ecological communities (Tambe et al. 2008). The idle waste management system caused instantaneous fear to RBW and surroundings in the Eastern Himalayas (O'Neill 2019). Besides, increasing vehicular pressure due to tourist movements may cause environmental degradation. Increasing population and economic development have encouraged more surface water demand along with rapid hydropower generation and infrastructure development in the hill. It has been estimated that the development of water-based tourism industry near the Gazaldoba reservoir will create more than 13.64 million liters per capita/day water demands. Additionally, the water-based tourism industry will also generate massive waste that significantly affects the reservoir and surroundings. Furthermore, the landuse land covers altered due to agriculture, urbanization, and construction of roads may cause deforestation and expose the mountain soil to erosion and vulnerable to landslides. The construction of roads dumped massive mucks/sediments into the river. These anthropogenic activities are the primary reason behind the sedimentation in RBWs.

8.7.2 Variations in Water Level

Reservoir operations have influenced the water level in the Tista basin, whereas water level in the reservoirs is seasonal and depends upon the intensity of dam operations. Many water-based ES, like flora and fauna diversity of the reservoir water (Leria and Cartonati 2008; Wantzen et al. 2008) and water-based recreations (Daugherty et al. 2011) are severely affected by reservoir-water-level fluctuations De Lima et al. (2017). The daily water-level fluctuations and variations in downstream environmental flow (e-flow) due to upstream regulations in the Tista river may affect the ES (Rudra 2018). Tista River is a transboundary river. Consequently, the geopolitics of quantitative water sharing (50–50 proportion proposal) between India and Bangladesh has virtually ignored the e-flow of the river which may also affect the lean season water-level of the reservoir-based ES in the basin. The Gazaldoba reservoir observed more than 5 m of water-level variation out of the highest 6 m deep reservoir (see Figure 8.2) while proposed 50–50 proportion water treaty may affect the water level of the reservoir.

8.8 Climate Change

The concentration of greenhouse gases (GHGs) like carbon dioxide (IPCC 2000: global mean CO_2–370 ppm, expected to mount 490–1260 ppm by 2100), methane, and nitrous oxide are prominently increasing, which primarily drives the global climate change, associated with "an increase in the global mean temperature of the earth's surface and increased variability in precipitation regimes with greater frequency of extreme events" (Gopal et al. 2010). The models and long-term trend analysis of temperature and precipitation (1971–2000) in Eastern Himalayan regions predicted that both are expected to go up with a significant amount in all seasons by the end of the twenty-first century (Gopal et al. 2010; ICIMOD 2010). Climate change will affect the glacial melt discharge in the Eastern Himalayan river basin. As a result, reducing streamflow into the reservoirs will adversely affect the food security of the region (Immerzeel et al. 2010). The temperature (minimum and maximum) in the basin increased between 1951 and 2010, while the snowfall reduced and overall precipitation projected (2100) to be decreased in the Tista basin (Sharma and Goyal 2020). Subsequently, there is a possibility in streamflow reduction in future. In the Brahmaputra river basin, more than 27% of the river flow of the total discharge was generated by the snow and glacial melt. Immerzeel et al. (2010) found that the summer and late spring meltwater discharge are eventually expected to reduce consistently and considerably after a period with a dramatic increase in streamflow due to acceleration in glacier melting. However, the reservoirs in the glacier-melted river basins are considered as the large buffers for future water supply against the climate-induced variabilities of water resources in the densely populated basin (Palmer et al. 2008). On the contrary, the uncontrolled dam operations and an unrealistic number of proposed dams may deteriorate the situation and lead to water stress conditions in the Himalaya.

8.9 Management and Conservation of RBWs

Across the world and specifically in the Eastern Himalaya, RBWs are providing essential ES, despite the constructions and regulations of reservoirs that have altered the ecology of the river. RBWs can play dual roles, either support the benefits to the needy people, or may bring destructions one step closer. Still, due to the reasons mentioned above in Section 8.1.1, the RBWs are not given much importance like the privilege of natural mainstream wetlands. In the global scale, India lacks an accurate and updated database of ES of RBWs; therefore, their capabilities are also attributed very sporadically and dispersedly. Hence, the first step toward managing RBWs should be the identification of RBW through proper classification and database generation, and further recognize it as a significant wetland system considering the essential ES it provides. With that comes the urgency to protect the existing reservoirs that provide important ES and manage the regulation policies of these reservoirs to get more ecological benefits and also avoid adverse impact due to the construction of new reservoirs in the sensitive biodiversity hotspot region like Himalaya. Therefore, the management strategies need to be formulated to protect the reservoir operations rather than only dam removal or destructions, as dam removal has several negative impacts on the ecology (Pizzuto 2002). India needs to put a brake on the new proposed hydro projects on the Himalaya. Instead, it should focus on the existing projects to enhance the benefits and reduce the adverse impact observed in the above sections. The ignored reality is that the Government has never thought beyond to follow up on the country's future energy requirements that may invent a replacement of the existing hydropower and can significantly reduce the need for so many new dams (World Wildlife Fund 2007; Pandit and Grumbine 2012). The significant issues like the stormwater and sediment flux can be controlled in the upland reservoirs by managing and decreasing regulation of reservoirs (Barros et al. 2011). The adverse impact can be mitigated by introducing flood-tolerant woody species in the reservoir surroundings which also can provide the necessary economic resources, such as wood, fruits, and medicinal oils and additionally enhancing the beauty of the landscape as a vital step for sustainable management (Willison et al. 2013). The renovation of natural flow regimes should be adopted by barrage and river management authorities, and technologies can be used to lessen the impact of dams on migratory species (Allen et al. 2010). Fish ladder or fish passes are suggested for mitigation and management of fish species movement along the obstructed stretches of the river across the upstream and downstream stretches in the Tista basin (Bhatt et al. 2017). Although such measures are impractical in many cases of the Himalayan river dams, the ex-situ conservation through the development of hatchery of native fish species was suggested by many Environmental Impact Assessment (EIA) reports in the Tista basin. Sustenance of the minimum e-flow in the downstream stretches of the dams is crucial for the riparian ecology, life and livelihood opportunities, and overall development of the aquatic ecosystem of the basin. For example, a detailed EIA study has suggested that the minimum e-flow of the Tista Stage IV project downstream should be at least 5 m^3/s. However, the complete implementation of EIAs is a rare event and always ignored due to unwillingness for further performance. Besides, the EIAs are not enough to address the extended impact of dams more than streamflow and sediment discharge modifications, but also their impact on biodiversity, species loss, and sociocultural profiles of the surroundings (Bhatt et al. 2012; Grumbien and Pandit 2013). Eventually, more demonstrations and protests against dam constructions are widespread in the Eastern Himalayas by the Indian civil society (Dupka et al. 2018). In this scenario, the EIA Law's approval to nearly 300 new dam proposals is clearly beyond the carrying capacity of the existing ecosystem that makes their function and reliability very doubtful and their capacity and intentions questionable (Grumbine and Pandit 2013). Surprisingly, the non-governmental organizations (NGOs) are more trustworthy and reliable nowadays rather than the regional governments (ICIMOD 2009, 2010) that miserably failed to manage and rely more on the NGOs to wait for their contributions (O'Neill 2019).

Finally, following the suggestion of Bhatt et al. (2017), the river ecosystem continuum approach should be integrated for the discussion of the transboundary water-sharing treaty between two countries. In addition to that, the importance of RBWs across the basin needs to be recognized for future policies and integrated basin management program.

8.10 Conclusion

Global population growth and energy and food nexus have promoted a large number of dam constructions that served multiple services by storing the flowing water into the dam-impounded reservoirs. Gradually, these reservoirs developed a new ecosystem with ageing and act as a wetland to provide ES in the surroundings like any other natural wetlands. Thus, human-made wetlands were formed due to the construction of small- to large-scale dams across the globe, which have been studied here as RBWs. Along with the multifaceted benefits of RBWs, the construction of dams and reservoir operations altered the geomorphology, hydrology, and ecology of the river basin. These effects can be seen in the river basins of Indian Himalayas as this region of the earth has a high density of hydropower dams. Specifically, the Eastern Himalayas, the repository of biodiversity, is going to be affected by multiple proposed dams. Conversely, the existing reservoirs have many benefits that are apparent and established from several studies. The Tista basin demonstrated the benefits and adverse effects of RBWs. This basin is tagged with the highest density of dams across the Himalayas where many small-scale reservoirs of the RoR cascading dams are providing multiple ES, like economic, provisional, ecological, recreational, and sociocultural services. However, the construction of 14 dams including a barrage in the basin has destroyed the large forest cover, replaced the native species, adversely affected flora and fauna, reduced the streamflow, and modified the sediment into the downstream. On the other hand, water-level variations, anthropogenic activities, and the threat of the climate change in the basin are the most significant challenges for the sustainable ES provided by the RBWs. The drastic water-level variations affect the ecology and blur the delineation of RBW boundaries, which is the principal disadvantage of reservoirs as a wetland in comparison with the natural wetland. Again, the proposed quantitative sharing of transboundary Tista water may affect the water level of the existing reservoirs ignoring the extensive ES of the Gazaldoba wetland. Increasing water demand, rising tourist pressure, and land-use land-cover alteration due to the process of urbanization in the basin remain critical anthropogenic stress. Additionally, the eastern part of the Himalayas is most vulnerable due to variabilities in precipitation, warming climate, and glacial melting. Thus, to prevent the impact of climate change, more water needs to be conserved by checking the increasing demand and through conservation of surface water into reservoirs. It is essential to manage the RBWs to protect the present ecosystem from enriching the species' richness and biodiversity of the basin. On the contrary, no more dams should be constructed in the Himalayan region to avoid the destruction of the existing ecology and minimize the regulation to maintain the environmental flow in the downstream and upstream reservoirs. To ensure these conservation and management strategies for sustaining the benefits from the RBWs, the Government should take the EIAs and carrying capacity assessment reports seriously by encouraging participation of civil societies and NGOs. Finally, the river ecosystem continuum approach should be integrated considering the protection of the overall ecosystem of the RBWs for further discussion on transboundary water sharing.

Therefore, it can be concluded that the dual role of the human as creator and destroyer of RBWs can be converted to protector through the management of reservoirs and regulation of dam operations. Further, the ecosystem of the RBWs can be conserved by avoiding the construction of new dams in the Himalayas that essentially could transform the adverse effects into benefits.

References

Abedini, M., Md Said, M.A., and Ahmad, F. (2012). Effectiveness of check dam to control soil erosion in a tropical catchment (The Ulu Kinta Basin). *Catena* 97: 63–70. https://doi.org/10.1016/j.catena.2012.05.003.

Acharya, B.K. Sanders, N. J., Vijayan, L., and Chettri, B. (2011). Elevational gradients in bird diversity in the eastern himalaya: an evaluation of distribution patterns and their underlying mechanisms. *PLoS One* 6 (12) https://doi.org/10.1371/journal.pone.0029097.

Agostinho, A.A., Pelicice, F.M., and Gomes, L. (2008). Dams and the fish fauna of the Neotropical region : impacts and management related to diversity and fisheries. *Brazilian Journal of Biology* 68 (4, Suppl): 1119–1132.

Agostinho, A.A., Thomaz, S.M., and Gomes, L. (2004). Threats for biodiversity in the floodplain of the Upper Paraná River: effects of hydrological regulation by dams. *Ecohydrology and Hydrobiology* 4 (3): 267–280.

Ahlers, R., Budds, J., Joshi, D., Merme, V., and Zwarteveen, M. (2015). Framing hydropower as green energy: assessing drivers, risks and tensions in the Eastern Himalayas. *Earth System Dynamics* 6: 195–204. https://doi.org/10.5194/esd-6-195-2015.

Allen, D.J., Molur, S., and Daniel, B.A. (2010). *The Status and Distribution of Freshwater Biodiversity in the Eastern Himalaya*. Cambridge, UK and Gland, Switzerland: IUCN.

Barros, N., Cole, J. J., Tranvik, L. J. et al. (2011). Carbon emission from hydroelectric reservoirs linked to reservoir age and latitude. *Nature Geoscience* 4 (September): 593–597. https://doi.org/10.1038/ngeo1211.

Bhatt, J.P., Manish, K., and Pandit, M.K. (2012). Elevational gradients in fish diversity in the Himalaya: water discharge is the key driver of distribution patterns. *PLoS One* 7 (9) https://doi.org/10.1371/journal.pone.0046237.

Bhatt, J.P., Tiwari, S., and Pandit, M.K. (2017). Environmental impact assessment of river valley projects in upper Teesta basin of Eastern Himalaya with special reference to fish conservation: a review. *Impact Assessment and Project Appraisal* 5517: 1–11. https://doi.org/10.1080/14615517.2017.1354642.

Brandt, S.A. (2000). Classification of geomorphological effects downstream of dams. *Catena* 40: 375–401.

Gopal, B., Shilpakar, R.L., Sharma, E. (2010). *Climate Change Impact and Vulnerability Functions and Services of Wetlands in the Eastern Himalayas: Impacts of Climate Change'*. ICIMOD.

Bro, A.S., Moran, E., and Calvi, M.F. (2018). Market participation in the age of big dams: the Belo Monte hydroelectric dam and its impact on rural Agrarian households. *Sustainability (Switzerland)* 10 (5) https://doi.org/10.3390/su10051592.

Buechler, S., Sen, D., Khandekar, N., and Scott, C.A. (2016). Re-linking governance of energy with livelihoods and irrigation in Uttarakhand, India. *Water (Switzerland)* 8 (10): 1–22. https://doi.org/10.3390/w8100437.

Ceregato, S.A. and Petrere, M. Jr. (2003). Comparação financeira entre as pescarias artesanais no complexo de Urubupungá, no médio rio Paraná (Brasil). *Brazilian Journal of Biology* 63 (4): 673–682.

Cheng, C.T. and Chau, K.W. (2004). Flood control management system for reservoirs. *Environmental Modelling and Software* 19 (12): 1141–1150. https://doi.org/10.1016/j.envsoft.2003.12.004.

Datta, T. (2011). Human interference and avifaunal diversity of two wetlands of Jalpaiguri, West Bengal, India. *Journal of Threatened Taxa* 3 (12): 2253–2262.

Daugherty, D.J., Buckmeier, D.L., and Kokkanti, P.K. (2011). Sensitivity of recreational access to reservoir water level variation: an approach to identify future access needs. *North*

American Journal of Fisheries Management 31: 63–69. https://doi.org/10.1080/02755947.2011.559846.

De Lima, F.T., Reynalte-tataje, D.A., and Zaniboni-filho, E. (2017). Effects of reservoirs water level variations on fish recruitment. *Neotropical Ichthyology* 15 (3): 1–10. https://doi.org/10.1590/1982-0224-20160084.

Department of Tourism, Government of Sikkim, Gangtok. (2017). Tourism and Civil Aviation Department. http://www.sikkimtourism.gov.in/Webforms/General/DepartmentStakeholders/DeptAwards.aspx (accessed March 2019).

Dharmadhikary, S. (2008). Mountains of concrete: dam BuilDing in the himalayas. International Rivers, USA.

Downing, J. A., Cole, J. J., Duarte, C. M. et al. (2006). The global abundance and size distribution of lakes, ponds, and impoundments. *Limnology and Oceanography* 51 (5): 2388–2397.

Fernando, C.H. (1984). Reservoirs and lakes of southeast Asia (Oriental region). In: *Lakes and Reservoirs. Ecosystems of the World*, vol. 23 (ed. F.B. Taub), 411–446. Amsterdam: Elsevier Science Publishers.

Fu, B.J., Wu, B.F., Lü, Y.H. et al. (2010). Three Gorges project: efforts and challenges for the environment. *Progress in Physical Geography* 34: 741–754.

Garg, J.K., Singh, T.S., and Murthy, T.V.R. (1998). 'Wetlands of India', Project Report: RSAM/SAC/RESA/PR/01/98. Space Applications Centre (ISRO), Ahmedabad, pp. 239.

Gehrke, P.C., Gilligan, D.M., and Barwick, M. (2002). Changes in fish communities of the Shoalhaven River 20 Years after construction of Tallowa Dam, Australia. *River Research and Applications* 286: 265–286. https://doi.org/10.1002/rra.669.

Ghosh, K. and Singh, R. (2012). Reservoir based Wetland ecosystem services, Gazaldoba Reservoir, West Bengal, India. *Proceeding: International Wetland Symposium (IWS)*, Pokhara, Nepal, Conservation and Sustainable Use of Wetlands in Nepal, ISBN: 978-9937-2-5938-5, pp. 236:133-142.

Government of India (GOI) (2008). *Hydropower Policy*. New Delhi: Ministry of Power http://powermin.nic.in/whats_new/pdf/new_hydro_policy.pdf.

Grill, G., Lehner, B., Lumsdon, A. E. et al. (2015). An index-based framework for assessing patterns and trends in river fragmentation and flow regulation by global dams at multiple scales. *Environmental Research Letters* 10 https://doi.org/10.1088/1748-9326/10/1/015001.

Grumbine, R.E. and Pandit, M.K. (2013). Threats from India's Himalaya Dams. *Science* 339 (January 2013): 1–3. https://doi.org/10.1126/science.1227211.

Grumbine, R. and Xu, J. (2011). Mekong hydropower development. *Science* 332: 178–179.

Harrison, J. A., Maranger, R. J., Alexander, R. B. et al. (2009). The regional and global significance of nitrogen removal in lakes and reservoirs. *Biogeochemistry* 93: 143–157. https://doi.org/10.1007/s10533-008-9272-x.

Immerzeel, W.W., van Beek, L.P.H., and Bierkens, M.F.P. (2010). Climate change will affect the asian water towers. *Science* 328 (1382) https://doi.org/10.1126/science.1183188.

International Centre for Integrated Mountain Development (ICIMOD) (2009). *A Manual for an Inventory of Greater Himalayan Wetlands*. Kathmandu: Hill Side Press.

International Centre for Integrated Mountain Development (ICIMOD) (2010). *Functions and Services of Wetlands in the Eastern Himalayas: Impacts of Climate Change*. Kathmandu: Hill Side Press.

IPCC (2000). *Special Report on Emission Scenarios* (eds. Nakicenovic et al.). Cambridge: Cambridge University Press.

Kanade, R. and John, R. (2018). Topographical in fl uence on recent deforestation and degradation in the Sikkim Himalaya in India; Implications for conservation of East Himalayan broadleaf forest. *Applied Geography* 92: 85–93. https://doi.org/10.1016/j.apgeog.2018.02.004.

Wantzen, K.M., Rothhaupt, K.-O., Mortl, M. et al. (2008). Ecological effects of water-level fluctuations in lakes : an urgent issue. *Hydrobiologia* 613: 1–4. https://doi.org/10.1007/s10750-008-9466-1.

Khagram, S. (2004). *Dams and Development: Transnational Struggles for Water and Power.* New York: Cornell University Press.

Lauerwald, R., Regnier, P., Figueiredo, V. et al. (2019). Natural lakes are a minor global source of N 2 O to the atmosphere. *Global Biogeochemical Cycles* https://doi.org/10.1029/2019GB006261.

Lehner, B. and Döll, P. (2004). Development and validation of a global database of lakes, reservoirs and wetlands. *Journal of Hydrology* 296: 1–22.

Lehner, B., Liermann, C. R., Revenga, C. et al. (2011). High-resolution mapping of the world's reservoirs and dams for sustainable river-flow management. *Front Ecol Environ, The Ecological Society of America* 9 (9): 494–502. https://doi.org/10.1890/100125.

Leira, M. and Cantonati, M. (2008). Effects of water-level fluctuations on lakes: an annotated bibliography. *Hydrobiologia* 613: 171–184. https://doi.org/10.1007/s10750-008-9465-2.

López-Moreno, J.I., Vicente-Serrano, S.M., Begueria, S. et al. (2009). Dam effects on droughts magnitude and duration in a transboundary basin: the lower river tagus, pain and Portugal. *Water Resources Research* 45 (2): 1–13. https://doi.org/10.1029/2008WR007198.

Maavara, T., Dürr, H.H., and Van Cappellen, P. (2014). Worldwide retention of nutrient silicon by river damming: from sparse dataset to global estimate. *Global Biogeochemical Cycles*: 842–855. https://doi.org/10.1002/2014GB004875.

Maavara, T., Parsons, C. T., Ridenour, C. et al. (2015). Global phosphorus retention by river damming. *PNAS* 112 (51): 15603–15608. https://doi.org/10.1073/pnas.1511797112.

Maavara, T., Lauerwald, R., Regnier, P., and Cappellen, P. Van (2017). Global perturbation of organic carbon cycling by river damming. *Nature Communications* 8 (15347): 1–10. https://doi.org/10.1038/ncomms15347.

Macklin, P.A. et al. (2018). Carbon dioxide dynamics in a lake and a reservoir on a tropical island (Bali, Indonesia). *PLoS One* 13 (6): 1–20. https://doi.org/10.1371/journal.pone.0198678.

Mitsch, W.J. and Gosselink, J.G. (1986). *Wetlands*, 539. New York: Van Nostrand Reinhold.

Mitsch, W.J. and Gosselink, J.G. (2007). *Wetands*, 4e. New York: Wiley.

Mittermeier, R.A., Myers, N., Thomsen, J.B. et al. (1998). Biodiversity hotspots and major tropical wilderness areas: approaches to setting conservation priorities. *Conservation Biology* 12: 516–520.

Mittermeier, R.A., Roblies-Gil, P., Hoffmann, M. et al. (2004). *Hotspots Revisited – Earth's Biologically Richest and Most Endangered Terrestrial Ecoregions.* Mexico City: CEMEX.

MOEF (1990). *Wetlands of India – A Directory*. New Delhi: Ministry of Environment & Forests.

Mullick, R.A., Perret, S.R., and Babel, M.S. (2013). Benefit functions for instream water uses – a case of the Teesta River, Bangladesh. *Journal of Applied Water Engineering and Research* 1 (2): 118–128. https://doi.org/10.1080/23249676.2013.875490.

NACSE (Northwest Alliance for Computation Science & Engineering) (2011). *Transboundary Freshwater Dispute Database*. Corvallis: Oregon State University http://ocid.nacse.org/tfdd/index.php.

New, T. and Xie, Z.Q. (2008). Impacts of large dams on riparian vegetation: applying global experience to the case of China's Three Gorges Dam. *Biodiversity and Conservation* 17: 3149–3163.

Nilsson, C., Reidy, C. A., Dynesius, M., and Revenga, C. (2005). Fragmentation and flow regulation of the world's large river systems. *Science* 308 (405) https://doi.org/10.1126/science.1107887.

O'Neill, A.R. (2019). Evaluating high-altitude Ramsar wetlands in the Eastern Himalayas. *Global Ecology and Conservation* 20 https://doi.org/10.1016/j.gecco.2019.e00715.

Palmer, M. A., Liermann, C. A. R., Nilsson, C. et al. (2008). Climate change and the world's river basins : anticipating management options. *Frontiers in Ecology and the Environment* 6 (2): 81–89. https://doi.org/10.1890/060148.

Pandit, M.K. (2009). Other factors at work in the melting Himalaya: follow-up to Xu et al. *Conservation Biology* 23: 1346–1347.

Pandit, M.K. and Grumbine, R.E. (2012). Potential effects of ongoing and proposed hydropower development on terrestrial biological diversity in the Indian Himalaya. *Conservation Biology* 26 (6): 1061–1071. https://doi.org/10.1111/j.1523-1739.2012.01918.x.

Panigrahy, S., Murthy, T. V. R., Patel, J. G., and Singh, T. S. (2012). Wetlands of India : inventory and assessment at 1 : 50,000 scale using geospatial techniques. *Current Science* 102 (6): 852–856.

Patten, B.C., Jorgeusen, S.E., Dumont, H.J. et al. (1990). *Wetlands and Continental Shallow Water Bodies*, vol. I, 759. The Hague: SPB Academic Publishing.

Pizzuto, J. (2002). Effects of dam removal on river form and process. *BioScience* 52 (8): 683–691.

Ramsar Convention Secretariat. (2016). An introduction to the Ramsar convention on Wetlands. *Handbook 1 – International Cooperation on Wetlands*, 5th Edition, 108.

Rao, K.L. (1975). *India's Water Wealth*, 255. New Delhi: Orient Longman.

RCS (2006). *'The Ramsar Convention Manual: A Guide to the Convention on Wetlands', (Ramsar, Iran, 1971)*, 4e. Gland (Switzerland): Ramsar Convention Secretariat.

Richter, B., Matthews, R., Harrison, D., and Wigington, R. (2003). Ecologically sustainable water management: managing river flows for ecological integrity. *Ecological Applications* 13: 206–224.

Rudra, K. (2018). Rivers of the Ganga- Brahmaputra- Meghna Delta. *Springer International Publishing AG part of Springer Nature* https://doi.org/10.1007/978-3-319-76544-0.

Gopal, B. and Sah, M. (1995). Inventory and classification of wetlands in India. *Vegetatio* 118: 39–48.

Seth, P. (2016). "Gajaldoba for awild week". https://www.flickr.com/photos/23985194@N06/24707558260/in/photostream/

Sharma, A. and Goyal, M.K. (2020). Assessment of the changes in precipitation and temperature in Teesta River Basin in Indian Himalayan region under climate change. *Atmospheric Research* 231 (August 2019): 104670. https://doi.org/10.1016/j.atmosres.2019.104670.

Sharma, K.P. (1985). Impact of river basin development projects in India with reference to fisheries development. *Proceedings of the International Seminar on Environmental Impact Assessment of Water Resources Projects*, University of Roorkee, Roorkee, India, 1, pp. 25–32.

Singh, B.P. (2001). The Eastern Himalaya Wetlands, forests and beliefs. *Economic and Political Weekly* 36 (31): 2968–2975.

Smakhtin, V. and Anputhas, M. (2006). An assessment of environmental flow requirements of Indian river basins. International Water Management Institute Research Report, Colombo, Sri Lanka, 107.

Stanley, E.H. and Doyle, M.W. (2003). Trading off: the ecological effects of dam removal. *Frontiers in Ecology and the Environment* 1 (1): 15–22.

Stone, R. (2008). Three gorges dam: into the unknown. *Science* 321: 628–632.

Tambe, S., Ghose, D., and Arrawatia, M.L. (2008). Designing a participatory policy framework for the conservation of lakes in the Sikkim Himalaya. *Proceedings of Taal 2007: the 12th World Lake Conference* (Sengupta, M. and Dalwani, R., eds.), pp. 2056e2060. http://www.moef.nic.in/sites/default/files/nlcp/Indian%20Case%20Studies/Q-44.pdf.

The Asia Foundation. (2013). Political economy analysis of the Teesta River Basin. Report.

WCD (World Commission on Dams) (2000). *Dams and Development: A Framework for Decision Making*. London, UK: Earthscan.

Wiejaczka, Ł., Prokop, P., Kozłowski, R., and Sarkar, S. (2018). Reservoir's impact on the water chemistry of the Teesta River mountain course (Darjeeling Himalaya). *Ecological Chemistry and Engineering S* 25 (1): 73–88. https://doi.org/10.1515/eces-2018-0005.

Wijesundara, C. and Dayawansa, N. (2011). Construction of large dams and their impact on cultural landscape: a study in victoria reservoir and the surrounding area. *Tropical Agricultural Research* 22 (2): 211. https://doi.org/10.4038/tar.v22i2.2830.

Willison, J.H.M., Li, R., and Yuan, X. (2013). Conservation and ecofriendly utilization of wetlands associated with the Three Gorges Reservoir. *Environmental Science and Pollution Research* 20: 6907–6916. https://doi.org/10.1007/s11356-012-1438-3.

Wu, J., Huang, J., Han, X. et al. (2004). The Three Gorges Dam: an ecological perspective. *Frontiers in Ecology and the Environment* 2: 241–248.

WWF (World Wildlife Fund). (2007). Climate solutions. WWFs vision for 2050. WWF International, Gland, Switzerland.

Yuan., L. H. Jiang., W. G., Luo., Z. L., He, X. H., and Liu, Y. H. (2014). Analysis of wetland change in the Songhua River Basin from 1995 to 2008. *IOP Conference Series: Earth and Environmental Science* 17 (012125) https://doi.org/10.1088/1755-1315/17/1/012125.

Zarfl, C., Lumsdon, A.E., Berlekamp, J. et al. (2015). A global boom in hydropower dam construction. *Aquatic Sciences* 77 (1): 161–170. https://doi.org/10.1007/s00027-014-0377-0.

Zwieten, P.A.M., Béné, C., Kolding, J., Brummett, R., Valbo-Jørgensen, J. eds. (2011). Review of tropical reservoirs and their fisheries – the cases of Lake Nasser, Lake Volta and Indo-Gangetic Basin reservoirs. FAO Fisheries and Aquaculture Technical Paper, Rome, (557), pp. 148.

9

Spatiotemporal Evaluation of Causes and Consequences of Wetland Degradation

Vijay Kumar[1], *Deeksha Ranjan*[1], *and Khushaboo Verma*[2]

[1] *Department of Applied Sciences & Humanities, Faculty of Engineering & Technology, Rama University, Kanpur, Uttar Pradesh, India*
[2] *Faculty of Education, Banaras Hindu University, Varanasi, Uttar Pradesh, India*

9.1 Introduction

Wetlands are the most productive and complex ecosystem on the earth where water is the primary factor that controls the environment and associated plant and animal life. It is present in the region inundated with water temporarily or permanently. Here the water table is present at or near the surface.

Due to the water inundation, the wetlands are characterized by flooding or saturated soil as well as an oxygen-deficient environment. Such an environment develops special potential in several organisms to adopt the existing conditions and, hence, leads to greater biodiversity.

The term "wetland" was first used by the US Fish and Wildlife Service (USFWS) for the waterfowl habitat in the United States. Various ecologists and government officials have defined the term "wetland". There is no standard definition of wetlands in existence; however, Ramsar Convention, an intergovernmental treaty signed in Ramsar, Iran in 1971 has defined it which is the most widely accepted. According to Article 1.1 of the convention, wetlands are:

> areas of marsh, fen, peatland, or water, whether natural or artificial, permanent or temporary, with water that is static or flowing, fresh, brackish, or salt, including areas of marine water the depth of which at low tide does not exceed six meters.

9.2 Classification of Wetlands

There are different classification systems of wetlands. According to the Classification of Wetland and Deepwater Habitat of the United States, 1979 (USFWS), there are five major types of wetlands:

Wetlands Conservation: Current Challenges and Future Strategies, First Edition.
Edited by Sanjeev Sharma and Pardeep Singh.

Marine: It includes coastal wetlands such as coastal lagoons, rocky shores, seagrass beds, and coral reefs

Estuarine: It includes tidal marshes, mudflats, and mangrove swamps, etc.

Lacustrine: It includes wetlands associated with lakes

Riverine: It includes wetlands along rivers and streams

Palustrine: It is marshy in nature and contains marshes, swamp, and bogs

According to the Ramsar system of classification, there are 42 different types of wetlands which are grouped under three categories:

i) Marine/Coastal Wetlands
ii) Inland Wetlands
iii) Man-made wetlands

Marine/Coastal Wetlands

A) **Permanent shallow marine waters:** These are less than 6 m deep at low tide, such as sea bays and straits.
B) **Marine subtidal aquatic beds:** Such as kelp beds, sea-grass beds, and tropical marine meadows.
C) **Coral reefs.**
D) **Rocky marine shores:** Such as rocky offshore islands, sea cliffs, etc.
E) **Sand, shingle, or pebble shores:** Such as sand bars, spits, and sandy islets; includes dune systems and humid dune slacks.
F) **Estuarine waters:** These include permanent waters of estuaries and estuarine systems of deltas.
G) **Intertidal mud, sand, or salt flats.**
H) **Intertidal marshes:** Such as salt marshes, salt meadows, tidal brackish, and freshwater marshes.
I) **Intertidal forested wetlands:** Such as mangrove swamps and tidal freshwater swamp forests.
J) **Coastal brackish/saline lagoons:** These include brackish to saline lagoons with at least one relatively narrow connection to the sea.
K) **Coastal freshwater lagoons:** Such as freshwater delta lagoons.

Zk(a). Karst and other subterranean hydrological systems: It may be marine/coastal.

Inland Wetlands

L) **Permanent inland deltas.**
M) **Permanent rivers/streams/creeks:** Such as waterfalls.
N) **Seasonal/intermittent/irregular rivers/streams/creeks.**
O) **Permanent freshwater lakes (over 8 ha):** Such as large oxbow lakes.
P) **Seasonal/intermittent freshwater lakes (over 8 ha):** Such as floodplain lakes.
Q) **Permanent saline/brackish/alkaline lakes.**
R) **Seasonal/intermittent saline/brackish/alkaline lakes and flats.**
Sp. Permanent saline/brackish/alkaline marshes/pools
Ss. Seasonal/intermittent saline/brackish/alkaline marshes/pools

Tp. Permanent freshwater marshes/pools: It is characterized by ponds (below 8 ha), marshes, and swamps on inorganic soils; with emergent vegetation water-logged for at least most of the growing season.
Ts. Seasonal/intermittent freshwater marshes/pools on inorganic soils: Such as sloughs, potholes, seasonally flooded meadows, sedge marshes, etc.
U. Non-forested peatlands: Such as shrubs or open bogs, swamps, and fens.
Va. Alpine wetlands: Such as alpine meadows, and temporary waters from snowmelt.
Vt. Tundra wetlands: Such as tundra pools, and temporary waters from snowmelt.
W. Shrub-dominated wetlands: Such as shrub swamps, shrub-dominated freshwater marshes, etc.
Xf. Freshwater, tree-dominated wetlands: Such as freshwater swamp forests, seasonally flooded forests, wooded swamps, etc.
Xp. Forested peatlands: Such as peat swamp forests.
Y. Freshwater springs.
Zg. Geothermal wetlands.
Zk(b). Karst and other subterranean hydrological systems.

Human-made Wetlands

1) **Aquaculture:** Such as fish/shrimp ponds.
2) **Ponds:** Such as farm ponds, stock ponds, and small tanks (generally below 8 ha).
3) **Irrigated land:** Such as irrigation channels and rice fields.
4) **Seasonally flooded agricultural land.**
5) **Salt exploitation sites:** Such as salt pans, salines, etc.
6) **Water storage areas:** Such as reservoirs/barrages/dams/impoundments (generally over 8 ha).
7) **Excavations:** Such as gravel/brick/clay pits; borrow pits, mining pools.
8) **Wastewater treatment areas:** Such as sewage farms, settling ponds, oxidation basins, etc.
9) **Canals and drainage channels, ditches.**

Zk(c). Karst and other subterranean hydrological systems.

9.3 Causes of and Consequence of Wetland Degradation

There are two causes of wetland degradation:

1) Natural causes
2) Anthropogenic causes

9.3.1 Natural Causes

Natural forces such as storm surges, barrier islands, flooding, salinization, climate change etc., are the potential causes of the loss of wetlands. These forces have degraded the coastal wetlands of New Jersey, Louisiana, Barbataria Bay, etc. (Table 9.1).

Table 9.1 Major natural causes of wetland loss.

Natural cause(s)	Example	Source
Storms surge	Coastal wetlands in New Jersey	Narayan et al. (2017)
Disintegration of barrier islands	Wetland loss in Barataria Bay	Fitzgerald et al. (2007)
Flooding and salinization	Wetlands of coastal Louisiana	Gaugh and Grace (1998)
Herbivory	Loss of coastal Louisiana Marsh	Shaffer et al. (2015), Johnson and Foote (1997), Randall and Foote (2005)
Climate change	High-latitude wetland and Alpine zones	Government of Australia (2019)
Major shifts in a river's course	Mississippi River	Neill and Deegan (1986)

9.3.1.1 Storms Surge

The storms surge generated due to hurricane is one of the natural causes of wetland degradation. The degradation of wetlands occurs due to erosion, deposition and marsh salinization (Narayan et al. 2017). The storm surge causes the penetration of inland farther which leads to the displacement of animals and disruption in the food supply. Hauser et al. (2015) have investigated the effect of hurricane sandy storm surge on the coastal wetlands in New Jersey. They noticed that moderate flooding and marsh dieback were the most common type of damage. They also find out that saline marshes and herbaceous wetlands were the most degraded wetlands (Hauser et al. 2015).

9.3.1.2 Disintegration of Barrier Islands

The barrier islands act as a buffer that reduces the force of the ocean current and thus protects the estuaries and wetlands. The disintegration of these islands reduces the habitat provided by the wetlands because of exposure to wave action tidal surge, saltwater intrusion, etc. The cumulative effect of all such activities causes the disintegration of wetlands. The loss in wetlands also occurs due to an increase in sea level. The wetland loss in Barataria Bay was the outcome of an increase in the sea level in the Barataria barrier system of the Mississippi River delta plain (FitzGerald et al. 2007).

9.3.1.3 Flooding and Salinization

Gough and Grace (1998) have investigated the effect of flood, salinity, and herbivory on the coastal plant communities. They observed that the increased salinity reduced the species number and biomass. They found an interesting result where the plants from higher salinity sites were consumed selectively (Gaugh and Grace 1998).

9.3.1.4 Herbivory

Shaffer et al. (2015) showed that the loss of coastal Louisiana marsh was due to the nutria herbivory and waterfowl herbivory. They observed that aggressive nutria control resulted in the recovery of herbaceous vegetation (Shaffer et al. 2015). Johnson and Foote (1997) also investigated that the herbivory has a strong influence on vegetation present in the wetland (Johnson and Foote 1997). In a study, conducted by Randall and Foote (2005) managed an impoundment in rapidly eroding marsh of coastal Louisiana to enhance plant production. They did not observe any significant annual production of *Spartina* between the managed and unmanaged areas, whereas the annual production of *Schoenoplectus* was greater in the unmanaged marsh. They concluded that the management study was not effective in increasing plant production in rapidly eroding marshes of coastal Louisiana. They observed that the annual aboveground production of *Schoenoplectus* reduced under nutria herbivory, whereas it did not show any effect on the aboveground production of *Spartina*. It was found that the stem of the *Spartina* increased in the absence of herbivory. They concluded that the change in plant structure due to nutria herbivory may degrade the wetland (Randall and Foote 2005).

9.3.1.5 Climate Change

Wetlands are among the most vulnerable ecosystems to climate change. The global temperature is increasing because numerous anthropogenic activities have led to an increase in the greenhouse gases which are responsible for the absorption and emission of radiant energy. The global increase in temperature due to such activities has disturbed the natural ecosystem of wetland by affecting the flora and fauna. The wetlands such as coral reefs, mangroves, and swamps, and the wetland at high latitude and alpine zones are at risk. The inland freshwater wetlands are much prone to be affected by the global increase in temperature. According to the Australian government, the wetlands which are highly modified or degraded may be more sensitive and less resilient to climate change (Government of Australia 2019).

9.3.1.6 Major Shifts in a River's Course

The shift in the Mississippi River caused the erosion and degradation of the delta. Thus the delta running along the sea is eroded by the marine coastal processes leading to the formation of barrier islands. These barrier islands keep separating from the mainland coast by the bays and lagoons and thus affect the habitat of wetland by restricting the flow of water into it (Neill and Deegan 1986).

9.3.2 Anthropogenic Causes of Wetland Loss

According to the Millennium Ecosystem Assessment, there are six major anthropogenic causes of wetland loss (MEA 2005), which are as follows:

i) Infrastructure development
ii) Conversion of land
iii) Water withdrawal
iv) Eutrophication and pollution

v) Overharvesting and overexploitation
vi) Introduction of invasive species

9.3.2.1 Infrastructure Development

The wetlands are degraded at an alarming rate because of the development of several infrastructures around the wetland habitat. The infrastructure such as a levee is built for the purpose of impounding water for agriculture or aquaculture and restrict the movement of water from rivers and tides into the wetlands. It also alters the transportation of sediments and nutrients on the ocean coast and thus alters the balance of salt water and fresh water (Galatowitsch 2018). Earlier, Lake Chilika was a complex shallow marine, brackish, and freshwater wetlands which were a natural habitat for diverse fishery and endangered species such as *Irrawaddy* dolphin and wintering ground for several migratory birds. The degradation of these wetlands resulted from the development of infrastructure (Kumar et al. 2020).

The development of the floodplain by the filling of wetlands and removal of vegetation from wetlands are also responsible for the wetland loss. The development of floodplain seeks to install artificial stream-stabilizing devices such as rip-rap and bulk-heads which aims to check the natural meandering of the rivers. Such activities affect the wetlands by creating new wetlands and diminishing the existing ones (Table 9.2).

The development of the roads/railway tracks along the river valley led to the narrowing of the floodplain and destabilizing of the river. These structures also affect the drainage system from upland to floodplains where most of them are built on areas that once were wetlands.

9.3.2.2 Land Conversion

Various processes such as agricultural production, construction, or peat mining require the draining of water from the adjacent wetlands. The draining of water from the wetlands is achieved by cutting ditches into the ground which collect the water and transport it. Such activities lower the water table and affect the wetlands (Oslund et al. 2010).

Table 9.2 Major natural causes of wetland loss.

Anthropogenic causes	Example	Source
Infrastructure development	Levees for impounding water for agriculture or aquaculture	Galatowitsch (2018)
Land conversion	Prairie Pothole Region of Minnesota	Oslund et al. (2010)
Water withdrawal	Azraq Oasis Wetland of Jordan	Jawarneh and Biradar (2017)
Eutrophication and pollution	Fresh Water Marshes of South Florida	Liston et al. (2008)
Overharvesting and overexploitation	High-Latitude Wetland and Alpine Zones	
Introduction of invasive species	Crayfish in a small shallow lake, Chozas	Rodriguez et al. (2005)

9.3.2.3 Water Withdrawal

The activities like deliberate diversion of water from the stream, withdrawal of water, and ground-water pumping change the natural wetland regime. Such effects become severe in dryland agricultural regions; for example, the Azraq Oasis wetland present in Jordan is a unique ecosystem in the arid region. Here several freshwater lakes and marshes are present which receive a very minute quantity of spring discharge because of the aquifers which have been depleted. This wetland supports great biological, socioeconomic, and cultural values. The development of agriculture is highly dependent on the well-drilling and dam-building which has extracted the water in a huge amount. The huge extraction of water converted the freshwater Oasis wetland into highly saline water (Jawarneh and Biradar 2017).

9.3.2.4 Eutrophication and Pollution

Wetlands are the sink of the nutrients which absorb a great amount of nutrients especially in the form of fertilizers, sewage, animals waste, etc. But the increase in nutrient beyond the limit can be a threat to the wetlands. Deegan et al. (2012) have conducted a nine-year nutrient-enrichment study and investigated that eutrophication (nutrient enrichment) has become a global problem for coastal ecosystems. They concluded that eutrophication is a potential cause of loss of salt marsh (Deegan et al. 2012).

Liston et al. (2008) have conducted a study in freshwater marshes of South Florida to show the effect of increased phosphorus (eutrophication) on the macroinvertebrate and periphyton. After two years of phosphorus loading, they observed that the density of the microinvertebrate was 2–16 times higher in periphyton mat than the benthic flock. They found that the biomass of the periphyton decreased with the increase in phosphorus enrichment. As the phosphorus enrichment was increased from intermediate to high, the periphyton was completely absent. The growth of macroinvertebrate was associated with the growth of periphyton. They concluded that as the enrichment of phosphorus increased, the density of the macroinvertebrate also increased until the mats of periphyton were lost. Thereafter, the density of macroinvertebrate started to decrease because of loss of habitat (Liston et al. 2008).

9.3.2.5 Overharvesting and Overexploitation

The wetlands are the most productive ecosystem and, hence, a great source of crop production, fisheries and sand mining etc. The overexploitation and overconsumption of these resources from wetlands led to its degradation.

9.3.2.6 Introduction of Invasive Species

The introduction of invasive species such as water hyacinth (*Eichornia crassipes*) and Salvinia (*Salvinia molesta*) into the ecosystem of native species compete with the native species and check the waterways. Rodriguez et al. (2005) have shown that the introduction of crayfish in a small shallow lake, Chozas turned the clear water lake rich in various plants, invertebrates, amphibians, and birds, to turbid water lake deficient in the mentioned flora and fauna. Earlier to this, they have also performed an experiment by excluding the crayfish which confirmed that the crayfish acts as a herbivorous organism that consumes a macrophyte and establishes the trophic cascade effect on the wetland. They concluded that the crayfish had a main role in the destruction of submerged plants (Rodriguez et al. 2005). Burlakova et al. (2009) have confirmed the potential effect of *Pomacea insularum* (apple snail) on the aquatic ecosystem by investigating its feeding rate under laboratory condition. They concluded that the rate of feeding of *Pomacea insularum* on the native species was greater than the invasive plants (Burlakova et al. 2009).

9.3.2.7 Others

Grazing Overgrazing is one of the serious causes of wetland degradation which led to the compaction of the soil, removal of the vegetation and destabilization of the stream bank. Livestock usually spend more time in wetlands because of the great source of forage and water which become the cause of degradation of wetlands.

Agriculture Generally, the wetlands are flat areas that are enriched with organic soil. Such circumstances make it a highly productive land after draining. Therefore, numerous wetlands are drained and converted into agricultural lands, thus degrading the wetlands.

Mining

Hydrologic Alterations The hydrological conditions are the main characteristics of a wetland that remains saturated with water either permanently or temporarily. An alteration in hydrologic condition involves either the removal of water from the wetlands or raising the level of the wetlands. Once these changes occur in the characters, functions and values of the wetlands are lost. These alterations can take place in the following ways:

Dumping The present condition of the wetland has become threatened because of the dumping of the municipal solid wastes also which buries the hydric soil. Such practices effectively lower the water table which leads to the degradation of wetland because of the loss of hydrophytic plants.

Dredging and Draining The wetlands are the reservoirs of productive soil, sand, etc. The removal of these materials from the riverbed lowers the water table of the surrounding area and thus the wetlands adjacent to such an area are degraded. The draining is also one of the reasons for wetland degradation by the formation of ditches into the ground which collect and transport the water and dry up the nearby wetlands.

Diversion of Flow The diversion of the water flow around the wetlands causes the lowering of the water table and plays a potential role in wetland degradation.

Devegetation The vegetations of the wetlands play an important role in the wetland ecosystem by maintaining the water level and also by altering the soil erosion. They provide a habitat for the wildlife and involve in soil and water chemistry.

9.4 Consequences of Wetland Loss

Loss of wetland results in numerous problems such as loss of biodiversity, loss of soil nutrients, decrease in water level, loss of habitat, salinization, displacement of populations, water-borne diseases, and other adverse effects on the natural ecosystem (Kumsa 2015) as given in Table 9.3.

9.4.1 Loss of Biodiversity

Wetlands possess huge biodiversity but extensive anthropogenic activities such as land conversion and drainage have threatened the wetlands. These enhanced activities have put biodiversity in great danger. Verones et al. (2013) have studied the impact of water consumption on the biodiversity of wetlands of international importance. They assessed the impact on the richness of waterbirds, mammals, nonresidential birds, amphibians, and reptiles. They concluded that the effect of water consumption was the greatest on waterbirds (Verones et al. 2013). Findlay and Bourdages have studied that the construction of road may result in significant loss of biodiversity due to habitat fragmentation, species invasion, and increase in wildlife access by humans (Findlay and Bourdages 2000).

Table 9.3 Consequences of wetland loss.

Major consequences	Example	Source
Loss of biodiversity	Huge water consumption in Kenya and	Verones et al. (2013)
Decrease in water level	Lake Alemaya	Ghermandi et al. (2008)
Loss of habitat	Georgian Bay in Lake Huron	Fracz and Fraser (2013)
Climate change	Development of the aquaculture pond	Macleod et al. (2011)
Emission of greenhouse gases	The feed and the wastes added to aquaculture	Hue et al. (2012)
Erosion of river delta	Mississippi delta	Martinez et al. (2009)

9.4.2 Decrease in Water Level

Wetlands are the greatest water reservoirs that help in maintaining the water level by storing it. The loss of wetlands results in the diminishing of the water level. Ghermandi et al. (2008) have noticed that Lake Alemaya of South-Eastern completely dried up in 2004 because of the excessive abstraction of water.

9.4.3 Loss of Habitat

The wetlands provide a suitable habitat for a huge number of biological species including fishes, mammals, birds, amphibians, etc., because they have nutrient-rich soil, huge water resources, and suitable environmental conditions. The loss of wetlands due to numerous anthropogenic activities such as land conversion, overgrazing, the building of roads, and other infrastructures, etc., results in the loss of habitat. Fracz and Fraser (2013) have estimated the quantity of the fish habitat lost due to the declining water level. They noticed that the wetland habitat gets disconnected because of the declined water level in coastal wetlands of eastern Georgian Bay in Lake Huron (Fracz and Fraser 2013).

9.4.4 Climate Change

The wetlands serve as a great sink of carbon. The development of the aquaculture ponds by clearing the mangroves and subsequently excavating them may oxidize more than 1400-tonne carbon per hectare. According to an estimate, 70-tonne carbon per hectare per year would enter into the atmosphere if half of the carbon is oxidized in 10 years. Such a huge amount of carbon released into the atmosphere is about 50 times the carbon sequestration rate which led to an increase in the global atmospheric temperature. If such activity is continued for a few decades, it results in climate change which, in turn, increases the sea level due to the melting of polar ice (Macleod et al. 2011).

9.4.5 Emission of Greenhouse Gases

Aquaculture also acts as a source of greenhouse gases such as N_2O and CO_2. The feed and the wastes added to it are decomposed by microorganisms and this leads to the emission of N_2O and CO_2 which are greenhouse gases. Thus, the conversion of coastal wetlands into aquaculture becomes the source of greenhouse gases (Hue et al. 2012). Pendleton also supported this fact in his study and stated that the conversion of wetlands into non-wetland structures such as aquaculture. He estimated that these structures are responsible for the emission of 0.04–0.28 Pg carbon per year (Pendleton et al. 2012).

According to the NASA Earth Observatory 2013, the flood has caused great destruction in southeastern Asia such as Bangladesh, Thailand, the Philippines, and Indonesia. These regions suffer from the flood every year. The constant flood forced the people to migrate out of these regions due to the reduction in fishery production because of the reduced mangrove area (Li et al. 2018).

9.4.6 Erosion of River Delta

Loss of wetlands also results in the loss of river delta due to erosion. Martinez et al. (2009) have studied that the loss of wetlands has resulted in the loss of the Mississippi delta due to erosion (Martinez et al. 2009). Du et al. (2016) also studied that the Yangtze River delta has changed morphologically due to the combined action of reduced sediment supply, sand mining, and erosion (Du et al. 2016).

References

Burlakova, L.E., Karatayev, A.Y., Padilla, D.K. et al. (2009). Wetland restoration and invasive species: apple snail (Pomacea insularum) feeding on native and invasive aquatic plants. *Restoration Ecology 17* (3): 433–440.

Deegan, L.A., Johnson, D.S., Warren, R.S. et al. (2012). Coastal eutrophication as a driver of salt marsh loss. *Nature 490* (7420): 388–392.

Du, J.L., Yang, S.L., and Feng, H. (2016). Recent human impacts on the morphological evolution of the Yangtze River delta foreland: a review and new perspectives. *Estuarine, Coastal and Shelf Science 181*: 160–169.

Findlay, C.S. and Bourdages, J. (2000). Response time of wetland biodiversity to road construction on adjacent lands. *Conservation Biology 14* (1): 86–94.

Fishery Pendleton, L., Donato, D.C., Murray, B.C. et al. (2012). Estimating global "blue carbon" emissions from conversion and degradation of vegetated coastal ecosystems. *PloS one* 7 (9): e43542.

FitzGerald, D., Kulp, M., Hughes, Z. et al. (2007). Impacts of rising sea level to backbarrier wetlands, tidal inlets, and barrier islands: Barataria Coast, Louisiana. In: *Coastal Sediments' 07* (eds. N.C. Kraus and J.D. Rosati), 1179–1192. ASCE Library.

Fracz, A. and Chow-Fraser, P. (2013). Impacts of declining water levels on the quantity of fish habitat in coastal wetlands of eastern Georgian Bay, Lake Huron. *Hydrobiologia 702* (1): 151–169.

Galatowitsch, S.M. (2018). Natural and anthropogenic drivers of wetland change. In: *The Wetland Book II: Distribution, Description, and Conservation* (eds. C.M. Finlayson, G.R. Milton, R.C. Prentice and N.C. Davidson), 359–367. Springer Nature.

Ghermandi, A., van den Bergh, J.C., Brander, L.M. et al. (2008). *The Economic Value of Wetland Conservation and Creation: A Meta-Analysis*. SSRN.

Gough, L. and Grace, J.B. (1998). Effects of flooding, salinity and herbivory on coastal plant communities, Louisiana, United States. *Oecologia 117* (4): 527–535.

Hauser, S., Meixler, M.S., and Laba, M. (2015). Quantification of impacts and ecosystem services loss in New Jersey coastal wetlands due to Hurricane Sandy storm surge. *Wetlands 35* (6): 1137–1148.

Hu, Z., Lee, J.W., Chandran, K. et al. (2012). Nitrous oxide (N_2O) emission from aquaculture: a review. *Environmental science & technology 46* (12): 6470–6480.

Jawarneh, R.N. and Biradar, C.M. (2017). Decadal national land cover database for Jordan at 30 m resolution. *Arabian Journal of Geosciences 10* (22): 483.

Johnson, L.A. and Foote, A.L. (1997). Vertebrate herbivory in managed coastal wetlands: a manipulative experiment. *Aquatic Botany 59* (1-2): 17–32.

Kumar, R., Pattnaik, A.K., and Finlayson, C.M. (2020). Ecosystem services: implications for managing Chilika. In: *Ecology, Conservation, and Restoration of Chilika Lagoon, India* (eds. C.M. Finlayson, G. Rastogi, D. Mishra and A. Pattnaik), 63–94. Cham: Springer.

Kumsa, A. (2015). *GIS and Remote Sensing based analysis of population and environmental change: The case of Jarmet wetland and its surrounding environments in Western Ethiopia*. Doctoral dissertation, Thesis report, Addis Ababa University, Addis Ababa.

Li, X., Bellerby, R., Craft, C., and Widney, S.E. (2018). Coastal wetland loss, consequences, and challenges for restoration. *Anthropocene Coasts 1* (1): 1–15.

Liston, S.E., Newman, S., and Trexler, J.C. (2008). Macroinvertebrate community response to eutrophication in an oligotrophic wetland: an in situ mesocosm experiment. *Wetlands 28* (3): 686–694.

Martinez, L., O'Brien, S., Bethel, M. et al. (2009). Louisiana Barrier Island Comprehensive Monitoring Program (BICM) Volume 2: Shoreline Changes and Barrier Island Land Loss 1800's-2005.

Mcleod, E., Chmura, G.L., Bouillon, S. et al. (2011). A blueprint for blue carbon: toward an improved understanding of the role of vegetated coastal habitats in sequestering CO2. *Frontiers in Ecology and the Environment 9* (10): 552–560.

Narayan, S., Beck, M.W., Wilson, P. et al. (2017). The value of coastal wetlands for flood damage reduction in the northeastern USA. *Scientific reports 7* (1): 1–12.

Neill, C. and Deegan, L.A. (1986). The effect of Mississippi River delta lobe development on the habitat composition and diversity of Louisiana coastal wetlands. *American Midland Naturalist* 116: 296–303.

Oslund, F.T., Johnson, R.R., and Hertel, D.R. (2010). Assessing wetland changes in the Prairie Pothole Region of Minnesota from 1980 to 2007. *Journal of Fish and Wildlife Management 1* (2): 131–135.

Prasad, S.N., Ramachandra, T.V., Ahalya, N. et al. (2002). Conservation of wetlands of India-a review. *Tropical Ecology 43* (1): 173–186.

Randall, L.A.J. and Foote, A.L. (2005). Effects of managed impoundments and herbivory on wetland plant production and stand structure. *Wetlands 25* (1): 38–50.

Rodríguez, C.F., Bécares, E., Fernández-Aláez, M., and Fernández-Aláez, C. (2005). Loss of diversity and degradation of wetlands as a result of introducing exotic crayfish. *Biological invasions 7* (1): 75.

Shaffer, G.P., Day, J.W., Hunter, R.G. et al. (2015). System response, nutria herbivory, and vegetation recovery of a wetland receiving secondarily-treated effluent in coastal Louisiana. *Ecological engineering 79*: 120–131.

Verones, F., Saner, D., Pfister, S. et al. (2013). Effects of consumptive water use on biodiversity in wetlands of international importance. *Environmental science & technology 47* (21): 12248–12257.

MEA (2005). Ecosystems and Human Well-Being: wetlands and water synthesis.

10

The Status of Current Knowledge, Distribution, and Conservation Challenges of Wetland Ecosystems in Kashmir Himalaya, India

Shahid Ahmad Dar[1], Sami Ullah Bhat[1], and Irfan Rashid[2]

[1] *Department of Environmental Science, School of Earth and Environmental Sciences, University of Kashmir, Hazratbal Srinagar, Jammu and Kashmir, India*
[2] *Department of Geoinformatics, School of Earth and Environmental Sciences, University of Kashmir, Hazratbal Srinagar, Jammu and Kashmir, India*

10.1 Introduction

Wetlands are "areas of marsh, fen, peatland, or water, whether natural or artificial, permanent or temporary, with water that is static or flowing, fresh, brackish, or salt, including areas of marine water the depth of which at low tide does not exceed six meters" (Ramsar 1971). They are the most important, highly diversified, sensitive, adaptive, and productive systems of the Earth's surface (Turner et al. 2000; Ghermandi et al. 2008; Keddy et al. 2009), offering many vital services to the human population (Prasad et al. 2002) like fisheries, agricultural production, fodder, water supplies, habitat for biodiversity, and purification of water resources (Raich and Schlesinger 1992; Whitehouse et al. 2008). Besides, they provide an important function as storage basins in the hydrological cycle and act as huge reservoirs for offsetting the impacts of climate change by absorbing large amounts of atmospheric CO_2 (Turner et al. 2000; Junk et al. 2013). Apart from this, they support vast diversity according to their geographical setting, dominant species, water regime, soil chemistry, and sediment features (Bassi et al. 2014).

Despite the provision of ecological services and functions by wetlands, they have been degraded, got vanished and are modified strongly worldwide (Junk et al. 2013; Reis et al. 2017). As per estimates, there had been a decline of 64–71% in the area of wetlands worldwide during the twentieth century (Davidson 2014). Worldwide, the dependence of societies on wetland ecosystems for resources, agriculture, water, dam construction, and other uses has created tremendous pressures on these ecologically sensitive areas (Molur et al. 2011). As a consequence, many species dependent on wetlands including 37% mammal species, 21% bird species, and 20% fish species are either threatened or extinct (MEA 2005). This massive deterioration in the function and decline in the area of wetland ecosystems all over the world is extensively recognized and widely endorsed and has therefore led to the actions for their conservation and restoration (Gardner et al. 2015).

Wetlands Conservation: Current Challenges and Future Strategies, First Edition.
Edited by Sanjeev Sharma and Pardeep Singh.

The convention on wetlands of international importance, the Ramsar Convention (www.ramsar.org), is a treaty agreed in 1971 with the aim of identifying the wetlands having high importance and endorsing their protection. However, actions to guarantee the protection of wetlands has not always been effective. Despite the creation of a network of Ramsar sites, the degradation and loss of wetland ecosystems continue all over the world.

Kashmir Himalaya in the Indian Himalayan region has a rich natural heritage of wetland ecosystems that are famous as waterfowl habitats (Habib 2014). Besides having scientific and recreational importance, these freshwater ecosystems are harboring natural biodiversity and have been playing a great role in the economic and sociocultural activities of the valley since ancient times (Kaul and Pandit 1980; Pandit 1982). However, over the last few decades, the rapidly expanding human population, large-scale land use/land cover changes, burgeoning development projects and improper use of watersheds have all caused a substantial decline in wetland resources of the valley. Large-scale losses of wetland areas have resulted due to conversion for various industrial, agricultural, and urban developments. These have led to hydrological perturbations and pollution and their effects. Over the years, the reckless use of pesticides, fertilizers, and large-scale deforestation in the catchment areas has led to excessive nutrient, silt, and sediment loadings to the wetlands. This has resulted in the reduction of the water depth of the wetlands together with their colonization by emergent species of aquatic macrophytes.

It has been found that the management of wetlands has received inadequate attention in the national water sector agenda, the valley of Kashmir being no exception. However, there has been a little policy push up after formulation of Wetland Rules, 2017 regarding the protection of wetlands. This has lead to the formation of wetland authorities entrusted to look after the protection and management of wetlands. Further, the majority of research on wetland management in Kashmir Himalaya relates to the limnological aspects and ecological/environmental economics of wetland management. But, the physical (hydrological and land-use changes in the catchment) and socioeconomic processes leading to limnological changes have not been explored substantially. This chapter therefore provides special insights into the wetland wealth of Kashmir Himalaya in terms of their geographic distribution and extent, ecosystem services they provide, and various stresses they are exposed to. The chapter also discusses the efforts in the management of these fragile ecosystems, identifies the institutional vacuum and suggests priority areas where immediate attention is required in order to formulate better conservation strategies for these productive systems.

10.2 Wetlands Over North-Western Kashmir Himalaya

10.2.1 Current Status

Kashmir Himalaya is a large intermountain valley in the northwestern part of Indian Himalaya between Zanskar and Pir-Panjal range. Kashmir Himalaya by its unique geographical position, varied terrain, and temperate climate has, no wonder, a wealth of wetland ecosystems (Figure 10.1). Presently, there are 755 natural wetlands covering an area of 42 663 ha, holding approximately 2.67% of the total area of the Kashmir Valley (National Wetland Atlas 2010). Occurring from the valley floor/mid-latitudes to very high altitudes, they harbor huge diversity of aquatic flora and fauna. The Kashmir Himalayan wetlands are the major

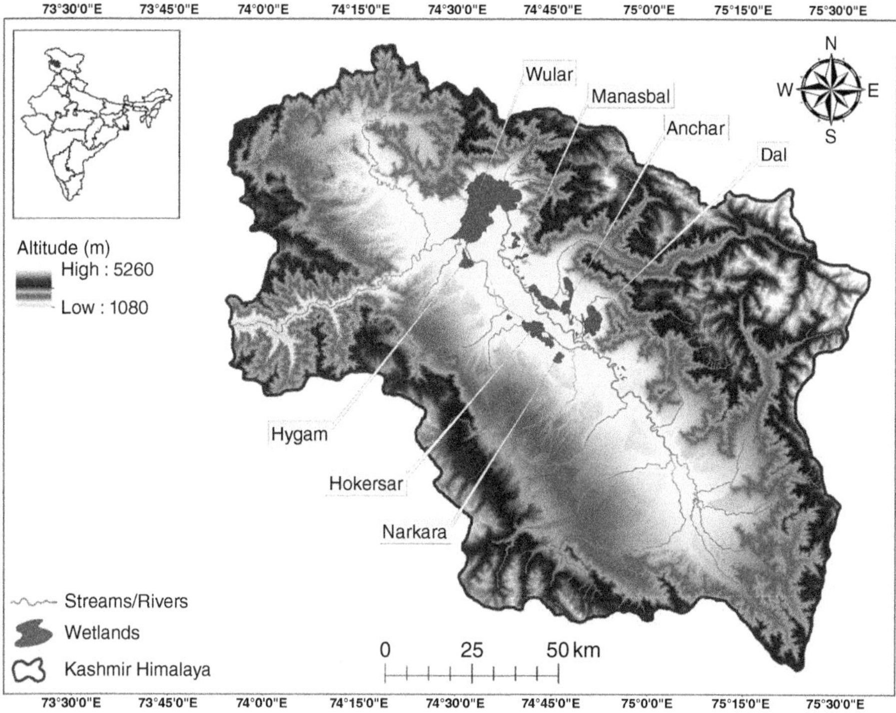

Figure 10.1 Major Wetland ecosystems of Kashmir Himalaya. Background data: Advanced Spaceborne Thermal Emission Reflection Radiometer Digital Elevation Model (ASTER DEM).

sources of livelihoods for a large portion of the population. They are of vital importance keeping in view the delivery of a variety of ecosystem services and economic goods (Dar et al. 2020a). The wetlands of this region provide a critical function of buffering floodwaters and moderating high pollution and nutrient loads from various river discharges and drainages entering into them. They support rich biological diversity and provide habitats to millions of migratory birds from Central Asian flyway (Pandit 1991; Mushtaq and Pandey 2014).

Wetlands in the Kashmir Valley show multiple signs of deterioration (Khan 2015; Dar et al. 2020b) due to water pollution (Najar and Khan 2012) and population growth (Kuchay and Bhat 2014). The wetlands are being treated as wastelands by both the public and policymakers as large quantities of domestic sewage, agricultural waste, plastics, and other municipal solid wastes are being dumped in these water bodies (Shah et al. 2015; Showqi et al. 2018). Large-scale land transformations, urbanization, rapidly increasing human population and deforestation in the catchment areas have caused a substantial decline in the areal extent of wetlands over this region (Rashid and Naseem 2008; Rashid and Aneaus 2019). Approximately, 3510.07 ha of wetland area have been lost from 1992 to 2015 due to encroachments, siltation, pollution, and conversion of wetland areas into non-wetland areas (Alam et al. 2019). These changes have further led to the decline of hydrological flows to wetlands, thereby seriously impacting their health. Large-scale deforestation, soil erosion, and reduced river flows have resulted in the degradation of wetland ecosystems over this region (Romshoo and Rashid 2014).

Various international wetland organizations such as Ramsar Convention have been developed for protection and management of wetlands but these provide protection to few wetlands because only a few numbers of selected wetlands as "Ramsar sites" have received considerable attention by means of technical and financial support while the remaining unattended ones are in an abandoned state. Though the Government of Jammu and Kashmir (J&K) has established the Department of Wildlife Protection J&K, various bodies like Lakes and Waterways Development Authority (LAWDA), various Developmental Authorities like Manasbal Development Authority (MDA), Wular Development Authority (WDA), Wular Conservation and Management Authority (WUCMA), etc., for management of lakes and wetlands, they have been unable to arrest the degradation and encroachment of wetland areas (Bhan and Trisal 2016). From the last 3–4 decades, approximately ` 10 billion have been pumped and invested in the restoration and management of famous wetlands of the valley (Dal, Manasbal, and Wular). Regrettably, the success of these investments has been ambiguous and often challenged as these sensitive ecosystems are not monitored systematically. The problems of wetlands in the Kashmir Valley are not solved fully as the planners and managers use certain inefficient restoration practices, keeping the root problems alive. Having said this, it is argued that the management and protection of wetland ecosystems in Kashmir Himalaya is crucial for their wise use and sustainable development. It is pertinent to mention here that the wetland management has received both insufficient attention and implementation in the national water management program (Bassi et al. 2014), with the valley of Kashmir being no exception. However, we expect a change in the management priority of wetlands after the constitution of Wetland authority in J&K.

Wetlands in Kashmir Himalaya depict a range of trophic status, like oligotrophic, eutrophic, and some mostly concentrated in the Srinagar city are in the hypertrophic state (Dar et al. 2020b), while others are in the progression toward eutrophication (Khan 2008). Though the process of eutrophication is driven naturally corresponding to the ageing of the lakes, the trophic state of wetlands in Kashmir Himalaya is largely driven by cultural eutrophication (Tahseen and Balkhi 2018; Shah et al. 2019a). Cultural eutrophication of water bodies over this region is a recent phenomenon of the past 2–3 decades concurring with visible developmental progress in the wetland catchment areas (Yaqoob et al. 2008; Khan et al. 2014; Ganaie et al. 2015). Although there has not been ample advancement in industrial development in Kashmir Himalaya, the key factors contributing to cultural eutrophication of lakes and wetlands are unplanned urban development, rapid increase in human population, land system changes, encroachments, fertilizer runoff, and soil erosion from the catchment areas (Shah et al. 2019a).

10.2.2 Wetland Classification

In Kashmir Himalaya, wetlands include streams, lakes, marshes, rivers, and swampy areas. The Kashmir Himalayan wetlands are broadly classified into two categories: high-altitude wetlands and mid-altitude wetlands (Zutshi 1989). There are some clear and crystalline high-altitude wetlands (HAWs) above 3000 m in height and some highly polluted wetlands in mid-altitudes coexisting in the Kashmir Valley. The brief characteristics of some noted wetlands of Kashmir Himalaya are presented in Table 10.1. On the basis of altitude, the Kashmir Himalayan wetlands can be classified into two categories:

Table 10.1 Characteristics of some noted wetlands of Kashmir Himalaya.

S. No.	Name	Coordinates	Altitude (m-asl)	Area (ha)	River basin	District	Protection status
1.	Ahansar	34°13′43.81″N 74°39′36.96″E	1585	20.2	Jhelum	Ganderbal	Unprotected
2.	Anchar	34°08′33.47″N 74°47′01.02″E	1582	690.5	Sindh, Jhelum	Srinagar	Unprotected
3.	Bod Sar/ Chatlam	34°00′47.43″N 74°56′14.21″E	1587	852	Jhelum	Pulwama	Conservation reserve (CR)
4.	Bod Sar	33°50′27.92″N 74°25′40.29″E	3958	46.67	Jhelum	Budgam	Unprotected
5.	Brari Nambal	34°05′12.78″N 74°48′53.60″E	1585	43.7	Jhelum	Srinagar	LAWDA
6.	Dal	34°07′16.50″N 74°51′37.26″E	1581	2400	Jhelum	Srinagar	LAWDA
7.	Freshkoori/ Phashkuri	33°59′58.80″N 74°55′42.31″E	1591	341	Jhelum	Pulwama	CR
8.	Gadsar	34°25′18.40″N 75°03′26.76″E	3624	39.84	Neelum (Pakistan)	Ganderbal	Unprotected
9.	Gangabal	34°25′54.97″N 74°55′22.64″E	3612	162.4	Sindh	Ganderbal	Unprotected
10.	Hokersar	34°06′10.52″N 74°42′42.46″E	1588	1375	Jhelum	Budgam/ Srinagar	Ramsar site/ CR
11.	Hygam	34°14′22.17″N 74°31′25.77″E	1576	700	Jhelum	Baramulla	CR
12.	Khanpur Sar	34°12′21.51″N 74°40′31.90″E	1583	36.3	Jhelum	Ganderbal	Unprotected

(*Continued*)

Table 10.1 (Continued)

S. No.	Name	Coordinates	Altitude (m-asl)	Area (ha)	River basin	District	Protection status
13.	Khushalsar	34°06′51.00″N 74°47′ 58.32″E	1583	109.6	Jhelum	Srinagar	Unprotected
14.	Kishansar	34°23′47.37″N 75°06′10.31″E	3819	31.94	Neelum (Pakistan)	Ganderbal	Unprotected
15.	Konsar nag	33°30′34.32″N 74°46′11.56″E	3819	152.6	—	Pulwama	Unprotected
16.	Krentchoo-Chandhara	33°59′44.26″N 74°56′10.84″E	1586	128	Jhelum	Pulwama	CR
17.	Malgam	34°16′31.88″N 74°39′06.07″E	1581	450	Jhelum	Bandipora	CR
18.	Manibugh	34°00′48.08″N 74°55′21.25″E	1588	106	Jhelum	Pulwama	CR
19.	Marsar	34°08′36.61″N 75°06′36.44″E	3849	43.73	Dagwan, Jhelum	Anantnag	Unprotected
20.	Manasbal	34°14′54.30″N 74°40′16.57″E	1580	334.6	Jhelum	Ganderbal	MDA
21.	Mirgund	34°07′19.69″N 74°38′11.87″E	1582	400	Jhelum	Baramulla	CR
22.	Nambli Narkara	34°02′19.04″N 74°46′05.76″E	1585	446	Doodhganga	Budgam	Unprotected
23.	Naranbagh	34°11′23.44″N 74°40′47.72″E	1582	9.7	Jhelum	Ganderbal	Unprotected
24.	Nilnag	33°51′22″N 74°41′36″E	2111	7.8	Endorheic	Budgam	Unprotected
25.	Nundkol	34°25′06.74″N 74°56′06.57″E	3505	37.4	Sindh	Ganderbal	Unprotected

26.	Shallabug	34°09′08.36″N 74°44′24.21″E	1580	1600	Sindh, Jhelum	Ganderbal	CR
27.	Sheikh Sar	34°13′46.85″N 74°38′22.09″E	1586	31.38	Jhelum	Bandipora	Unprotected
28.	Sheshnag	34°05′38.14″N 75°29′49.55″E	3584	53.45	Lidder	Anantnag	Unprotected
29.	Sonasar	34°04′05.76″N 75°28′29.14″E	3736	17.26	Lidder	Anantnag	Unprotected
30.	Tarsar	34°08′23.83″N 75°08′49.02″E	3490	84.84	Jhelum	Anantnag	Unprotected
31.	Vishansar	34°23′16.67″N 75°07′06.15″E	3949	47.25	Neelum	Ganderbal	Unprotected
32.	Waskura	34°13′06.82″N 74°39′39.54″E	1582	21	Jhelum	Ganderbal	Unprotected
33.	Wular	34°21′20.80″N 74°32′38.42″E	1576	8671	Jhelum	Bandipora	Ramsar site

Source: Data generated from Environmental Systems Research Institute Basemap, ancillary data from Department of Wildlife Protection J&K Govt. © John Wiley & Sons.

10.2.2.1 High Altitude Wetlands (HAWs)

HAWs are those waterbodies that lie above 3000 m altitude (Chatterjee et al. 2010). The HAWs cover an area of approximately 2020 ha (Panigrahy et al. 2010). Their position in the upper reaches of major tributaries makes these wetland ecosystems important globally. The HAWs are created and sustained by glacier and snow melting. These include meadows, mountain streams, marshes, and glacial lakes. The HAWs are clear, deep, mostly oligotrophic, and devoid of macrophytic vegetation with little or no anthropogenic pressures. Most of the HAWs in Kashmir Himalaya lie in Hindu Kush Himalayan and Pir Panjal range and act as huge water reservoirs for various river basins such as Jhelum and Sindh. Some notable HAWs are Gangabal, Gadsar, Marsar, Nundkol, Sheshnag, and Tarsar, etc. There are many other HAWs that have high ecological, economic, and cultural importance, but most of these are yet to be studied and, therefore, little information exists.

10.2.2.2 Mid-Altitude Wetlands (MAWs)

MAWs are those waterbodies that lie in the Kashmir Valley between 1500 and 3000 m altitude. These are mostly shallow having high nutrient and pollution load and mostly in eutrophic to hypereutrophic state. This category includes lakes, marshes, rivers, and lakeshores covering an area of 40643 ha. The MAWs have a characteristic zonation of various plant and biotic communities. The MAWs are characterized by the shallow water region along the peripheral margins where the emergents (*Typha sagitarria*, *Phragmites australis*, and *Phragmites communis*) stand in water with their greater portions above the surface of the water. Beyond this is the zone where there are free-floating plants and submerged aquatic macrophytes. A typical characteristic of most of the MAWs is the occurrence of floating gardens (*raads*) which are wide and long tracts of land created artificially using filling materials for growing vegetables and other crops. The MAWs play a key role in the hydrography of the Jhelum river basin. Some of the noted wetlands in the mid-altitudes are Anchar, Dal, Hokersar, Manasbal, and Wular.

10.2.3 Wetland Distribution and Extent in Kashmir Himalaya

In Kashmir Himalaya, the natural wetlands cover a total area of 42 663 ha, comprising approximately 2.67% of the total area of the Kashmir Valley (National Wetland Atlas 2010) (Table 10.2). These wetlands lie along the flood plains of River Jhelum which drains the mountain slopes forming terraces through cut and fill processes, traverses sluggishly through a large part of the valley from South to North and leaves it through a gorge in the north-west to enter Pakistan where it joins river Indus at Muzaffargarh (Sharma et al. 2012). The present extent and distribution of wetlands in the Kashmir Valley no longer concur with that which previously existed. The wetlands have attained major reductions in both the area and ecological functions. Wetland destruction is going on unabated in most parts of the valley and large areas of wetlands have been destroyed at many places at a tremendous rate. In the Kashmir Valley, there are various assessments on the distribution and area coverage of wetlands owing to different methods adopted for the assessment of wetlands. The latest inventory on wetlands of Kashmir Himalaya is the National Wetland Atlas 2010, prepared by Space Application Centre (SAC) (accessible at http://www.moef.nic.in/downloads/public-information/NWIA_Jammu_and_Kashmir_Atlas.pdf), wetland

Table 10.2 District-wise area statistics of wetlands in Kashmir Himalaya.

District profile		Wetland category					
		Lakes/ ponds	High-altitude wetlands	Riverine wetlands	River/ stream	Wetlands (< 2.25 ha)	Total
Anantnag and Kulgam	Number	—	69	15	11	23	118
	Area (ha)	—	1026	273	5553	23	6875
Baramulla and Bandipora	Number	2	38	29	13	15	97
	Area (ha)	11273	448	1478	3146	15	16360
Budgam	Number	—	11	9	12	48	80
	Area (ha)	—	150	1932	1272	48	3402
Kupwara	Number	18	—	2	5	70	95
	Area (ha)	96	—	6	2212	70	2384
Pulwama and Shopian	Number	—	2	7	5	251	266
	Area (ha)	—	4	347	2956	251	3561
Srinagar and Ganderbal	Number	14	29	25	7	23	99
	Area (ha)	2194	392	5457	2012	23	10081

Source: Data compiled from National Wetland Atlas: Jammu and Kashmir (2010), Ministry of Environment and Forests, Govt. of India. © John Wiley & Sons.

assessment was carried for the entire Kashmir Himalaya at a scale of 1 : 5000 and a total of 755 natural wetlands were recognized and mapped.

Kashmir Himalaya was once having a huge sea of wetlands but during the last five decades, the landscape has experienced frequent variations in the extent and functioning of wetlands in response to several economic and infrastructural developmental activities. In the mid-altitudes of Kashmir Himalaya, a rapidly growing human population has resulted in the transformation of wetland areas into non-wetland uses (Nengroo et al. 2017; Ganaie et al. 2020). Estimates suggest that 3510.07 ha of the area of wetlands have been lost to other land uses between 1992 and 2015 (Alam et al. 2019). This is related to high population growth which increased from 1 712 964 persons in 1951 to 6 888 475 persons in 2011 (Census 2011). The prevailing loss of wetland ecosystems in Kashmir during the last three to five decades is regarded as the main reason for intensifying the flood deluge in the Kashmir Valley during September 2014 (Romshoo et al. 2018). Srinagar, the capital city lost saviors (Wetlands), the entire Srinagar city was flooded for more than three weeks with water levels going up to 30 feet after J&K received more than 21-inch rainfall during September 2014 (Romshoo et al. 2018). The damage caused by the 2014 floods would have been comparatively less provided its wetlands had been in an interconnected fashion and ecologically healthy state. With the continuous expansion and spreading of the Srinagar city, most of the wetlands that used to act as sponges during floods have been taken over by settlements and residential colonies (Rashid and Naseem 2008; Kuchay and Bhat 2014). The wetlands like *Arat*, *Gandakshah*, *KhanKhan*, and *Batamalum Nambal* were encroached and earth-filled and were used for the construction of built-up and settlements. A water channel *Nallah Mar* taking excess waters from *Brari Nambal* to *Khushalsar* wetland was earth-filled and converted into a motorable road (Dar et al. 2021). It will be pertinent to mention here that the city had lost 50% of the area of wetlands between 1911 and 2004 due to urbanization (Rashid and Naseem 2008). Besides urbanization and encroachments, lack of wetland policy implementation, environmental literacy, and passive citizen science aspect specifically at the regional level is also a major cause for the degradation of wetlands over this region (Dar et al. 2020a). Owing to projected climate change and anthropogenic pressures, the wetlands become the most stressed resources in Kashmir Himalaya. The wetlands reveal significant changes in morphology and loss of area owing to anthropogenic pressures, encroachments, growing demands for agricultural land, and changes in water regimes and depleting streamflows (Romshoo and Rashid 2014).

10.3 Wetland Functions and Values

Wetlands are amongst the most productive ecosystems of the earth's surface (Ghermandi et al. 2008; Keddy et al. 2009). Because of the rich biodiversity, broad food webs, the support and the functions they perform, they are distinctively designated as "biological supermarkets" and "kidneys of the landscape" (Barber et al. 2001; Guo et al. 2017). In addition, wetlands supply a broad assortment of services and goods and acquire a range of characteristics of values and goods for societies (Raich and Schlesinger 1992; Whitehouse et al. 2008). They provide important provisioning, regulating, supporting, and cultural services that make great fiscal value from their potential, direct, or indirect use (MEA 2005). Some of the functions and services provided by wetlands are described below.

10.3.1 Regulatory functions

10.3.1.1 Regulation of Global Climate

The regulation of global climate is the most important regulatory function of wetlands. Wetland soils and plants act as sinks for large fractions of atmospheric carbon dioxide, therefore helping in combating climate change (Kraiem 2002; MEA 2005). Among all terrestrial ecosystems, wetlands have the highest carbon content supported by their resistance to droughts and high availability of nutrients (Pant et al. 2003). The physical and biogeochemical cycles of Kashmir Himalayan wetlands help stabilize the climatic conditions of the valley (Dar et al. 2020a; Shah et al. 2020). The benefits provided by the regulatory function help in the maintenance of a favorable temperate climate, which is essential for crop productivity, human well-being, and cultural and recreational activities.

10.3.1.2 Groundwater Recharge and Discharge

Wetland ecosystems in the Kashmir Himalayan region bring a range of hydrological services that help in the recharge and discharge of groundwater. The recharge-discharge function is an important but complicated part of wetland hydrology. The groundwater recharge function of wetlands is recognized as a valuable aspect of their overall role in the landscape. Groundwater discharge maintains a high water table in wetlands, whereas recharge to the underlying aquifers replenishes groundwater supplies.

10.3.1.3 Water Purification

Wetland ecosystems perform an important role in the purification of water resources by the reduction of high nutrient levels and sediment loads (Kumar et al. 2017). Wetland ecosystems act as filters for certain kinds of wastes and soluble contaminants. The absorbent and storing ability depends on the vegetation cover and sediment characteristics of wetland ecosystems. The physical, chemical and biological processes in wetlands immobilize and transform a wide range of environmental contaminants and nutrients which in excess cause severe eutrophication and pollution levels (WISA 2007). Heavy metals, pesticides and industrial wastes are bound to soil, sediments and aquatic macrophytes and therefore rendered more or less inert.

10.3.1.4 Natural Hazard and Flood Control

Wetland ecosystems lessen the damaging nature of floods and stormwater events by acting as buffers, absorbing huge volume of excess waters, thereby averting potentially catastrophic effects of storms and flash floods (Kumar 2018). In Kashmir Himalaya, the wetlands are mostly concentrated in the stretch of river Jhelum from Pampore to Bandipora where the major tributaries of river Jhelum empty their waters into the main Jhelum river/channel. In the Kashmir Valley, the wetlands have a remarkable setup as they are dotted along the right and left bank floodplains of the main Jhelum river/channel. Their presence from south to north of the Kashmir Valley prevents floodwaters from entering uptown areas. Along the entire valley, they act as storage basins by absorbing excess waters during peak flows, thereby providing buffering services to the whole Kashmir Valley from destructive floods.

10.3.1.5 Sediment Retention

In Kashmir Himalaya, the large-scale deforestation in the catchment areas brings excessive silt and sediment loads to the wetlands. Aquatic plants and macrophytes growing along the lakes and wetlands contribute greatly to the reduction of soil erosion and enable sedimentation of soil particles (WISA 2014). This function of soil retention is largely due to the physical features of wetland plants. The leaves and stems of free-floating, submersed, and roots of emergent macrophytes hold the soil particles, thus preventing the pollution of freshwater streams and rivers. The binding effect of wetland vegetation helps in the stabilization of banks and shores and also in the accretion of sediments, thus counteracting the forces of erosion and subsidence.

10.3.2 Provisioning Functions

10.3.2.1 Food Resources

The wetland ecosystems are an important source of eatable animals and plants (Horwitz et al. 2012). People in Kashmir Himalaya use fertile wetland soils as agricultural grounds for growing crops and vegetables. The main crops grown are pumpkins, turnips, cucumbers, haak, tomatoes, beans, peas, sweet potatoes, and onions. The people living in the peripheries have been engaged in various occupations such as the harvesting of weeds (green fodder), harvesting of lotus rootstocks (*Nadru*), fishing, and nuts of water chestnut (*Singhara*) (Abubakr et al. 2011).

10.3.2.2 Raw Materials

Wetlands over the Kashmir Valley flourish in various natural goods and products. They provide useful materials such as reeds, grasses, and wood for making furniture and construction purposes (Bhat et al. 2017). The wood makes the framework of chairs and tables, furniture, and for making baskets while the vegetation, especially leaves and stems of *Typha* sp., and grasses provide materials for mats and thatching.

10.3.2.3 Medicinal Resources

Aquatic macrophytes such as *Typha angustata*, *Nelumbo nucifera*, *Phragmites communis*, *Trapa natans, and Nymphea alba* harvested from the valley wetlands are important sources of medicines used in traditional system (Abubakr et al. 2011; Fatima et al. 2018). The therapeutic uses of *Typha angustata* include treatment of nose bleeds, hematemesis, hematuria, uterine bleeding, dysmenorrheal, and abscesses. The rootstocks of the plant are used as astringent and the underground stem is used as tonic, febrifuge, and diuretic (Lavnert 1981). *Nelumbo nucifera* is used as astringent, cardiotonic, febrifuge, emollient, and diuretic. It was used in the treatment of diarrhea, tissue inflammation, and homeostasis. The rhizome extract was used as antidiabetic and anti-inflammatory properties due to the presence of asteroidal triterpenoid. Leaves were used as an effective drug for hematemesis, epistaxis, hemoptysis, hematuria, and metrorrhagia. Flowers were used to treat diarrhea, cholera, fever, and hyperdipsia (Paudel and Panth 2015). The rhizomes of *Phragmites communis* contain silicic acid and 5.6% sugar. Therapeutically, the plant parts are used as analgesic, antispasmodic, and hypotensive (Duke 1998). Fruits of *Trapa natans* are sweet, astringent, cooling, diuretic, and tonic. They are used in dyspepsia, haemorrhages, diarrhoea, and dysentery. The presence of saponins, glycerides, phenolic compounds, flavonoids, and phytosterols makes it an important medicinal ingredient in many ayurvedic preparations. Therapeutic uses also include making liniments for the cure of rheumatism. Therapeutic uses of *Nymphea alba* include its astringent and antiseptic properties besides treatment in bronchial congestion. The rhizome contains alkaloids, viz. nymphacine, nupharine, glycosides, and tannins. The general calming and sedative effects on the nervous system make the plant species useful in the treatment of insomnia, anxiety, and other disorders where nervous agitation is a factor.

10.3.3 Cultural Functions

10.3.3.1 Tourism, Aesthetics, and Recreation

The wetlands of the Kashmir Valley are famous worldwide as tourist hotspots for nature lovers. The wetlands like Dal, Manasbal, and Wular provide pleasant sights and invariably attract tourists for refreshment, relaxation, and recreation (Aslam et al. 2018). Most of the people enjoy the additional attractions associated with wetlands like boating, fishing, and watching wildlife and migratory birds.

10.3.3.2 Scientific and Educational Information

Wetlands provide limitless prospects for nature studies, education, and serve as natural laboratories for scientific investigations leading to the writing of several research articles each year (De Groot et al. 2002). Nowadays, wetland soils are used as key components for assessing the status and recovery after restoration projects are implemented. The macrophytes growing in wetlands are increasingly utilized for assessing the potential to remediate heavy metals. Wetland ecosystems also function as reference centers for monitoring changes in the environment like changes in global climate, floods, and glacier melting.

10.3.4 Supporting Functions

10.3.4.1 Biodiversity Habitats

Wetlands in Kashmir Himalaya provide overwintering resorts to millions of migratory birds migrating from the Palearctic region and breeding and nesting grounds to a host of other birds in summer (Pandit 1991). The aquatic habitats support 23 species of fish fauna and 117 species of macrophytes, belonging to 69 genera and 42 families. A multitudinal number of more than 300 algal species belonging to phytoplankton, periphyton, and phytobenthic communities have been registered from wetland biotopes. The rich biodiversity of wetlands is also indicated by 150 species of zooplankton, 79 species of meiobenthos, 17 species of macrobenthos, and 45 species of macrofauna (Pandit 2008).

10.3.4.2 Nutrient Cycling

Wetland soils serve as long-term sinks for nutrients and as integrators of environmental and ecological conditions of wetlands. Wetland ecosystems facilitate the functional characteristics of the cycling of nutrient materials at regional scales. The biota in the wetland soils bring decomposition of organic matter and help in liberating nutrients for the growth of plants and bring the gaseous exchange to the atmospheric compartment of the biosphere (Reddy et al. 2010).

10.3.5 Economic Values

Wetlands in Kashmir Himalaya form an important economic resource by regulating the production of various ecosystem goods. The natural products harvested or used by humans from wetlands such as fish, fruits, forage, nuts, birds, fodder, and timber have high economic values. Similarly, various food resources such as *Nadru* and vegetables produced in wetlands provide high economic returns. Alone in Dal Lake, returns of harvesting of lotus rootstocks (*Nadru*) were estimated at 57.17 US $ per kanal of land. Similarly, returns from vegetable production per kanal were estimated at 97.19 US $. Annual earnings of food resources are 4000 US $ from Dal Lake (Wani et al. 2013) and 3944165 US $ from Wular Lake (Kaul et al. 2016). The annual returns from hotels and houseboats are 84.39 US $ alone from the Dal ecosystem. The estimated economic values of wood resources (willows) from Wular Lake are 2.4 million US $.

10.4 Drivers of Wetland Degradation

The wetlands of Kashmir Himalaya face a multiple facet of problems which require adequate attention. Some of these threats are general to every wetland while the others are peculiar to the particular wetland. Besides, the threats are so interrelated to each other that a solution to one problem is not possible in isolation. Coupled with this comes the

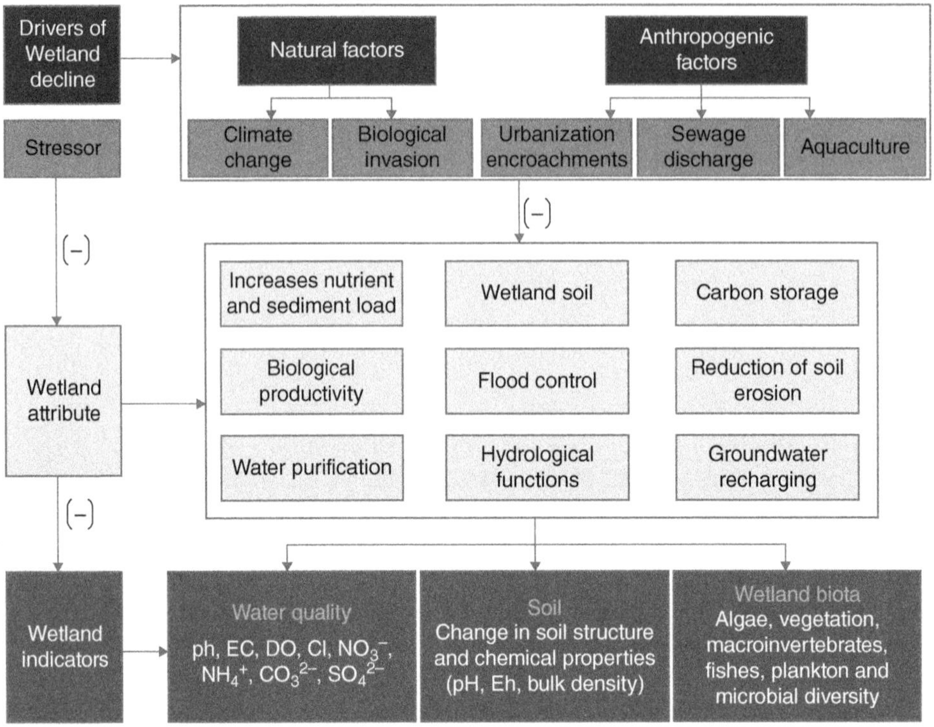

Figure 10.2 Schematic showing the effects of natural and anthropogenic drivers on wetland structure and functions.

socioeconomic influence since the livelihood of the large population is reliant on various resources from the wetlands for its survival. The main threats which result in the degradation of wetland resources are human activities. These include filling of wetland areas, encroachments, land system changes, discharge of sewage, and other wastes. All these stressors adversely affect the physical, chemical, and biological properties of wetlands and processes become stressed (Figure 10.2).

Globally, the drivers that degrade wetland ecosystems are grouped into six classes which include land use land cover changes, urbanization, alterations in hydrological regimes, degradation of water quality, overexploitation of resources and introduction of exotic species (Galatowitsch 2018). The lives of people of the Kashmir Valley are intricately connected with water bodies as they have been living close to them from times immemorial. Over the years, the lakes and wetlands in Kashmir have been used as open access resources which any consumer could attempt to exploit to one's fullest. The impact of human activities on lakes and wetlands is both direct and indirect. Some of the threats faced by wetlands are described below.

10.4.1 Land System Changes

Land system changes are the major threats to wetlands (Phethi and Gumbo 2019). These stressors have altered the hydrological balances of wetlands resulting in drylands which over a period of time are used for the construction of residential colonies (Rashid et al. 2017;

Rashid and Aneaus 2019). In Kashmir Himalaya, the wetlands have been taken over by concrete surfaces. The tremendous anthropogenic pressures have translated into the loss of wetland areas since the neighboring settlements proliferate and expand toward the wetland areas (Bhat et al. 2019; Dar et al. 2020a) (Figure 10.3a).

10.4.2 Pollution

Most of the wetlands of Kashmir Himalaya are targets to inflows of municipal solid wastes, domestic sewage, and fecal matter. Besides, fertilizer runoff from agriculture fields aggravates a load of nutrients and pollutants to the wetlands (Showqi et al. 2018). This ultimately leads to the reduction of oxygen levels, eutrophication, and degradation of water quality. The discharge of untreated fecal matter from houseboats in Dal Lake and Nigeen Lake (Parvez and Bhat 2014; Dar et al. 2020b), domestic sewage, and the effluents over the whole region have led to pollution and high levels of nutrients in these water bodies, which has not only destroyed the aesthetic beauty of the wetlands but also the hydrobiological setup. The reckless and unplanned developments around the wetlands had not only caused deterioration of the chemical quality of water but also turned the wetland peripheries into easily amenable solid waste-dumping sites. The plastics and other solid wastes are directly thrown into the wetlands choking some of the important waterways (Figure 10.3b).

Figure 10.3 Threats perceived by wetlands in Kashmir Himalaya: (a) Built-up encroaching in the wetland area; (b) Wetland choked by plastics and other solid wastes; (c) Excessive weed infestation by *Azolla sp.*; (d) Illegal hunting of migratory waterfowl going unabated in wetlands of Kashmir Himalaya, (photograph taken be first author at Waskura Wetland, Ganderbal, J&K).

10.4.3 Floating Agriculture

One of the major direct human impacts on the wetland ecosystems of Kashmir Himalaya has been the result of the creation of floating gardens of different sizes, shapes, and thickness. The floating gardens during low water levels get fixed to the bottom and result in permanent marsh islands. The marsh islands are used by local dwellers for agricultural cultivation who also grow some willow plantations there. People use fertilizers in floating gardens that result in high nutrient loadings to wetlands; also the willow plantations result in altered wetland hydrology. Overall, the impact of floating gardens is grave and severe as within a couple of years, these result in permanent drylands which are used for construction of houses.

10.4.4 Siltation

Siltation is perhaps the most common problem faced by every wetland in Kashmir Himalaya. The inflows of water into the wetlands bring a huge amount of silt and sediments. Large amounts of silt have resulted in the reduction of the water depth of wetlands. The impact is further aggravated by the deforestation in the upper catchments of wetlands (Alam et al. 2011; Romshoo et al. 2018). During 1960s, in Hokersar wetland, *Myriophyllum spicatum*, *Potamogeton lucens*, and *Hydrilla verticellata* formed the dominant vegetation of the wetland. The excessive silt and sediments brought by running waters from higher reaches as a result of deforestation in the catchment areas have brought a shift in the macrophytic vegetation community of the wetland. Nowadays, the wetland is infested by *Nymphoides peltatum* and *Trapa natans* which are able to withstand low depth and turbid water conditions.

10.4.5 Roads and Railways

The construction of roads has fragmented the wetlands of Kashmir Himalaya. In Dal Lake, a road connecting Nishat to Saida Kadal was constructed, which has virtually fragmented the water body. Also, the foreshore road on the north-eastern side of Dal Lake was constructed by filling a large area of the water body. The Khushalsar wetland was bifurcated into two parts – northern Gilsar and southern Khushalsar by road construction. The Hokersar wetland is also fragmented by road construction. A large area of Narkara wetland was separated by the construction of a railway line. These roads have fragmented the water bodies, which has impacted hydrology and has helped in the rapid colonization of the wetland areas.

10.4.6 Plantations

The proliferation of the willow and poplar plantations within the wetlands of Kashmir Himalaya has hindered the hydrological flows, adversely affecting the wetland functioning. The dense willow plantations are documented as grave threats to wetlands (Final Draft Report Wular Lake 2014) as they cause a variety of harmful ecological and morphological changes to wetlands and other aquatic ecosystems (Poppe et al. 2006; Dar et al. 2020a).

10.4.7 Overexploitation

The wetland resources of Kashmir Himalaya are generally rich in fish, fodder, and birds. There is thus a temptation to exploit these resources (Khan and Ali 2013). Unfortunately, many times, this takes the form of indiscriminate fishing and not even sparing the young stock, resulting in the fall of fish landings. The major undesirable practices are the use of very small nets, whereby even the very young fish are caught.

10.4.8 Weed Infestation

Alien invasive species are widespread in almost all valley lakes and wetlands (Shah and Reshi 2014). The high nutrient inputs and climate change impacts have resulted in widespread changes in aquatic vegetation. Nowadays, excessive growth of *Azolla* sp. has affected almost all water bodies of Kashmir Himalaya and is creating numerous environmental problems (Figure 10.3c). In Dal Lake, it has reached nuisance levels replacing *Salvinia sp.* It has caused a decrease in light penetration and dissolved oxygen content of the Hokersar wetland besides competing with other macrophytic species within the wetland (Uheda et al. 1999).

10.4.9 Hunting and Poaching

The wetlands of Kashmir Himalaya remain swarmed with thousands of migratory birds every day. The birds go to the peculiar wetlands where they find the environment friendly and peaceful with abundant food supplies. During the last few decades, the hunting and poaching of migratory birds have gone unabated in the lakes and wetlands across Kashmir Himalaya. The hunters shoot hundreds of migratory birds every day from almost all the wetlands in the valley and fetch hefty amounts of money from their sale (Figure 10.3d). The changes occurring in these environments have resulted in the reduced arrival of migratory birds (Pandit 1991).

10.4.10 Land Reclamation

Reclamation of wetland areas is a common practice observed in the wetlands of Kashmir Himalaya. The reclamation of wetland areas often results due to high population growth (Kuchay and Bhat 2014). Large-scale reclamation of wetland areas for agricultural, road-building and housing needs has caused widespread shrinkage of wetland areas over this region (Pandit 1991; Rashid and Naseem 2008). In some cases, parts of wetland areas are used by locals for aquatic-agriculture produce. These areas are drained by putting different types of bunds and filled using different materials to increase the area for agricultural purposes.

10.5 Wetland Conservation in Kashmir Himalaya

The wetland ecosystems of Kashmir Himalaya are under remarkable stress due to pollution, encroachments, urbanization, hydrological alterations, exotic species, and climate change (Farooq and Muslim 2014; Alam et al. 2019; Dar et al. 2020a,b). The Department of

Table 10.3 Status of wetland conservation in Kashmir Himalaya.

S. No.	Particulars	Wetlands
1.	Total number of wetlands in Kashmir Himalaya	755
2.	Number of wetlands designated as Ramsar sites	2
3.	Wetlands designated as conservation reserves	9
4.	Wetlands protected by LAWDA and MDA	3
5.	Overall number of wetlands in conservation	13[a]
6.	Percent under conservation	1.72

[a] Hokersar both a Ramsar site and a Conservation reserve.
Source: Data compiled from National Wetland Atlas: Jammu and Kashmir (2010), Ministry of Environment and Forests, Govt. of India; Department of Wildlife Protection J&K. © John Wiley & Sons.

Wildlife Protection in the Union Territory (UT) of J&K is the head authority for managing the wetland ecosystems over Kashmir Himalaya. However, recently in 2020, Department of Forest, Ecology and Environment has been designated as Nodal Department for Wetland authority of J&K. Though a number of Himalayan wetlands are marked under the list of protected areas, further extension of the policies to include all the wetlands in Kashmir Himalaya is the need of the hour. In Kashmir Himalaya, two wetlands – Hokersar in Budgam and Srinagar – and Wular Lake in Bandipora and Baramulla districts are designated Ramsar sites (http://www.ramsar.org/wetland/india) and a number of other wetlands are designated conservation reserves (CR) (Table 10.1). However, to curb the encroachments around the wetlands designated as CR, the Wildlife Warden Wetlands Division (J&K), forest department and revenue department have taken up the step of demarcation of the current wetland extents as per revenue records. The overall status of wetland protection and conservation is very poor as only 1.7% of the wetlands are protected while other wetlands hardly figure in any conservation and management plans (Table 10.3).

10.5.1 Legal Framework

Though the Department of Wildlife Protection J&K is the nodal authority under whose jurisdiction come all the wetland resources of Kashmir Himalaya. However, they are obliquely managed by a number of other departments like Forest Department of J&K, various Developmental Authorities, like LAWDA, MDA, WDA, WUCMA, etc., and also come under the legislative framework like The Jammu and Kashmir Kahcharai Act, 1954; The Water Prevention and Control of Pollution Act, 1974; Jammu and Kashmir Wildlife Protection Act, 1978; Jammu and Kashmir State Forest Corporation Act, 1978; Forest Conservation Act, 1980; The Environment Protection Act, 1986; Biological Diversity Act, 2002; LAWDA under the State Development Act, 1970; and The Jammu and Kashmir Reorganization Act, 2019. Provisions underneath these regulatory measures vary from the protection of the quality of water and notification of environmentally sensitive zones for paying toward conservation, management, and enhancement of the biological diversity of the water bodies of J&K UT.

10.5.2 Conservation Challenges

Daunting challenges characterize the wetland ecosystems of Kashmir Himalaya as economic, environmental, and sociocultural changes are vigorously impacting the health of wetland resources over this region, thereby hampering the ecological services, livelihoods, and, ultimately, sustainability. New challenges for wetlands are the growing human population, unplanned urbanization, and climate change effects, which all impact the health of wetland ecosystems by way of increasing the pollution loads and declining streamflows (MEA 2005; Sharma et al. 2009). The necessity of a coordinated and integrated approach for the conservation of physical, chemical, and biological characteristics of Kashmir Himalayan wetlands has been felt by many researchers both in the past and at present (Pandit 1991; Shah et al. 2019b; Dar et al. 2020b). It has been largely recognised that the mandate of some departments is overlapping when the question of conservation of wetlands comes at the forefront like Department of Forest, Wildlife, and LAWDA etc. This ultimately leads to situation wherein at times it becomes increasingly difficult to pinpoint and identify the grey areas in the wetland conservation. Despite the creation of Ramsar sites and CRs, the Department of Wildlife Protection J&K has still to overcome shooting challenges to promote the conservation of physicochemical and biological characteristics and reducing the encroachments of Kashmir wetlands. The challenges should be included in policymaking and integrating these as the central sermon toward the conservation, management, and sustainable development of wetland ecosystems.

10.5.3 Conservation Strategies

Although the Government has established several conservation agencies like the Department of Wildlife Protection J&K, LAWDA, various developmental authorities, like MDA, WDA, and WUCMA, etc., for the management of lakes and wetlands over this region, they have failed in the protection of these important reserves. To sustainably manage the wetland ecosystems, a comprehensive monitoring program is the need of the hour (Table 10.4).

10.5.4 Knowledge Gaps

After an extensive survey of the literature, it is revealed that in Kashmir Himalaya, wetlands are still the most ignored ecosystems as far as conservation and research studies are concerned. The management approaches for the conservation of wetlands are not built on site specifications, but they are of perspective nature and do not address the actual degradation causes on the ground. There is a lack of implementation of restoration strategies for the protection of degraded wetlands. For addressing various drivers of wetland decline, management programs demand continuous research efforts. However, this has failed to happen for most of the wetlands. In Kashmir Himalaya, most of the research is available on some select wetlands (Anchar, Dal, Hokersar, Manasbal, and Wular) focused on limnological aspects with very limited knowledge on functional aspects like impacts of changing climate and land use the land cover on hydrological processes and relationship of hydrological

Table 10.4 Steps for the implementation of a comprehensive wetland management/monitoring program.

Wetland area	Recommended steps
Policy/ programmatic	Implement a region-wide wetland-monitoring program Establish a strategy to comprehensively assess wetlands using the existing programs Develop a funding agency to support monitoring programs Support periodic (after every three years) evaluation of the effectiveness and progress of wetland management program
Mapping	• A physical survey imperative to mapping and delimitation of the historical physical dimensions of wetlands using remote sensing (RS) and geographical information system (GIS) is an urgent necessity. • Demonstrating new wetland mapping standards for updating the wetland inventory as a base map for tracking changes in physical extents.
Resource inventories	• A detailed resource inventory of each selected wetland is a must for any effective conservation action.
Perspective plan	• Drawing up of a perspective plan setting out priorities for the next 5–10 years. • Adhoc and annual action plan taken up should be ultimately orchestrated with the perspective plan.
Wetland authority	• Development of region-wide wetland management authority • There should be an enduring organizational structure for protection and wise use of wetland • Creation of district-wise steering committee to provide coordination on approaches and strategies for wetland monitoring and assessment
People's involvement	• The best way to protect the wetlands is not only by legislation but by educating the people on the importance of the wetland and what they stand to lose if these ecosystems perish. • An awareness-building program should therefore receive top priority

regimes with ecosystem services of wetlands. Research on the rest of the remaining wetlands in Kashmir Himalaya lacks both in limnological and functional aspects.

Regarding the HAWs, the SAC has done a great job as far as inventorying of these wetlands is concerned but these HAWs are the most ignored ecosystems in research, conservation, and management plans. The changes that occurred in these pristine ecosystems have not been reported over a period of time, nor other HAWs have been studied. Yet, there has been also no effort to scientifically examine the biota and the relationship of organisms with abiotic factors in HAWs, nor is there any scientific study to document the natural and anthropogenic pressures they are subjected to. The impacts of climate change on HAWs are also unknown. Yet very little is known about climate change triggering glacier melt and erratic precipitation events on the hydrological regimes, aquatic biota, and resource availability of HAWs in Kashmir Himalaya. To address these predictive modeling and multi-criteria decision analysis approaches should be carried to evaluate likely impacts on HAWs. Future research efforts are needed to systematically survey HAWs' biodiversity and monitor their responses to global change.

10.6 Conclusion

The wetlands in Kashmir Himalaya are an integral part of the hydrological cycle and contribute significantly to the human well-being in the region. Despite their economic, ecological, and cultural significance, they have been put under tremendous pressures and threats due to land system changes mainly related to urbanization, infrastructure development, and agricultural intensification. These challenges are vigorously impacting the health of wetlands over this region, thereby hampering the ecological services, livelihoods, and sustenance of these ecosystems. Unless more research followed by immediate conservation and restoration strategies are not initiated, these important ecosystems which are essential capitals for thousands of people in the region are continuously going to get depleted in the imminent future.

Acknowledgments

The first author highly acknowledges the Senior Research Fellowship award by the University Grants Commission-Maulana Azad National Fellowship (UGC-MANF) during the study period. The authors would also like to acknowledge the support of the Department of Environmental Science, University of Kashmir, Srinagar.

References

Abubakr, A., Nahvi, A., and Kundangar, M.R.D. (2011). Ecological studies and uses of valued aquatic plants in Kashmir wetlands. *Nature, Environment and Pollution Technology 10* (1): 115–118.

Alam, A., Rashid, S.M., Bhat, M.S., and Sheikh, A.H. (2011). Impact of land use/land cover dynamics on Himalayan wetland ecosystem. *Journal of Experimental Sciences 2* (3): 60–64.

Alam, A., Bhat, M.S., and Maheen, M. (2019). Using Landsat satellite data for assessing the land use and land cover change in Kashmir valley. *GeoJournal*: 1–15.

Aslam, J., Ameta, S.K., Aslam, R., and Iqbal, O. (2018). Tourism in Kashmir Valley: growth, environmental impacts and sustainability. *International Journal of Pure & Applied Biosciences 6* (6): 1095–1106. https://doi.org/10.18782/2320-7051.7210.

Barber, L.B., Leenheer, J.A., Noyes, T.I., and Stiles, E.A. (2001). Nature and transformation of dissolved organic matter in treatment wetlands. *Environmental Science & Technology 35* (24): 4805–4816.

Bassi, N., Kumar, M.D., Sharma, A., and Pardha-Saradhi, P. (2014). Status of wetlands in India: a review of extent, ecosystem benefits, threats and management strategies. *Journal of Hydrology: Regional Studies 2*: 1–19.

Bhan, M. and Trisal, N. (2016). Fluid landscapes, sovereign nature: conservation and counterinsurgency in Indian-controlled Kashmir. *Critique of Anthropology 37* (1): 67–92. https://doi.org/10.1177/0308275X16671786.

Bhat, G.M., Islam, M.A., Malik, A.R. et al. (2017). Willow industries in Kashmir valley: present dynamics and future prospects. *Rashtriya Krishi 12* (1): 79–83.

Bhat, M.S., Ahmad, B., Alam, A. et al. (2019). Flood hazard assessment of the Kashmir Valley using historical hydrology. *Journal of Flood Risk Management 12*: e12521.

Chatterjee, A., Blom, E., Gujja, B. et al. (2010). WWF initiatives to study the impact of climate change on Himalayan high-altitude wetlands (HAWs). *Mountain Research and Development 30* (1): 42–52.

Dar, S.A., Bhat, S.U., Rashid, I., and Dar, S.A. (2020a). Current status of Wetlands in Srinagar City: threats, management strategies, and future perspectives. *Frontiers in Environmental Science* **7**: 199.

Dar, S.A., Bhat, S.U., Aneaus, S., and Rashid, I. (2020b). A geospatial approach for limnological characterization of Nigeen Lake, Kashmir Himalaya. *Environmental Monitoring and Assessment 192* (2): 1–18.

Dar, S.A., Rashid, I., and Bhat, S.U. (2021). Land system transformations govern the trophic status of an urban wetland ecosystem: perspectives from remote sensing and water quality analysis. *Land Degradation & Development*: 1–18. https://doi.org/10.1002/ldr.3924

Davidson, N.C. (2014). How much wetland has the world lost? Long-term and recent trends in global wetland area. *Marine and Freshwater Research 65* (10): 934–941.

De Groot, R.S., Wilson, M.A., and Boumans, R.M. (2002). A typology for the classification, description and valuation of ecosystem functions, goods and services. *Ecological Economics 41* (3): 393–408.

Duke, J.A. (1998). *Handbook of Energy Crops: Phragmites australis (Cav.) Trin. ex Steud.* West Lafayette, IN: Center for New Crops and Plants Products, Purdue University 5 p. http://www.hort.purdue.edu/newcrop/duke_energy/phragmites_australis.html.

Farooq, M. and Muslim, M. (2014). Dynamics and forecasting of population growth and urban expansion in Srinagar city-a geospatial approach. *The International Archives of Photogrammetry, Remote Sensing and Spatial Information Sciences 40* (8): 709.

Fatima, T., Iftikhar, F., and Hussain, S.Z. (2018). Ethno-medicinal and pharmacological activities of lotus rhizome. *Journal of Pharmaceutical Innovation 7* (4): 238–241.

Final Report on Environmental Impact Study of Revised Management Action Plan for Wular Lake. (2014). http://wular.org/wp-content/uploads/2019/07/Final%20Draft%20Report%20Wular%20Wular%20Lake%2025-04-2014.pdf

Galatowitsch, S.M. (2018). Natural and anthropogenic drivers of wetland change. In: *The Wetland Book* (eds. C. Finlayson, G. Milton, R. Prentice and N. Davidson), 359–367. Dordrecht: Springer.

Ganaie, M.A., Parveen, M., and Balkhi, M.H. (2015). Physico-chemical profile of three freshwater flood plain lakes of River Jhelum, Kashmir (India). *International Journal of Multidisciplinary Research and Development 2* (11): 527–532.

Ganaie, T.A., Jamal, S., and Ahmad, W.S. (2020). Changing land use/land cover patterns and growing human population in Wular catchment of Kashmir Valley, India. *GeoJournal*: 1–18.

Gardner, R.C., Barchiesi, S., Beltrame, C. et al. (2015). State of the World's wetlands and their services to people: a compilation of recent analyses. In: *Ramsar Briefing Note*, vol. **7**, 1–20. Gland, Switzerland: Ramsar Convention Secretariat.

Ghermandi, A., van den Bergh, J.C., Brander, L.M. Nunes, P.A.L.D. (2008). The economic value of wetland conservation and creation: a meta-analysis. Fondazione Eni Enrico Mattei Mattei working paper, Milan, Italy, 238, 1–25. doi:10.2139/ssrn.1273002.

Guo, M., Li, J., Sheng, C. et al. (2017). A review of wetland remote sensing. *Sensors 17* (4): 777.

Habib, M. (2014). Bird community structure and factors affecting the avifauna of Hokersar wetland Kashmir. *International Journal of Current Research* **6** (7): 7397–7403. https://www.journalcra.com/sites/default/files/issue-pdf/5817.pdf.

Horwitz, P., Finlayson, M., and Weinstein, P. (2012). *Healthy Wetlands, Healthy People: A Review of Wetlands and Human Health Interactions Geneva. Secretariat of the Ramsar Convention on Wetlands*. Switzerland: *The World Health Organisation*.

Junk, W.J., An, S., Finlayson, C.M. et al. (2013). Current state of knowledge regarding the world's wetlands and their future under global climate change: a synthesis. *Aquatic Sciences 75* (1): 151–167.

Kaul, V. and Pandit, A.K. (1980). Management of wetland ecosystems as wildlife habitats in Kashmir. In: *Proceedings of the International Seminar: Management of Environment* (ed. B. Patel), 31–53. Bombay, India: Bhabha Atomic Research Centre.

Kaul, R., Masoodi, A., Rasool, A. et al. (2016). Economic Feasibility of Willow Removal from Wular Lake, Jammu & Kashmir. Final Report Submitted to: GIZ/Ministry of Environment & Forests Government of India, 1–84.

Keddy, P.A., Fraser, L.H., Solomeshch, A.I. et al. (2009). Wet and wonderful: the world's largest wetlands are conservation priorities. *BioScience 59* (1): 39–51.

Khan, M.A. (2008). Chemical environment and nutrient fluxes in a Floodplain Wetland Ecosystem, Kashmir Himalayas, India. *Indian Forester 134* (4): 505–514.

Khan, M.A. (2015). Environmental characteristics and phytoplankton productivity of a shallow Ramsar-site floodplain in the western Himalaya. *Lakes & Reservoirs: Research & Management 20* (2): 69–76.

Khan, I. and Ali, M. (2013). Current status of the Fish Fauna of River Jhelum, Kashmir, J&K. *Open Access Scientific Reports* **2** (3): 1–3. https://doi.org/10.4172/scientificreports694.

Khan, M.U.Z., Ganaie, I.M., and Bhat, K.H. (2014). Limnological study of dal lake Srinagar J&K. *International Journal of Engineering Research and General Science 2* (6): 558–573.

Kraiem, H.A.B.I.B. (2002). Biophysical and socio-economic impacts of climate change on Wetlands in the Mediterranean. *Mediterranean Roundtable,* 1–39. https://www.uicnmed.org/web2007/CDCambio_climatico/contenido/F/PDF/CC_f4.pdf.

Kuchay, N.A. and Bhat, M.S. (2014). Urban sprawl of Srinagar city and its impact on wetlands-A spatio-temporal analysis. *International Journal of Environment and Bioenergy 9* (2): 122–129.

Kumar, R. (2018). Wular Lake, Kashmir. In: *The Wetland Book* (eds. C. Finlayson, G. Milton, R. Prentice and N. Davidson), 1705–1710. Dordrecht: Springer.

Kumar, R., Bhatt, J.R., and Goel, S. (2017). *Natural Capital of Wetlands*, 45. Wetlands International South Asia: New Delhi.

Lavnert, E. (1981). *Edible and Medicinal Plants*. Himalayan Publishers.

Millennium Ecosystem Assessment (2005). *Ecosystems and Human Well-Being: Wetlands and Water Synthesis*. Washington, D.C.: World Resources Institute.

Molur, S., Smith, K.G., Daniel, B.A., and Darwall, W.R.T. (2011). *The Status and Distribution of Freshwater Biodiversity in the Western Ghats, India*. IUCN; Zoo Outreach Organisation: Cambridge, UK and Gland, Switzerland; Coimbatore, India.

Mushtaq, F. and Pandey, A.C. (2014). Assessment of land use/land cover dynamics vis-à-vis hydrometeorological variability in Wular Lake environs Kashmir Valley, India using multitemporal satellite data. *Arabian Journal of Geosciences 7* (11): 4707–4715.

Najar, I.A. and Khan, A.B. (2012). Assessment of water quality and identification of pollution sources of three lakes in Kashmir, India, using multivariate analysis. *Environmental Earth Sciences 66* (8): 2367–2378.

National Wetland Atlas (2010). Jammu and Kashmir, SAC/RESA/AFEG/NWIA/ ATLAS/16/2010, Space Applications Centre, ISRO, Ahmedabad, India, 176p.

Nengroo, Z.A., Shah, A.H., and Bhat, M.S. (2017). Assessment of the impact of land use change on natural resource land of Srinagar Metropolitan Region of Kashmir Valley. *IOSR Journal of Humanities and Social Science 22* (7): 54–60.

Pandit, A.K. (1982). Looking out for the wildlife in Jammu and Kashmir Himalaya. *Environmental Awareness* **5**: 53–62.

Pandit, A.K. (1991). Conservation of wildlife resources in wetland ecosystems of Kashmir, India. *Journal of Environmental Management 33* (2): 143–154.

Pandit, A.K. (2008). Biodiversity of wetlands in Kashmir Himalaya. *Proceedings of the National Academy of Sciences, India – Section B: Biological Sciences* **78**: 29–51.

Panigrahy, S., Singh, T.S., Patel, J.G. et al. (2010). *National Wetland Atlas: Jammu and Kashmir, SAC/RESA/AFEG/NWIA/ATLAS/16/2010*. ISRO, Ahmedabad, India, 176 p. https://vedas.sac.gov.in/vedas/downloads/atlas/Wetlands/NWIA_Jammu_and_Kashmir_Atlas.pdf: Space Applications Centre.

Pant, H.K., Rechcigl, J.E., and Adjei, M.B. (2003). Carbon sequestration in wetlands: concept and estimation. *Food, Agriculture and Environment 1* (2): 308–313.

Parvez, S. and Bhat, S.U. (2014). Searching for water quality improvement in Dal lake, Srinagar, Kashmir. *Journal of Himalayan Ecology and Sustainable Development 9*: 51–64.

Paudel, K.R. and Panth, N. (2015). Phytochemical profile and biological activity of Nelumbo nucifera. *Evidence-based Complementary and Alternative Medicine* **789124**: 1–16.

Phethi, M.D. and Gumbo, J.R. (2019). Assessment of impact of land use change on the wetland in Makhitha village, Limpopo province, South Africa. *Jàmbá: Journal of Disaster Risk Studies 11* (2): 1–6.

Poppe, L., Rutherford, I., Price, P. et al. (2006). River and riparian land management: controlling willows along Australian rivers. Technical guideline No. 6. Land and Water Australia. Australian Govt, Canberra, 17.

Prasad, S.N., Ramachandra, T.V., Ahalya, N. et al. (2002). Conservation of wetlands of India-a review. *Tropical Ecology 43* (1): 173–186.

Raich, J.W. and Schlesinger, W.H. (1992). The global carbon dioxide flux in soil respiration and its relationship to vegetation and climate. *Tellus B 44* (2): 81–99.

Ramsar (1971). Convention on Wetlands of International Importance especially as Waterfowl Habitat. https://www.ramsar.org/sites/default/files/documents/library/scan_certified_e.pdf

Rashid, I. and Aneaus, S. (2019). High-resolution earth observation data for assessing the impact of land system changes on wetland health in Kashmir Himalaya, India. *Arabian Journal of Geosciences 12* (15): 453.

Rashid, H. and Naseem, G. (2008). Quantification of loss in spatial extent of lakes and wetlands in the suburbs of Srinagar city during last century using geospatial approach. *Paper presented at: Proceedings of Taal: The 12th World Lake Conference*, 653–658.

Rashid, I., Romshoo, S.A., Amin, M. et al. (2017). Linking human biophysical interactions with the trophic status of Dal Lake, Kashmir Himalaya, India. *Limnologica 62*: 84–96.

Reddy, K.R., DeLaune, R., and Craft, C.B. (2010). Nutrients in wetlands: Implications to water quality under changing climatic conditions. *Final Report submitted to US Environmental Protection Agency*.

Reis, V., Hermoso, V., Hamilton, S.K. et al. (2017). A global assessment of inland wetland conservation status. *BioScience 67* (6): 523–533.

Romshoo, S.A. and Rashid, I. (2014). Assessing the impacts of changing land cover and climate on Hokersar wetland in Indian Himalayas. *Arabian Journal of Geosciences 7* (1): 143–160.

Romshoo, S.A., Altaf, S., Rashid, I., and Dar, R.A. (2018). Climatic, geomorphic and anthropogenic drivers of the 2014 extreme flooding in the Jhelum basin of Kashmir, India. *Geomatics, Natural Hazards and Risk 9* (1): 224–248.

Shah, M.A. and Reshi, Z.A. (2014). Characterization of alien aquatic flora of Kashmir Himalaya: implications for invasion management. *Tropical Ecology 55* (2): 143–157.

Shah, J.A., Pandit, A.K., and Shah, G.M. (2015). Dynamics of key plant nutrients (N & P) in Hokersar, a typical wetland of Kashmir Himalaya, India. *Journal of Environmental Engineering and Ecological Science 4* (1): 1–7.

Shah, J.A., Pandit, A.K., and Shah, G.M. (2019a). Physico-chemical limnology of lakes in Kashmir Himalaya, India. *Science and Technology 12* (3): 149–156.

Shah, A.B., Reshi, Z.A., and Shah, M.A. (2019b). Conspectus of Aquatic Macrophytic Flora of Jammu & Kashmir State, India. *International Journal of Scientific Research in Biological Sciences 6* (1): 167–176.

Shah, M.A., Shah, A.B., and Reshi, Z.A. (2020). Aquatic ecosystems of Jammu and Kashmir State. *Biodiversity of the Himalaya: Jammu and Kashmir State, Topics in Biodiversity and Conservation*, p. 18

Sharma, E., Chettri, N., Tse-Ring, K. et al. (2009). *Climate Change Impacts and Vulnerability in the Eastern Himalayas*. Kathmandu, Nepal: International Centre for Integrated Mountain Development (ICIMOD).

Sharma, V., Mishra, V.D., and Joshi, P.K. (2012). Snow cover variation and streamflow simulation in a snow-fed river basin of the Northwest Himalaya. *Journal of Mountain Science 9* (6): 853–868.

Showqi, I., Lone, F.A., Mehmood, M.A. et al. (2018). Assessment of some macro nutrients to determine the nutritional status of Anchar Lake of Kashmir Himalaya. *International Journal of Theoretical & Applied Sciences, Special Issue on Environmental Contaminants and Management 10* (1): 46–50.

Tahseen, A.A. and Balkhi, M.H. (2018). Trophic status assessment of Anchar Lake, Kashmir. *Journal of Pharmacognosy and Phytochemistry 7* (4): 1154–1156.

Turner, R.K., Van Den Bergh, J.C., Söderqvist, T. et al. (2000). Ecological-economic analysis of wetlands: scientific integration for management and policy. *Ecological Economics 35* (1): 7–23.

Uheda, E., Kitoh, S., and Shiomi, N. (1999). Response of six Azolla species to transient high-temperature stress. *Aquatic Botany 64* (1): 87–92.

Wani, M.H., Baba, S.H., Yousuf, S. et al. (2013). Economic valuation and sustainability of Dal lake ecosystem in Jammu and Kashmir. *Knowledge systems of societies for adaptation and mitigation of impacts of climate change* (pp. 95–118). Springer, Berlin, Heidelberg.

Wetlands International South Asia (WISA) (2007). *Comprehensive Management Action Plan for Wular Lake, Kashmir*. New Delhi: Wetlands International South Asia.

Wetlands International South Asia (WISA) (2014). Sarovar, the newsletter of wetlands International-South Asia. *Sarovar Newsletter*, vol. III, 1–30.

Whitehouse, N.J., Langdon, P.G., Bustin, R., and Galsworthy, S. (2008). Fossil insects and ecosystem dynamics in wetlands: implications for biodiversity and conservation. *Biodiversity and Conservation 17* (9): 2055–2078.

Yaqoob, K.U., Pandit, A.K., and Wani, S.A. (2008). Comparative physicochemical Limnology of three lakes of Kashmir Himalaya. *Proceedings of Taal 2007: The World Lake Conference* (ed. M. Sengupta and R. Dalwani), vol. 1927, 1922–1927.

Zutshi, D.P. (1989). Eutrophication of lakes. In: *Perspectives in Aquatic Biology* (ed. R.D. Khulbe), 87–98. New Delhi: Papyrus Publishing House.

11

Heavy Metal Pollution in Coastal Environment and Its Remediation Using Mangroves

An Eco-sustainable Approach

Jitendra Kumar Singh, Amit Kumar Yadav, Shalini Gupta, and Reeta Verma

School of Environment and Sustainable Development, Central University of Gujarat, Gandhinagar, Gujarat, India

11.1 Introduction

The coastal marine environment pollution is a consequence of indiscriminate discharge of untreated/partially treated industrial and sewage wastewater. Environmental pollution leads to undesirable alterations in physical, chemical, and biological characteristics of natural resources such as water, air, and soil which have harmful health effects on humans and other living organisms or prohibit their survival activities (Markert 1993; Krishna and Manickam 2017). According to numerous research studies, environmental pollutants from anthropogenic sources are recognized to possess adverse values and the ability to deteriorate the ecological integrity of the coastal environment. Among coastal ecosystems, the mangrove ecosystem is as rich in biological diversity as the tropical rainforest (Swaminathan 1991). Developmental projects categorized under anthropogenic activities affect the ecological functions and lead to destruction and depletion of coastal resources, mangrove habitats, ecosystem processes, and loss of biodiversity (Vijay et al. 2005). Industrial units and towns located around the coastal areas discharge effluents and municipal sewage into the sea. Entry of various pollutants into the coastal environment takes place by rivers, diffuse runoff and point sources (industry and urban). Land contributes to approximately 80% of ocean pollution; virtually anthropogenic activities solely affect the adjoining water quality and environment (Prabhahar et al. 2012). Among the existing water pollutants, metals are considered seriously prioritized due to their environmental persistence and ability to concentrate in organisms at different trophic levels. Rapid industrialization and urbanization have resulted in elevated levels of heavy metals in the biosphere (Lu et al. 2004). These metals are harmful to humans and animals and cause serious problems in the food chain owing to their bioaccumulation potential in living organisms. Increased accumulation of heavy metals above the normal level in substrates could have unfavorable environmental impacts (Paz-Alberto et al. 2014). Some heavy metals may reach dangerously higher levels due to developmental accomplishments: mining, automobile traffic, smelting, manufacturing, and agricultural wastes (Öncel et al. 2000). Heavy metal pollution is of serious concern at local, regional, and global scales.

Wetlands Conservation: Current Challenges and Future Strategies, First Edition.
Edited by Sanjeev Sharma and Pardeep Singh.

Elimination of heavy metals from coastal waste waters is of global concern because heavy metals cause serious problems in aquatic ecosystems owing to their long persistence and toxicity to aquatic organisms even at a lower concentration. Increased concentration of heavy metals may inhibit photosynthesis and respiration in mangroves, ultimately leading to death (Melville and Burchett 2002). Contamination of coastal water and soil area by heavy metals is of critical importance due to its impending adverse biological and ecological effects. Hence, it is very essential to monitor heavy metal pollution by taking suitable managerial measures to protect the valuable mangrove resources. Remediation approaches are being recognized to alleviate the heavy metal pollution in mangrove ecosystem. Remediation approaches may comprise dilution techniques, chemical stabilization, soil washing, thermal desorption, and phytoremediation (FFTC 2009). One of the most important mangrove capabilities is the ability of absorption and holding heavy metals in sediments which prevent the transportation of metal pollutants and purify water by absorbing impurities in coastal areas.

11.2 Pollution in Mangrove Habitats: A Global Concern

Mangroves are an indigenous species and a major contributor to the coastal environment. Pollution in mangrove habitat is a serious environmental concern with substantial consequences on coastal ecosystem health. Mangrove plant community can propagate in a wide range of harsh environmental conditions; unique adaptive features such as salt-excreting leaves, gas-exchanging pneumatophores, and viviparous propagules are present to cope up with harsh conditions (Primavera et al. 2004). Ecological resources such as mangrove resources have been deteriorated by anthropogenic undertakings and uncontrolled discharge of environmental pollutants (Noegrohati and Hadi 2008; Dsikowitzky et al. 2011).

Heavy metals are among one of the most widely found contaminants in mangrove ecosystems, primarily released from various anthropogenic activities such as shipping, dredging, oil spills, urban wastewater discharges, thermal pollution (hot-water outflows), agrochemicals, nutrient pollution (including sewage), heavy metals, pesticides, and petroleum products. The ecological functions and stabilization of mangrove ecosystems are affected by organic contaminants and heavy metals. These contaminants also cause harmful adverse effects to associated flora and fauna. Mangrove habitats are often contaminated and impacted with oil residues and petroleum hydrocarbons because of their distribution proximity transporting routes (Hoff et al. 2002), and oil pollution severely damages mangrove ecosystems (Marmiroli et al. 2006). The natural concentration of heavy metals in the environment is of less concern due to its low concentration and slow release rate; progress in the mining industry and agriculture practices has accelerated the release of metals into ecosystems, subsequently degrading the environment and ecosystem health. Increasing human pressure for domestic need and the development of industries are virtually destroyed large areas of virgin mangroves and are considered to be among the highly threatened habitats around the globe.

In recent years, the mangrove environment is being polluted with various contaminants by wastes and effluents from industrial sources. Wastes generated from industrial and domestic sources along with habitat destruction have a substantial influence on the coastal environments (Moore et al. 2004). Anthropogenic activities have resulted in deleterious effects on soil/water quality and generating pressure on the mangrove ecosystem functions; consequently resulting in the degradation of water quality and biodiversity, loss of critical habitats, and an overall decrease in the quality of life of local inhabitants (Herrera-Silveira and Morales-Ojeda 2009). Mangroves are highly resistant to heavy metal toxicity; however, accumulation of heavy metals in the mangrove sediment and water can have biological and ecological effects and the biotic components habituating in the ecosystem are susceptible to the deleterious effects of heavy metals (Zheng et al. 1997). Coastal environmental conditions facilitate mangroves to attain high adaptability and resistance to environmental stresses, such as high salinity, and metal and nutrient concentration.

11.3 Heavy Metal Cycling in the Mangrove Ecosystem

Heavy metal persistence in the mangrove ecosystem is a grave environmental threat worldwide. Heavy metals are naturally found in the environment or are produced by anthropogenic procedures and are considered as byproducts (Harvey 2011). The biogeochemical cycle of heavy metal found in coastal environments is complex due to the occurrence of multiple organic and inorganic metal compounds. The sources of heavy metal contamination in the coastal mangrove ecosystem including water and sediments may be natural and/or anthropogenic. Trace metals in the mangrove environment are primarily from tidal activities, rainfall, and land runoff. Mangrove ecosystems were affected by solid wastes and wastewaters, which are important sources of trace metals (Nriagu and Pacyna 1988). Heavy metals along with other factors affect the natural water and sediment quality significantly in estuaries and adjoining coastal environments, directly influencing the biotic components of the system (Nobi et al. 2010). Mangrove ecosystems are characterized by unique properties: they exhibit anaerobic conditions, and they are rich in organic matter and sulfide. Mangrove ecosystem is crucial for its participation in the biogeochemical cycles near the coast, specifically for the supply of nutrients and organic carbon to tropical coastal oceans (Dittmar et al. 2006; Bouillon et al. 2008).

Mangrove wetlands are constructive ecosystems that assist in the regulation of various dynamic processes, viz. biological cycles, water quality maintenance, nutrient movement, and support for food chains. Highly productive mangrove ecosystems (2500 mg $cm^{-2}d^{-1}$) with 1603 g/m^2 $year^{-1}$ rates of litter production could provide a sedimentary sink for trace metals (Ghosh et al. 1990; Bunt 1992; Gonneea et al. 2004). Most of the heavy metals accumulate in the biological environment in different ways via the nutrient chain. Mangrove ecosystems are under continuous threat of heavy metal contamination due to their long persistence in nature and constant anthropogenic inputs.

Heavy metals are nonbiodegradable in nature and have a tendency of bioaccumulation and biomagnification to hazardous levels in a food chain (Pan and Wang 2012). In mangrove environments, the cycling of heavy metals is a severe problem (Marchand et al. 2006; Pekey 2006) due to their general features of toxicity, persistence, and bioaccumulation capacity. Metals are resistant to biological or chemical degradation. Therefore, these can be locally accumulated or transported over large dimensions. Mangrove ecosystem may act as a sink or a source of heavy metals in coastal environments because of their variable physical and chemical properties (Harbison 1986). Usually, in oceans, the concentration of heavy metals is very low (Ash and Stone 2003). However, due to inputs from river systems, the concentration of heavy metals in coastal waters may be much higher (Morillo et al. 2004). Currently, a study on various aspects of heavy metal and mangroves as its bioindicators is being focused on. Identification of principal pathways responsible for the metals' transfer along with their significance and the prospective utilization of mangroves in coastal aquatic ecosystems as bioindicators of heavy metal contamination is being studied.

11.4 Heavy Metal Transport, Uptake, and Release

The main sources of metals are effluents and wastes from industries, urban runoff, sewage treatment plants, boating activity, runoff from agricultural lands containing pesticides, garbage dumps from domestic households, and mining operations (Tam and Wong 2000; MacFarlane 2002; Cox and Preda 2005; Gonzalez-Mendoza et al. 2007). Incoming tides flood the mangroves with saline water twice per day, hence often referred to as tidal forests. During this tidal flushing, the exchange of various materials is facilitated such as detritus, nutrients, and pollutants with contiguous waters (Imelda and Chandrika 2000). Transportation of heavy metals to coastal areas is basically done by water or wind, whereby these can be deposited as sediments. Mangroves have a capacity to efficiently trap suspended material from the water column (Furukawa et al. 1997). Mangrove roots often act as a barrier, retaining most of the heavy metals, thus reducing the translocation of heavy metals to other plant parts. In the mangrove ecosystem, metals' uptake is performed by roots; released by leaves and exported via detritus. The export of an enormous amount of organic matter along with heavy metals has a distinguishable effect on the existing food webs in the coastal environment. Ramose e Silva et al. (2006) recommended mangrove trees as possible biochemical reactors, due to their physiological and biochemical processes as well as an active role in the decomposition of organic matter within the sediments and influencing the mobility of heavy metals. Mangrove leaf litter can act as a source of a bioavailable form of heavy metals to the environment though the amounts are low and the dispersion of leaf litter is rapid with the low rate of metal release (Saenger and McConchie 2004).

11.5 Bioavailability and Concentration of Heavy Metals in the Sediments

The biogeochemistry of mangrove sedimentary environments can maintain conditions favorable for the retention of anthropogenic inorganic contaminants, particularly heavy metals (Lacerda 1998; Silva et al. 2003). Mudflat and tidal creek sediments seem to have traced heavy metal distribution largely affected by contamination, whereas metal distribution appears to be prominently influenced by root-sediment interactions in mangrove forest sediments. Mangrove sediments can accumulate heavy metals significantly (Lacerda et al. 1988; Tam and Wong 2000) because of the high affinity of organic matter (OM) for these pollutants (Nissenbaum and Swaine 1976). Tam and Wong (1995) demonstrated that mangrove soils can efficiently immobilize waste water-borne phosphorus, heavy metals though inefficient in nitrogen retention. Mangrove soils have been reported as a major sink for heavy metals (Tam and Wong 1995; Gueiros et al. 2003; Okbah et al. 2005).

Mangrove sediments quality, generally slightly acidic in nature, is used to judge the environmental condition and potential biological effects. The anaerobic condition in the soil (i.e. oxygen-deficient soils) helps sulfate-reducing bacteria to produce hydrogen sulfide. The grey or black color characteristic of the soil is due to the reduction of ferric compounds to ferrous sulfides. In estuaries, sediments under anoxic condition, a considerable amount of metal contaminants are bound to sulfides. The concentrations of heavy metal in the top layers of mangrove sediments are greater as compared to the top layer sediments of mud-flat sand mostly bounded with oxidizable matter (Krupadam et al. 2003; Goutam and Ramanathan 2013). Wetland sediments usually act as a sink for metals and very high concentrations of metals in their reduced state may be present in the anoxic zone. Evaluation of sediment quality plays a vital role in the assessment of mangrove ecosystem health (Olajire et al. 2005). Sediment contaminated with heavy metals may assist in the long-term availability of metals, their mobility and transport into the environment. Sediment also acts as an indicator of overlying water quality and a useful tool in the assessment of environmental pollution (Prabhakar et al. 2012).

Distribution and transport of heavy metals in sediment are an intricate process and are influenced by water quality, native biota, and sediment type. Mainly three mechanisms, viz., physico-chemical adsorption, physical accumulation, and biological uptake are involved in determining the concentration of heavy metals in the sediments (Hart 1982). Concentration (μg/g dry wt.) of some heavy metals (Cu, Cd, Cr, Ni, Pb, Zn, and Mn) from mangrove sediments in different mangrove habitats is given in Table 11.1. The sediment quality guidelines (Table 11.2) have been used to evaluate the level of heavy metal pollution in the sediments along with its potential adverse effects on the sediment system. Surface sediments' contamination with metals could directly affect the seawater quality and subsequent potential consequences to the sensitive organisms at the lowest trophic levels of the food chain and ultimately to human health (Christophoridis et al. 2009).

The range of detected concentration level of Cu (0.22–845 μg/g), Pb (0.24–1950 μg/g), Cd (0.002–34.73 μg/g), Cr (1.02–617 μg/g), Mn (2.90–2290 μg/g), Ni (0.47–252.1 μg/g), Zn (0.53–2372 μg/g), As (8–40 μg/g), Co (0.6–58 μg/g), Hg (2–1810μg/g), and Sn (0.8–174 μg/g) in mangrove sediments from diverse locations (Figure 11.1).

11.6 Factors Affecting Heavy Metals in the Sediment

The distribution and accumulation of heavy metals' concentration are influenced by numerous factors, for instance, soil texture, the total amount of organic matter, salinity, Cl^-, SO_4^{2-}, HCO_3^-, mineralogical composition, physical transport, and depositional environment of the region may alter metal concentration in sediment as well as on its discharge (Trefry and Parsley 1976; Buccolieri et al. 2006; Marchand et al. 2006; Goutam and Ramanathan 2013). The contaminated site may also affect the heavy metals' concentration due to its substantial positive relationship with organic carbon content and sediments' particle size. In the geochemistry phase, grain size and surface-to-volume ratio are the main

Table 11.1 Heavy metal concentrations (μg/g dry wt.) in mangroves' coastal sediments (depth 0–30 cm).

Location	Cu	Pb	Cd	Cr	Mn	Ni	Zn	References
Sunderbans mangrove, India	11–59	17–34	0.12–0.21	27—87	—	11–43	28–108	Chatterjee et al. (2009)
Andaman Islands, India	80.8–87.9	3.9–5.4	0.8–1.52	12.6–20.3	29.2–134.4	7.04–12.01	12.2–23.02	Nobi et al. (2010)
Alibag, Maharashtra, India[a]	92.59	19.51	5.31	545	1020	63.29	78.18	Pahalawattaarachchi et al. (2009)
Vamleshwar, Gujarat, India[a]	—	73.6	0.73	—	—	—	8.1	Nirmal et al. (2011)
Goutami-Godavari estuary, India	40.2	46.7	9.00	1.73	429.0	20.7	—	Ray et al. (2006)
Kakinada bay, India	42.4	18.2	4.12	1.02	446.0	44.5	—	Ray et al. (2006)
Godavari estuary, India	47.8	55.8	10.9	2.20	1059.0	25.7	—	Ray et al. (2006)
Thane Creek, Mumbai, India	110	41	—	90	2290	101	100	Lina and Nayak (2012)
Ulhas Estuary, Mumbai, India	173	82	—	239	2077	190	180	Lina and Nayak (2012)
South-West coast, Kerala, India	10–845	1100–1950	—	—	158–1124	—	31–2372	Thomasand and Fernandez (1997)
Northern Karnataka coast, India	1.8–23.1	1.1–3.9	2.5–4.0	1.2–13.9	228.7–343.7	14.3–55.1	21.8–150.3	Shanmugaarasu et al. (2013)
Gulf of Mannar, India	57	16	0.16	177	305	24	73	Jonathan et al. (2010)
Pichavaram, India	132.3	143.8	34.74	617	801	252.1	106	Ranjan et al. (2008)
Pichavaram, India	32	11.2	6.96	141.2	701	62	89	Ramanathan et al. (1999)
South East Coast, India	506.2	32.36	6.58	194.83	373	38.61	126.8	Raj and Jayaprakash (2008)
Mahanadi Delta, India	17.9	34.7	4.4	—	—	32.2	98.3	Behera et al. (2013)
Galatea Panama	4	32.5	7.2	12.8	143	74	10.9	Guzman and Jimenez (1992)
Payardia Panama	4	33.3	7.5	10.0	228	91.8	16.1	Guzman and Jimenez (1992)
Toro Pointa Panama	4.9	38.0	6.6	13.7	294	82.4	19.9	Guzman and Jimenez (1992)
Punta Mala Bay Panama	56.3	78.2	<10	23.3	295	27.3	105	Defew et al. (2005)
Yingluo Bay, China	18.9	10.0	0.077	9. 27	89.5	14.6	46.6	Zheng et al. (1997)
Shenzhen Bay, China	38.3	28.7	0.136	7.97	537. 0	25.0	114.0	Lin et al. (1997)

Jiulong River Estuary, China	29.7	18.3	0.094	4.73	583.0	16.9	111.0	Zheng et al. (1996a,b)
MNRs of South China	10.3–30.9	25.6–86.4	0.07–0.39	28.5–86.6	—	—	24.8–87.0	Liu et al. (2014)
Nanshamangrove, South China	113	55.3	0.78	155	880	48.4	159	Wu et al. (2014)
Native Futian mangroves, China	88.8	72.0	0.94	96.2	—	44.6	358.5	He et al. (2014)
Exotic Futian mangroves, China	93	77	2.96	81	—	45	252	He et al. (2014)
Shenzhen, Guangdong	38.3	28.7	0.136	7.97	—	25	114	Lin et al. (1997)
Shenzhen, Guangdong	45.58	62.92	—	58.04	—	62.80	125.08	Wang et al. (2003a,b)
Mai Po, Hong Kong	48–87	69–220	1.1–1.4	10–75	—	25–95	130–349	Ong Che (1999)
Deep Bay, Hong Kong	80	80	3	40	—	30	240	Tam and Wong (2000)
Punta Piuta Costa Rica	9.8	25.6	6.0	19.8	525	99.0	11.4	Guzman and Jimenez (1992)
Punta Portete Costa Rica	8.4	34.5	7.3	22.6	268	102	14.7	Guzman and Jimenez (1992)
Conception Bay, New Caledonia[b]	0.22	—	—	1.22	2.90	1.11	1.35	Marchand et al. (2011)
Sungei Buloh Singapore	7.06	12.3	0.18	16.6	—	7.44	51.2	Cuong et al. (2005), Cuong and Obbard (2006)
S. Khatib Bongsu	32.00	30.98	0.266	32.07	—	11.65	120.23	Cuong et al. (2005)
Cienaga Grande Colombia	23.3	12.6	1.92	13.2	623	32.5	91.0	Perdomo et al. (1999)
Buffalo River Brazil	76.0	131	3.64	118	524	—	479	Kehrig et al. (2003)
Guanabara Bay Brazil	79.6–91.7	166–170	—	37.7–43.3	524	—	448–505	Kehrig et al. (2003)
North-West Coast of South America[c]	253.8	81.3	1.9	94.5	469.6	82.2	678.3	Cadena et al. (2014)
Brisbane River Australia	3.1–30.2	20.1–81.9	0.1–1.9	13.3–54.3	—	2.4–57.6	40.8–144	Mackey and Hodgkinson (1995)
Mazatlan Harbor Mexico	7.7–90.9	14–100	2.0	7.6–42.5	—	6.1–30.3	46.4–348	Soto-Jiménez and Páez-Osuna (2001)
Dongzhai Haibor, Hannan	18	19	0.11	40	—	—	57	Qiu et al. (2011)

(Continued)

Table 11.1 (Continued)

Location	Cu	Pb	Cd	Cr	Mn	Ni	Zn	References
Gwatr Gulf, Iranian Coasts	21.27	—	0.25	—	—	63.57	45.91	Pakzadtoochaei (2013)
Arabian Gulf, United Arab Emirates	5.3–29.4	13.2–49.8	3.1–6.9	8.28–18.9	—	14.8–109	4.6–22.4	Shriadah (1999)
Kharchang Bay, Iranian Coasts	8.69	—	0.22	—	—	39.45	37.83	Pakzadtoochaei (2013)
Sirik Azini Creek, Iran	26–87	32–68	24–31	—	—	55–102	63–152	Parvaresh et al. (2011)
Kuantan, Malaysia	32.79	44.41	—	—	117.73	—	—	Yunus et al. (2011)
Peninsular Malaysia	31.9	83.1	0.8	6.0	—	—	4.3	Nazli and Hashim (2010)
Cross River Estuary, Nigeria	24–34	9–25	-	20–27	—	15–30	140–189	Essien et al. (2009)
Chumphon mangrove sediment, Thailand	2.26–51.19	0.24–12.50	0.002–0.04	2.73–30.25	3.47–267.90	0.47–14.00	0.53–40.85	Pumijumnong and Uppadit (2012)
Samut Sakhon mangrove forest sediment, Thailand	2.26–51.19	0.24–12.36	0.04–0.07	—	—	—	0.53–40.85	Buajan and Pumijumnong (2010)

[a] Metal concentration in mgl
[b] Metal concentration in μmol/g
[c] Metal concentration in μg/g

Table 11.2 Sediment quality guidelines for heavy metals (μg/g) pollution.

Sediment quality guide	Cu	Pb	Cd	Cr	Mn	Ni	Zn	References
GCSC	60–125	100–400	3–8	75–100	1.5–3	50	70–400	Alloway (1990)
SQG non-polluted	<25	<40	—	<25	—	—	<90	Luo et al. (2010)
SQG moderately polluted	25–50	40–60	—	25–75	—	—	90–200	Luo et al. (2010)
SQG heavily polluted	>75	>60	—	>75	—	—	>200	Luo et al. (2010)
MSQGs	35	60	0.5	80	—	—	150	SAQSIQ (2002)
Earth's crust	50	14.8	0.11	—	—	80	75	Wedepohl (1995)
The continental crust	32	1.46	0.079	80	—	38	70	Gao et al. (1998)

MNRs: Mangrove nature reserves, GCSC: General critical soil concentration, SQG: Sediment quality guidelines of USEPA, MSQGs: Marine sediment quality guidelines.

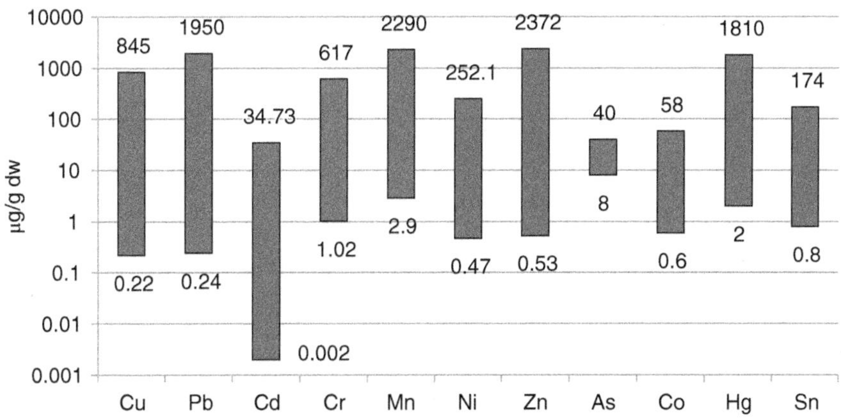

Figure 11.1 Level of heavy metal concentration (μg/g dw) in mangrove sediments.

influencing factors which may affect the heavy metal concentration (Maslennikova et al. 2012). Long (1992) suggested that the redox potential and sulfide concentration in sediments may intensely influence the amount of trace metals and their accessibility. Metal ion ligands' stability related to functional groups such as hydroxyl, amine, and carboxylic group associated with organic matter results in a significant association between organic matter and metal concentrations in the sediment (Riffaldi et al. 1983).

The biochemistry of heavy metals in sediment is controlled by a wide range of factors such as textural characteristics (particle size), dissolved or dry organic matter, and residue-binding characteristics, the kinetics of mobilization, depositional environments, and others. Variation in sediment conductivity and hydrogen ion concentration may lead to alteration in metal speciation and solubility, and accessibility can result in increased uptake into plants (Weis and Weis 2004). Subsequently, mangrove sediments assist to avert the heavy metals from entering the neighboring marine ecosystem.

11.7 Heavy Metal Accumulation in Mangrove Plants

Recently, bioaccumulation of metals in mangroves received considerable interest from researchers. Absorption and fixation of heavy metals by soil may cause remaining accumulation in the soil, and lead to overabsorption and accumulation of heavy metals in growing plants (Lian et al. 1999; RaviKumar et al. 2007). Each mangrove plant species imparts an explicit adaptation system, which may regulate their action toward heavy metals. Many mangrove species are able to propagate under the heavy metal-contaminated environment such as *Avicennia marina* and *Rhizophora stylosa* and have diverse bioaccumulation potential of different heavy metals (Table 11.3). These types of plants must have specific adaptations to survive there; they must be metallophytes, pseudo metallophytes, or hyperaccumulators. Sureshand Ravishankar (2004) reported that plants also consist of the ability of maximum bioaccumulation of different heavy metals by using phytochelatins and metallothioneins leading to complex formation with heavy metals and transferred into vacuoles. The levels of trace metals that bioaccumulate in mangroves vary seasonally and regionally (Lacerda and Abrao 1984). Tam and Wong (1997) proposed that the role of mangroves in holding heavy metals relies on plant age and biomass production. Mangrove roots play a key role as barrier for metal translocation. Maximum concentrations were found in roots as compared to the aerial parts of mangrove plants (MacFarlane and Burchett 2002). Lian et al. (1999) investigated that the varied concentrations of copper (2.1–7.8 μg/g), manganese (3.9–28 μg/g), zinc (5.7–60 μg/g), cadmium (0.014–0.057 μg/g), and lead (0.018–0.038 μg/g) in the propagules of *Ceriops tagal*, *Aegiceras corniculatum*, *Bruguiera sexangula*, *Sonneratia caseolaris*, *Rhizophora stylosa*, *Bruguiera gymnorrhiza*, *Kandelia candel*, *Avicennia marina*, *Sonneratia ovate*, and *Acanthus ilicifolius* mangrove species, respectively, in China (Figure 11.2).

The range of detected level of Cu, Pb, Cd, Cr, Mn, Ni, Zn, and Sn concentration in mangrove plants is 0.1–207, 0.02–225, 0.01–3.1, 0.2–347, 0.24–2472, 0.4–231.06, 0.03–1988, and 354.2–1769.8 μg/g dw, respectively, from multiple locations (Untawale et al. 1980; Peters et al. 1997; MacFarlane et al. 2007; Agoramoorthy et al. 2008; Lewis et al. 2011; Bayen 2012) are shown in Figure 11.3.

In mangrove ecosystems of China, variation in the concentration of different heavy metals such as Cu, Pb, Zn, Cd, Cr, Ni, Mn, Hg, and As was found in 15 mangrove species including *Sonneratia caseolaris*, *Sonneratia hainanensis*, *Sonneratia ovata*, *Sonneratia apetala*, *Kandelia candel*, *Bruguiera gymnorrhiza*, *Bruguiera sexangula*, *Aegicera corniculatum*, *Aegicera marina*, *Aegicera ilicifolius*, *Ceriops tagal*, *Rhizophora stylosa*, *Rhizophora apiculata*, *Excoecaria agallocha*, and *Lumnitzera apiculatal* ranged from less than 1 to 359, not determined (nd) to 113, 0.8 to 340, 0 to 1.8, 0.2 to 37.7, 0.2 to 50, 4 to 2,160, 0.02 to 4.3, and nd to 0.9 mg/kg dry weight, respectively (Zhang et al. 2014).

11.8 Heavy Metal Remediation Potential of Mangroves

Phytoremediation is a novel bioremediation technology or cost-effective unconventional technology in which plants are used for the cleanup or decontamination of harmful organic and inorganic pollutants from the terrestrial and aquatic ecosystem. Plants consist of natural ability to bioaccumulate, sequester, or detoxify contaminants in soils, water, or air

Table 11.3 List of mangrove plants known for heavy metal accumulation and phytoremediation.

Scientific name	Common name	Heavy metal	References
Avicennia alba	Api-api putih	Pb, Zn, and Cu	Panda et al. (2013)
Avicennia germinans	Black mangrove	Zn, Hg, Al, As, Cd, Cr, Cu, Fe, Pb, and Mg	Mejias et al. (2013)
Avicennia marina	Gray mangrove	Pb, Fe, Zn, Mn, Cd, and Cr	Lin et al. (1997), Shete et al. (2007), Kannappan et al. (2012), Paz-Alberto et al. (2014), Kathiresan et al. (2014)
Avicennia officinalis	Indian mangrove	Cd, Fe, Zn, Mn, Cu, and Pb	Sarangi et al. (2002), Kannappan et al. (2012)
Acanthus illicifolius	Holly mangrove	Fe, Cr, Pb, Zn, Cd, Ni, Cu, and Co	Thomas and Fernandez (1997), Akhand et al. (2012)
Aegiceras corniculatum	River mangrove	Zn, Cd, Cu, and Pb	Saenger and McConchie (2004)
Barringtonia racemosa	Powderpuff mangrove	Co, Fe, Zn, Mn, Cu, and Pb	Thomasand and Fernandez (1997)
Bruguiera sexangula	Oriental mangrove	Mn, Cd, Cu, Hg,Pb, and Zn	Lian et al. (1999), Gupta and Chakrabarti (2013)
Ceriops decandra	Yellow mangrove	Pb, Zn, and Cu	Panda et al. (2013)
Ceriops tagal	Tagal mangrove	Zn, Cd, Cu, and Pb	Saenger and McConchie (2004)
Excoecaria agallocha	Milky mangrove	Fe, Cr, Pb, Zn, Cd, Ni, Cu, and Co	Akhand et al. (2012)
Kandelia candel	Kandal-guriya	Cr	Rahman et al. (2009)
Laguncularia racemosa	White mangrove	Zn, Hg, Al, As, Cd, Cr, Cu, Fe, Pb, and Mg	Mejias et al. (2013)
Melaleuca cajuputi	Swamp tea tree	Cd and Zn	Mohd et al. (2013)
Rhizophora apiculata	Tall-stilt mangrove	Cu and Pb	Kamaruzzaman et al. (2009)
Rhizophora mangle	Red mangrove	Zn, Hg, Al, As, Cd, Cr, Cu, Fe, Pb, and Mg	Zheng et al. (1997), Mejias et al. (2013)
Rhizophora mucronata	Asiatic mangrove	Cd, Fe, Zn, Mn, Cu, and Pb	Panda et al. (2013), Sarangi et al. (2002), Kannappan et al. (2012)
Rhizophora stylosa	Stilted mangrove	Pb	Paz-Alberto et al. (2014)
Sonneratia alba	Sweet-scented apple mangrove	Pb	Paz-Alberto et al. (2014)
Sonneratia apetala	Sonneratia mangrove	Pb and Zn	Shete et al. (2007)
Sonneratia caseolaris	Crabapple mangrove	Fe, Cr, Pb, Zn, Cd, Ni, Cu, and Co	Akhand et al. (2012)

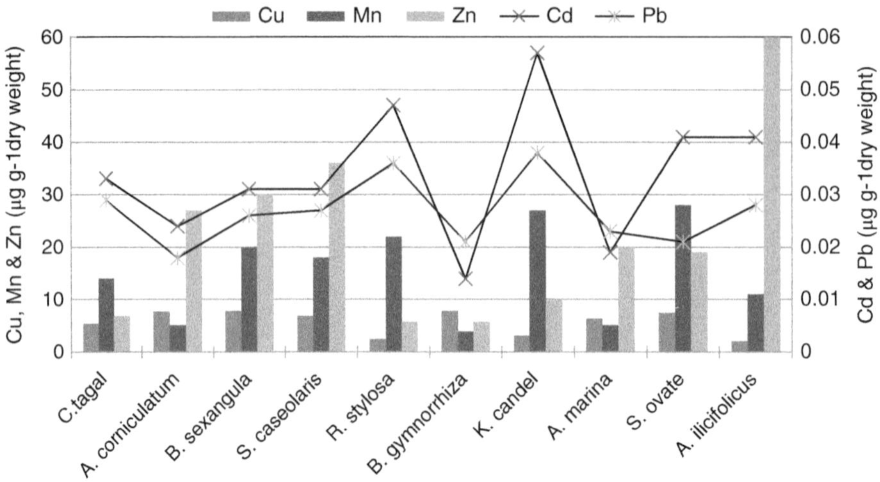

Figure 11.2 Contents of heavy metals (µg/g) in the propagules of 10 mangrove species.

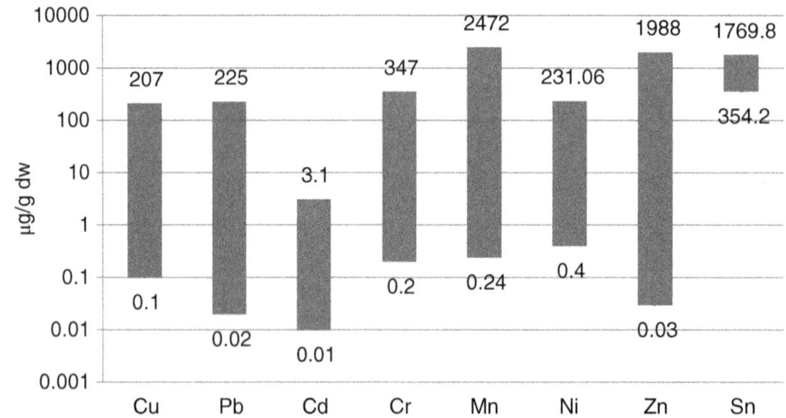

Figure 11.3 Range of detected level of heavy metal concentration (µg/g dw) in mangrove plants (all plants and all tissues).

through physical, chemical, or biological processes. Phytoextraction is the main process of phytoremediation in which plants tend to accumulate heavy metals from contaminated soils and translocate them to the aerial parts of a plant (Schwitzguébel 2001; Paz-Alberto et al. 2007, 2011, 2014; Uera et al. 2007). Recently, phytoremediation is an effective technique to reduce organic matter, nutrients and heavy metals' concentration in the polluted site and is also designated as green technology (Hadad et al. 2006; He et al. 2008; Valipour et al. 2009; Hongyu Wang et al. 2011; Huang et al. 2013). Plants possess various biochemical processes, like adsorption, transportation, translocation, hyperaccumulation, transformation, and mineralization, which may eventually remove toxic pollutants from contaminated sites (Meagher 2000). Mangroves play a key role in phytoremediation for removing heavy metals and other harmful pollutants in the wetland ecosystem (Zhang et al. 2011). The removal of heavy metals from contaminated sites is a recurring challenge, especially in developing countries. *Kandelia candel* plant reported for the removal of nutrients and different heavy metals from industrial wastewaters (Chu et al. 1998; MacFarlane et al. 2007). Several mangrove species such as *Almarina*, *Aegicers corniculatum*, *Bruguiera gymnorhiza*, *Sonneratia apetala*, Buch-Ham, known for bioaccumulation of nutrients and heavy metals (Boonsong et al. 2003).

Mangrove ecosystem is advantageous as it is a low-cost and high-efficiency wetland system for removal of toxic heavy metal pollutant from the marine environment and for the treatment of municipal wastewater effluent (Tam and Wong 1996; MacFarlane et al. 2003). It was found that *Avicennia schaueriana* (black mangrove) exhibits sensitivity to the oil-contaminated sediment signifying a promising mangrove species for hydrocarbon remediation (Moreira et al. 2010a,b).

The phytoremediation of total petroleum hydrocarbons (TPHs) using black mangrove showed approximately 89% removal from the mangrove sediment (Moreira et al. 2013). Ke et al. (2003) reported removal of pyrene from contaminated sediments using mangrove (*Kandelia candel* and *Bruguiera gymnorrhiza*) wetland systems. Saiyood et al. (2013) found that an evergreen mangrove tree, Bruguiera gymnorhiza, is tolerant to bisphenol A (BPA) and has the capability to remove BPA. Tables 11.3 and 11.4 show the list of mangrove species used for the remediation of heavy metals and organic contaminants (PAHs, TPHs, Pyrene, PCB, Bisphenol-A, and TDS), respectively.

Table 11.4 Accumulation of organic contaminants by mangrove plants.

Contaminant	Mangrove plant species	Removal site	Reference
Polycyclic aromatic hydrocarbons (PAHs) (mixed)	*Aegiceras corniculatum*	Sediment	Tam and Wong (2008)
PAHs	*Bruguiera gymnorrhiza*	Sediment	Wang et al. (2014)
Phenanthrene (PHE)	*Kandelia candel*	Sediment	Hong et al. (2009)
Fluoranthene (FLA)	*Kandelia candel*	Sediment	Hong et al. (2009)
Total petroleum hydrocarbons (TPHs)	*Avicennia schaueriana*	Sediment	Moreira et al. (2013)
	Rizophora mangle	Sediment	Moreira et al. (2011)
	Avicennia marina	Sediment	Sakineh et al. (2013)
Pyrene	*Kandelia candel*	Sediment	Ke et al. (2003)
	Bruguiera gymnorrhiza	Sediment	Ke et al. (2003)
Polychlorinated biphenyls (PCBs)	*Kandelia candel*	Sediment	Zhao et al. (2012)
	Aegiceras corniculatum	Sediment	Zhao et al. (2012)
Bisphenol A (BPA) and TDS	*Bruguiera gymnorrhiza*	Waste water	Saiyood et al. (2013)
Oil (Total hydrocarbons)	*Rizophora mangle*	Mangrove soils	Sanchez-Arias et al. (2013)
Organochlorine pesticides (OCPs)	*Avicennia marina*	Sediment	Shete et al. (2009)

11.9 Distribution of Heavy Metals in Different Plant Tissues of Mangrove Species

Circulation of heavy metals in leaves, stems, and roots may differ depending on the heavy metal concentration in the sediment, height of the plant, the types of heavy metals, and bioaccumulation tendency, biomass (Nirmal Kumar et al. 2011; Pakzadtoochaei 2013). Mangrove plant species of wetland distribute metals from root to aerial plant parts and tissues, then metals get accumulated in leaf tissue and stem tissue which differ according to the type of metal and its speciation. Transfer of metal to aerial parts of a plant is also dependent on the plant species, and diverse environmental conditions. *Avicennia marina* has the maximum affinity for Cu and Zn accumulation in fresh leaves compared to aged leaves. However, Pb & Cu accumulate in bark and wood compared to leaves of *Excoecaria agallocha* (Saenger and McConchie 2004). In *Rhizophora apiculata* seedlings, metal concentration declines from the root zone to stem and is followed by leaves (Moorthy and Kathiresan 1998). *Avicennia* mangrove species is found to oxidize the rhizosphere region, thus leading to sulfide reduction and enhance metal concentration in the exchangeable ionic metal (De Lacerda et al. 1993). Increased rate of Cd uptake and its bioavailability has been found in various wetland species under dry (oxidized) condition compared to flood (reduced) condition (Gambrell 1994). The concentrations of metals accumulated by different plant tissue parts of mangrove species are shown in Figure 11.4.

11.10 Application of Phytoremediation to Coastal Pollution Remediation

The United Nations Environment Program (UNEP 2003) supports phytoremediation as a sustainable tool for the treatment of environmental pollutants. Phytoremediation is considered eco-friendly and cost-effective technique. Some specific plant species possessed metal-accumulating tendency and removal of toxic metals from soil and water ecosystem (Rai 2008). Mangrove ecosystem creates a green environment by eliminating and converting harmful pollutants from waste water by sedimentation, filtration, microbial action, and absorption process (Wu et al. 2008; Wang et al. 2010; Pawar 2013). The potential of mangroves as phytoremediation plants from contaminated estuaries and coastal areas is an emerging research field due to its major advantage in the harvesting of plants for the abstraction of absorbed toxic metals that cannot be biodegraded. Potentially, halophyte provides an opportunity for phytoremediation application. Therefore, halophytes such as mangrove plant species are proposed as a relevant alternative and could be amended to deal with an adverse environmental condition. Consequently, holds pronounced potential for phytoremediation and removal of heavy metal-contaminated sites.

11.10.1 Phytoremediation Using Constructed Wetlands (CWs) Technology

The Constructed Wetlands (CWs) are constructed to employ the natural wetland function related to plants, soil system, and associated microbial population to clean the toxic pollutants in surface or waste water effluent (ITRC 2003; Jan Vymazal 2010; Patel Pratik and

Dharaiya Nishith 2014). The association and interaction among plant, water, soil, and microorganism play a key role in the ecoremediation concept. It acts as a biofilter, removing pollutants such as organic and inorganic pollutants (heavy metals) from the contaminated water. For the treatment of domestic wastewater, the first CW was used in Germany over 40 years ago in rural areas. Researchers have performed to the evaluation of hybrid CW's application for different wastewater treatments (Fenxia and Li 2009; Sayadi et al. 2012). Mangrove plant species might be used in CWs for the treatment of wastewater (Yang et al. 2008; Zhang et al. 2010).

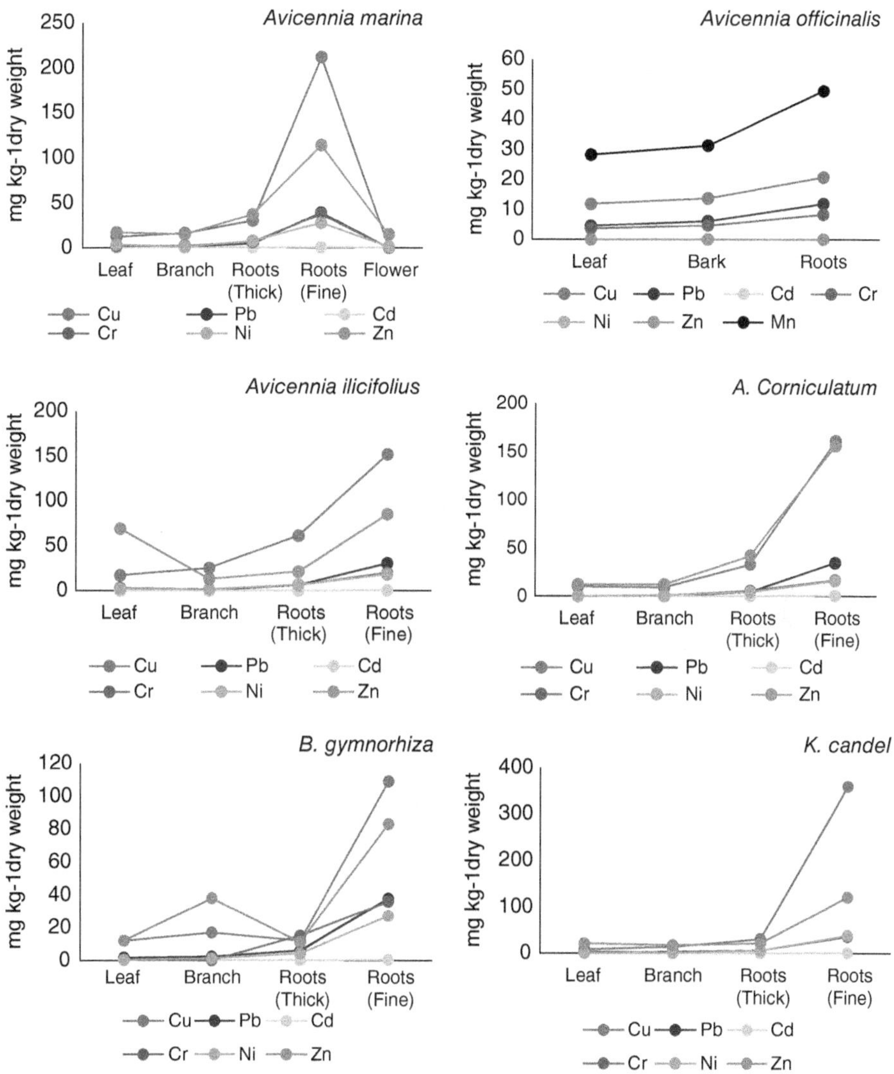

Figure 11.4 Scattering of the different trace metal concentrations in plant tissues of various mangrove species.

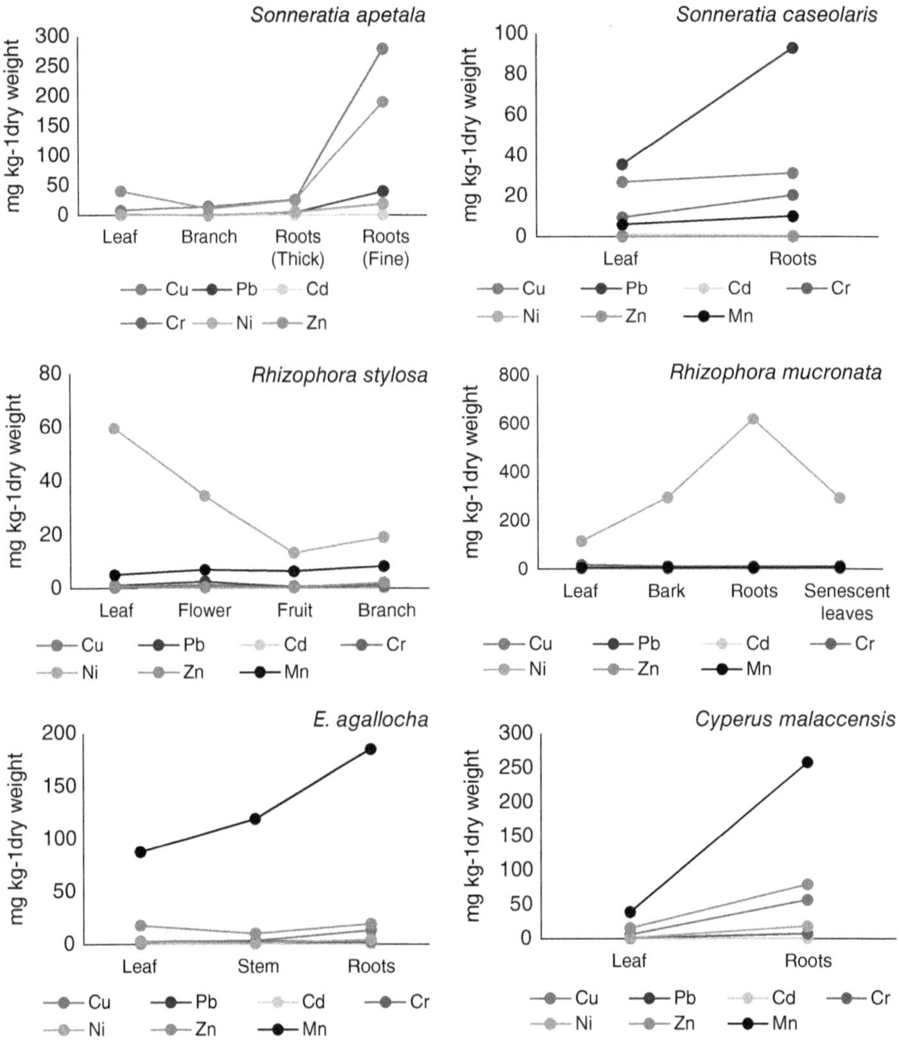

Figure 11.4 (Continued)

11.10.2 Phytoremediation Using Constructed Floating Bed

Phytoremediation using constructed floating bed technique was developed for the large-scale field remediation of pollutants from aquatic environments. Huang et al. (2013) has demonstrated the potential of salt-tolerant plant *Sesuvium portulacastrum* used in floating bed phytoremediation technique for effective treatment of organic pollutant percent in brackish water. Similarly, Zhang (2010) found that total nitrogen (TN) and total phosphorus (TP) concentrations efficiently reduced up to 62.2 and 51.4%, respectively, by using *S. portulacastrum* in floating bed technique with a mixture of sewage water with seawater in greenhouse. Hongyu Wang et al. (2011) observed that the organic matter from the polluted part of Guxin River in Hangzhou can be removed by the floating bed phytoremediation with *Eichornia crassipes* and *Lolium perenne* L. Phytoremediation technique of plants for the treatment of organic or inorganic contaminants in aquatic ecosystems may include plant uptake and microbial degradation at the root zone, supplemented with a sequence of complex physicochemical and biological progressions (Stottmeister et al. 2003).

11.11 Eco-remediation Technologies as Sustainable Natural Treatment Systems for Waste Water Treatment

Eco-remediation is the biotechnological application based on engineering fundamentals to recover and restore the terrestrial and aquatic ecosystem to deliver healthy atmosphere and environment for flora, fauna habitation, and human settlement (Kumari et al. 2020). Eco-remediation is basically a part of ecological engineering which can be defined as the application of ecological fundamentals for the management of ecosystem and design of sustainable ecosystem that assimilate human civilization with the environment for the betterment of both (Mitsch and Jorgensen 2003). Eco-restoration of heavy metal-contaminated land using wetland plant species tends to fix the toxic pollutant. Management of soil fertility, water purification, and clarification using microbes by ecological manipulation of lakes to elude eutrophication is one of the instances of ecosystem engineering among others. Eco-remediation technique for the treatment of waste stream and land not only helps in ecosystem restoration but also aids in the conservation of threatened biological communities.

Eco-remediation potential of mangrove ecosystem acts as an alternative, emerging elucidation for coastal tropical sites where environmental conditions are vital. Mangrove trees consist of high productivity and a natural tendency to absorb surplus nutrients and organic matter without any apparent functional or structural turbulences.

11.12 Conclusion and Future Prospects

The wetland ecosystem consisting of mangroves is exposed to a wide range of pollutants such as heavy metals, sewage wastes, pesticides, petroleum products, and other toxic chemical compounds. This paper concluded that transportation and conversion of toxic organic or inorganic compounds through the interaction of water, sediment, and soil, plant environment could be supportive for a scientific understanding of heavy metals' fate in the mangrove ecosystem. Mangroves are naturally growing plants across intertidal areas and are adapted to the contaminated coastal environment. Mangrove sediments are adept at fixing toxic heavy metals, and mangrove plant species own the capacity to uptake heavy metal contaminants via roots, stems, and leaves without any substantial, structural and functional modification. Mangrove plant species are able to minimize the heavy metal concentrations and other toxic pollutants in the coastal environment zone by their hyperaccumulation ability. Thus, mangroves have been recommended as an ideal plant for phytoremediation and a potential candidate for phytomining and potential bioindicators of coastal contamination as well. Hence, the plantation of potential mangrove species in the polluted coastal areas should be carried out in order to overcome the conflict of pollution load and for conservation and restoration of the mangrove ecosystem. Therefore, long-term in-situ field trails are required for the wide-range use of mangrove through phytostabilization as an effective technology of coastal pollution recovery. This technology needs further research on the mangrove ecosystem for contaminated water and soil treatment in coastal environment on a large scale.

References

Agoramoorthy, G., Chen, F.A., and Hsu, M.J. (2008). Threat of heavy metal pollution in halophytic and mangrove plants of Tamil Nadu, India. *Environmental Pollution 155* (2): 320–326.

Akhand, A., Chanda, A., Dutta, S. et al. (2012). Comparative study of heavy metáis in selected mangroves of Sundarban ecosystem, India. *Journal of Environmental Biology 33* (6).

Alloway, B.J. (1990). *Heavy Metals in Soils*, 37–40. NY USA: Willey.

Ash, C. and Stone, R. (2003). A question of dose. *Science 300* (5621): 925.

Bayen, S. (2012). Occurrence, bioavailability and toxic effects of trace metals and organic contaminants in mangrove ecosystems: a review. *Environment International 48*: 84–101.

Behera, B.C., Mishra, R.R., Patra, J.K. et al. (2013). Impact of heavy metals on bacterial communities from mangrove soils of the Mahanadi Delta (India). *Chemistry and Ecology 29* (7): 604–619.

Boonsong, K., Piyatiratitivorakul, S., and Patanaponpaiboon, P. (2003). Potential use of mangrove plantation as constructed wetland for municipal wastewater treatment. *Water Science and Technology 48* (5): 257–266.

Bouillon, S., Borges, A.V., Castañeda-Moya, E. et al. (2008). Mangrove production and carbon sinks: a revision of global budget estimates. *Global Biogeochemical Cycles 22* (2): 1–12.

Buajan, S. and Pumijumnong, N. (2010). Distribution of heavy metals in mangrove sediment at the Tha Chin Estuary, Samut Sakhon Province, Thailand. *Applied Environmental Research 32* (2): 61–77.

Buccolieri, A., Buccolieri, G., Cardellicchio, N. et al. (2006). Heavy metals in marine sediments of Taranto Gulf (Ionian Sea, southern Italy). *Marine Chemistry 99* (1-4): 227–235.

Bunt, J.S. (1992). How can fragile ecosystems best be conserved. In: *Use and Misuse of the Sea Floor* (eds. K.J. Hsu and J. Thiede), 229–242). (Dahlem Workshop reports environmental science report 11). Chichester: Wiley.

Cadena, J.C.F., Andrade, S., Silva-Coello, C.L., and De la Iglesia, R. (2014). Heavy metal concentration in mangrove surface sediments from the north-west coast of South America. *Marine Pollution Bulletin 82* (1-2): 221–226.

Chatterjee, M., Massolo, S., Sarkar, S.K. et al. (2009). An assessment of trace element contamination in intertidal sediment cores of Sunderban mangrove wetland, India for evaluating sediment quality guidelines. *Environmental Monitoring and Assessment 150* (1-4): 307.

Che, R.O. (1999). Concentration of 7 heavy metals in sediments and mangrove root samples from Mai Po, Hong Kong. *Marine Pollution Bulletin 39* (1-12): 269–279.

Christophoridis, C., Dedepsidis, D., and Fytianos, K. (2009). Occurrence and distribution of selected heavy metals in the surface sediments of Thermaikos Gulf, N. Greece. Assessment using pollution indicators. *Journal of Hazardous Materials 168* (2-3): 1082–1091.

Chu, H.Y., Chen, N.C., Yeung, M.C. et al. (1998). Tide-tank system simulating mangrove wetland for removal of nutrients and heavy metals from wastewater. *Water Science and Technology 38* (1): 361–368.

Cox, M.E. and Preda, M. (2005). Trace metal distribution within marine and estuarine sediments of western Moreton Bay, Queensland, Australia: relation to land use and setting. *Geographical Research 43* (2): 173–193.

Cuong, D.T. and Obbard, J.P. (2006). Metal speciation in coastal marine sediments from Singapore using a modified BCR-sequential extraction procedure. *Applied Geochemistry 21* (8): 1335–1346.

Cuong, D.T., Bayen, S., Wurl, O. et al. (2005). Heavy metal contamination in mangrove habitats of Singapore. *Marine Pollution Bulletin 50* (12): 1732–1738.

De Lacerda, L.D., Carvalho, C.E.V., Tanizaki, K.F. et al. (1993). The biogeochemistry and trace metals distribution of mangrove rhizospheres. *Biotropica*: 252–257.

Defew, L.H., Mair, J.M., and Guzman, H.M. (2005). An assessment of metal contamination in mangrove sediments and leaves from Punta Mala Bay, Pacific Panama. *Marine Pollution Bulletin 50* (5): 547–552.

Dittmar, T., Hertkorn, N., Kattner, G., and Lara, R.J. (2006). Mangroves, a major source of dissolved organic carbon to the oceans. *Global Biogeochemical Cycles 20* (1): 1–7.

Dsikowitzky, L., Nordhaus, I., Jennerjahn, T.C. et al. (2011). Anthropogenic organic contaminants in water, sediments and benthic organisms of the mangrove-fringed Segara Anakan Lagoon, Java, Indonesia. *Marine Pollution Bulletin 62* (4): 851–862.

Essien, J.P., Antai, S.P., and Olajire, A.A. (2009). Distribution, seasonal variations and ecotoxicological significance of heavy metals in sediments of cross river estuary mangrove swamp. *Water, Air, and Soil Pollution 197* (1-4): 91–105.

Fenxia, Y. and Li, Y. (2009). Enhancement of nitrogen removal in towery hybrid constructed wetland to treat domestic wastewater for small rural communities. *Ecological Engineering 35* (7): 1043–1050.

Food and Fertilizer Technology Center (FFTC). (2009). New solutions to soil pollution and distribution, bioavailability and management of heavy metals. http://www.agnet.org/library/ac/2007c/ (accessed 15 August 2011).

Furukawa, K., Wolanski, E., and Mueller, H. (1997). Currents and sediment transport in mangrove forests. *Estuarine, Coastal and Shelf Science 44*: 301–310.

Gambrell, R.P. (1994). Trace and toxic metals in wetlands – a review. *Journal of Environmental Quality 23* (5): 883–891.

Gao, S., Luo, T.C., Zhang, B.R. et al. (1998). Chemical composition of the continental crust as revealed by studies in East China. *Geochimica et Cosmochimica Acta 62* (11): 1959–1975.

Ghosh, P.B., Singh, B.N., Chakrabarty, C. et al. (1990). Mangrove Litter Production in a Tidal Creek of Lothian Island of Sundarbans, India. *Indian Journal of Marine Science*.

Gonneea, M.E., Paytan, A., and Herrera-Silveira, J.A. (2004). Tracing organic matter sources and carbon burial in mangrove sediments over the past 160 years. *Estuarine, Coastal and Shelf Science 61* (2): 211–227.

Gonzalez-Mendoza, D., Moreno, A.Q., and Zapata-Perez, O. (2007). Coordinated responses of phytochelatin synthase and metallothionein genes in black mangrove, Avicennia germinans, exposed to cadmium and copper. *Aquatic Toxicology 83* (4): 306–314.

Goutam, K. and Ramanathan, A.L. (2013). Geochemical assessment of heavy metal contamination in mangrove ecosystem: a Brief overview. *International Research Journal of Environmental Sciences 2* (3): 62–66.

Gueiros, B.B., Machado, W., Lisboa Filho, S.D., and Lacerda, L.D. (2003). Manganese behavior at the sediment-water interface in a mangrove dominated area in Sepetiba Bay, SE Brazil. *Journal of Coastal Research*: 550–559.

Gupta, S. and Chakrabarti, S.K. (2013). Effect of heavy metals on different anatomical structures of Bruguiera sexangula. *International Journal of Bio-resource and Stress Management 4* (4): 605–609.

Guzmán, H.M. and Jiménez, C.E. (1992). Contamination of coral reefs by heavy metals along the Caribbean coast of Central America (Costa Rica and Panama). *Marine Pollution Bulletin 24* (11): 554–561.

Hadad, H.R., Maine, M.A., and Bonetto, C.A. (2006). Macrophyte growth in a pilot-scale constructed wetland for industrial wastewater treatment. *Chemosphere 63* (10): 1744–1753.

Harbison, P.A.T. (1986). Mangrove muds – a sink and a source for trace metals. *Marine Pollution Bulletin 17* (6): 246–250.

Hart, B.T. (1982). Uptake of trace metals by sediments and suspended particulates: a review. In: *Sediment/Freshwater Interaction* (ed. P.G. Sly), 299–313. Dordrecht: Springer.

Harvey, J.K. (2011). Water encyclopedia – science issues, pollution sources: point and non-point. http://www.waterencyclopedia.com/Po-Re/Pollution-Sources-Point-and-Nonpoint.html.

He, P., Xu, S., Zhang, H. et al. (2008). Bioremediation efficiency in the removal of dissolved inorganic nutrients by the red seaweed, Porphyra yezoensis, cultivated in the open sea. *Water Research 42* (4-5): 1281–1289.

He, B., Li, R., Chai, M., and Qiu, G. (2014). Threat of heavy metal contamination in eight mangrove plants from the Futian mangrove forest, China. *Environmental Geochemistry and Health 36* (3): 467–476.

Herrera-Silveira, J.A. and Morales-Ojeda, S.M. (2009). Evaluation of the health status of a coastal ecosystem in southeast Mexico: assessment of water quality, phytoplankton and submerged aquatic vegetation. *Marine Pollution Bulletin 59* (1-3): 72–86.

Hoff, R., Hensel, P. et al. (2002). *Oil Spills in Mangroves, National Oceanic and Atmospheric Administration*. Washington: NOAA Ocean Service, Office of Response and Restoration.

Hong, Y.W., Yuan, D.X., Liao, D., and Liu, B.M. (2009). Accumulation of polycyclic aromatic hydrocarbons from contaminated soil by Kandelia candel. *Acta Oceanologica Sinica 28* (6): 24–29.

Huang, L., Zhuo, J., Guo, W. et al. (2013). Tracing organic matter removal in polluted coastal waters via floating bed phytoremediation. *Marine Pollution Bulletin 71* (1-2): 74–82.

Imelda, J. and Chandrika, V. (2000). Seasonal variations of sediment phenolics and aerobic hetrotrophs in mangrove swamps. *Indian Journal of Marine Sciences 29*: 52–56.

ITRC (2003). *Technical and Regulatory Guidance Document for Constructed Treatment Wetlands*. New Jersey: The Interstate Technology & Regulatory Council Wetlands Team.

Jonathan, M.P., Sarkar, S.K., Roy, P.D. et al. (2010). Acid leachable trace metals in sediment cores from Sunderban Mangrove Wetland, India: an approach towards regular monitoring. *Ecotoxicology 19* (2): 405–418.

Kamaruzzaman, B.Y., Ong, M.C., Jalal, K.C.A. et al. (2009). Accumulation of lead and copper in Rhizophora apiculata from Setiu mangrove forest, Terengganu, Malaysia. *Journal of Environmental Biology 30* (5): 821.

Kannappan, T., Shanmugavelu, M., and Karthikeyan, M.M. (2012). Concentration on heavy metals in sediments and mangroves from Manakudy Estuary (South West Coast of India). *European Journal of Biological Sciences 4*: 109–113.

Kathiresan, K., Saravanakumar, K., and Mullai, P. (2014). Bioaccumulation of trace elements by Avicennia marina. *Journal of Coastal Life Medicine 2* (11): 888–894.

Ke, L., Wang, W.Q., Wong, T.W. et al. (2003). Removal of pyrene from contaminated sediments by mangrove microcosms. *Chemosphere 51* (1): 25–34.

Kehrig, H.A., Pinto, F.N., Moreira, I., and Malm, O. (2003). Heavy metals and methylmercury in a tropical coastal estuary and a mangrove in Brazil. *Organic Geochemistry 34* (5): 661–669.

Krishna, I.M. and Manickam, V. (2017). *Environmental Management: Science and Engineering for Industry*. Oxford, UK: Butterworth-Heinemann.

Krupadam, R.J., Sarin, R., and Anjaneyulu, Y. (2003). Distribution of trace metals and organic matter in the sediments of Godavari estuary of Kakinada bay, East Coast of India. *Water, Air, and Soil Pollution 150* (1-4): 299–318.

Kumari, P., Singh, J.K., and Pathak, B. (2020). Potential contribution of multifunctional mangrove resources and its conservation. In: *Biotechnological Utilization of Mangrove Resources* (eds. J.K. Patra, R.R. Mishra and H. Thatoi), 1–26. Academic Press.

Lacerda, L.D. (1998). *Biogeochemistry of Trace Metals and Diffuse Pollution in Mangrove Ecosystems*, 63. Okinawa: International Society for Mangrove Ecosystems.

Lacerda, L.D. and Abrao, J.J. (1984). Metal bioaccumulation by mangrove and salt marsh intertidal sediment. *Revista Brasileira de Botânica* (7): 49–58.

Lacerda, L.D., Martinelli, L.A., Rezende, C.E. et al. (1988). The fate of trace metals in suspended matter in a mangrove creek during a tidal cycle. *Science of the Total Environment 75* (2-3): 169–180.

Lewis, M., Pryor, R., and Wilking, L. (2011). Fate and effects of anthropogenic chemicals in mangrove ecosystems: a review. *Environmental Pollution 159* (10): 2328–2346.

Lian, Y., Xu, J., Lin, P. et al. (1999). Five heavy metals in propagules of ten mangrove species of China. *Journal of Wood Science 45* (4): 343–347.

Lin, P., Wenjian, Z., and Zhenji, L. (1997). Distribution and accumulation of heavy metals in Avicennia marina community in Shenzhen, China. *Journal of Environmental Sciences (China) 9* (4): 472–479.

Lina, L.F. and Nayak, G.N. (2012). Heavy metals contamination in mudflat and mangrove sediments (Mumbai, India). *Chemistry and Ecology 28* (5): 435–455.

Liu, J., Wu, H., Feng, J. et al. (2014). Heavy metal contamination and ecological risk assessments in the sediments and zoobenthos of selected mangrove ecosystems, South China. *Catena 119*: 136–142.

Long, E.R. (1992). Ranges in chemical concentrations in sediments associated with adverse biological effects. *Marine Pollution Bulletin 24* (1): 38–45.

Lu, X., Kruatrachue, M., Pokethitiyook, P., and Homyok, K. (2004). Removal of cadmium and zinc by water hyacinth, Eichhornia crassipes. *Science Asia 30* (93): 103.

Luo, W., Lu, Y., Wang, T. et al. (2010). Ecological risk assessment of arsenic and metals in sediments of coastal areas of northern Bohai and Yellow Seas, China. *Ambio 39* (5-6): 367–375.

MacFarlane, G.R. (2002). Leaf biochemical parameters in Avicennia marina (Forsk.) Vierh as potential biomarkers of heavy metal stress in estuarine ecosystems. *Marine Pollution Bulletin 44* (3): 244–256.

MacFarlane, G.R. and Burchett, M.D. (2002). Toxicity, growth and accumulation relationships of copper, lead and zinc in the grey mangrove Avicennia marina (Forsk.) Vierh. *Marine Environmental Research 54* (1): 65–84.

MacFarlane, G.R., Pulkownik, A., and Burchett, M.D. (2003). Accumulation and distribution of heavy metals in the grey mangrove, Avicennia marina (Forsk.) Vierh.: biological indication potential. *Environmental Pollution 123* (1): 139–151.

MacFarlane, G.R., Koller, C.E., and Blomberg, S.P. (2007). Accumulation and partitioning of heavy metals in mangroves: a synthesis of field-based studies. *Chemosphere 69* (9): 1454–1464.

Mackey, A.P. and Hodgkinson, M.C. (1995). Concentrations and spatial distribution of trace metals in mangrove sediments from the Brisbane River, Australia. *Environmental Pollution 90* (2): 181–186.

Marchand, C., Lallier-Vergès, E., Baltzer, F. et al. (2006). Heavy metals distribution in mangrove sediments along the mobile coastline of French Guiana. *Marine Chemistry 98* (1): 1–17.

Marchand, C., Allenbach, M., and Lallier-Vergès, E. (2011). Relationships between heavy metals distribution and organic matter cycling in mangrove sediments (Conception Bay, New Caledonia). *Geoderma 160* (3-4): 444–456.

Markert, B. (1993). Plant as biomonitors/Indicators for heavy metals in the terrestrial environment. Weinheim VCH. Press: 670. White poplar (populus alba) as a biomonitor of trace elements in contaminated riparion forest. *Environmental Pollution 132*: 145–155.

Marmiroli, N., Marmiroli, M., and Maestri, E. (2006). Phytoremediation and phytotechnologies: a review for the present and the future. In: *Soil and Water Pollution Monitoring, Protection and Remediation* (eds. I. Twardowska, H.E. Allen and M.H. Haggblom), 403–416. Netherland: *Springer*.

Maslennikova, S., Larina, N., and Larin, S. (2012). The effect of sediment grain size on heavy metal content. *Lakes Reservoirs and Ponds 6* (1): 43–54.

Meagher, R.B. (2000). Phytoremediation of toxic elemental and organic pollutants. *Current Opinion in Plant Biology 3* (2): 153–162.

Mejías, C.L., Musa, J.C., and Otero, J. (2013). Exploratory evaluation of retranslocation and bioconcentration of heavy metals in three species of mangrove at Las Cucharillas marsh, Puerto Rico. *Journal of Tropical Life Science 3* (1): 14–22.

Melville, F. and Burchett, M. (2002). Genetic variation in Avicennia marina in three estuaries of Sydney (Australia) and implications for rehabilitation and management. *Marine Pollution Bulletin 44* (6): 469–479.

Mitsch, W.J. and Jørgensen, S.E. (2003). Ecological engineering: a field whose time has come. *Ecological Engineering 20* (5): 363–377.

Mohd, S.N., Majid, N.M., Shazili, N.A.M., and Abdu, A. (2013). Assessment of Melaleuca cajuputi as heavy metals phytoremediator for sewage sludge contaminated soil. *American Journal of Applied Sciences* 10 (9): 1087.

Moore, M.N., Depledge, M.H., Readman, J.W., and Leonard, D.P. (2004). An integrated biomarker-based strategy for ecotoxicological evaluation of risk in environmental management. *Mutation Research, Fundamental and Molecular Mechanisms of Mutagenesis 552* (1-2): 247–268.

Moorthy, P. and Kathiresan, K. (1998). UV-B induced alterations in composition of thylakoid membrane and amino acids in leaves of Rhizophora apiculata Blume. *Photosynthetica 35* (3): 321–328.

Moreira, I.T.A, Freitas, P.F., Nascimento, R.S.A. et al. (2010a). Selection of species plant mangrove for assessment phytoremediation of contaminated sediments by in oil and derivatives. *Rio Oil & Gas Expo and Conference, IBP2899_10.*

Moreira, I.T.A, Martins, C.M.S., Oliveira, M.C. et al. (2010b). Detailed and confirmatory investigation in an area with contaminated sediment, in order for the application of phytoremediation. *XII Workshop and Congress Geochemistry Organic American Latin Association (ALAGO).*

Moreira, I.T., Oliveira, O.M., Triguis, J.A. et al. (2011). Phytoremediation using Rizophora mangle L. in mangrove sediments contaminated by persistent total petroleum hydrocarbons (TPH's). *Microchemical Journal 99* (2): 376–382.

Moreira, I.T., Oliveira, O.M., Triguis, J.A. et al. (2013). Phytoremediation in mangrove sediments impacted by persistent total petroleum hydrocarbons (TPH's) using Avicennia schaueriana. *Marine Pollution Bulletin 67* (1-2): 130–136.

Morillo, J., Usero, J., and Gracia, I. (2004). Heavy metal distribution in marine sediments from the southwest coast of Spain. *Chemosphere 55* (3): 431–442.

Nazli, M.F. and Hashim, N.R. (2010). Heavy metal concentrations in an important mangrove species, Sonneratia caseolaris, in Peninsular Malaysia. *Environment Asia 3* (Special Issue): 50–55.

Nirmal Kurmar, J.I., Sajish, R.P., Nirmal Kurmar, R. et al. (2011). An assessment of the accumulation potential of Pb, Zn and Cd by Avicennia marina (Forssk.) Vierh. in Vamleshwar mangroves, Gujarat, India. *Notulae Scientia Biologicae 3* (1): 36–40.

Nissenbaum, A. and Swaine, D.J. (1976). Organic matter-metal interactions in recent sediments: the role of humic substances. *Geochimica et Cosmochimica Acta 40* (7): 809–816.

Nobi, E.P., Dilipan, E., Thangaradjou, T. et al. (2010). Geochemical and geo-statistical assessment of heavy metal concentration in the sediments of different coastal ecosystems of Andaman Islands, India. *Estuarine, Coastal and Shelf Science 87* (2): 253–264.

Noegrohati, S. and Hadi, S. (2008). Fate and behavior of organochlorine pesticides in the Indonesian tropical climate: a study in the Segara Anakan estuarine ecosystem. *CLEAN–Soil, Air, Water 36* (9): 767–774.

Nriagu, J.O. and Pacyna, J.M. (1988). Quantitative assessment of worldwide contamination of air, water and soils by trace metals. *Nature 333* (6169): 134–139.

Okbah, M.A., Shata, M.A., and Shridah, M.A. (2005). Geochemical forms of trace metals in mangrove sediments – Red Sea (Egypt). *Chemistry and Ecology 21*: 23–36.

Olajire, A.A., Altenburger, R., Küster, E., and Brack, W. (2005). Chemical and ecotoxicological assessment of polycyclic aromatic hydrocarbon – contaminated sediments of the Niger Delta, Southern Nigeria. *Science of the Total Environment 340* (1-3): 123–136.

Öncel, I., Keleş, Y., and Üstün, A.S. (2000). Interactive effects of temperature and heavy metal stress on the growth and some biochemical compounds in wheat seedlings. *Environmental Pollution 107* (3): 315–320.

Pahalawattaarachchi, V., Purushothamam, C.S., and Vennila, A. (2009). Metal phytoremediation potential of Rhizophora mucronata (Lam.). *Indian Journal of Marine Sciences 38* (2): 178–183.

Pakzadtoochaei, S. (2013). Comparing the concentration of heavy metals (Cd, Cu, Ni and Zn) in the sediment and different tissues of Avicenna marina in Iranian coasts of the Oman sea. *International Research Journal of Applied and Basic Sciences 4* (3): 561–569.

Pan, K. and Wang, W.X. (2012). Trace metal contamination in estuarine and coastal environments in China. *Science of the Total Environment 421*: 3–16.

Panda, S.S., Chaturvedi, N., Dhal, N.K., and Rout, N.C. (2013). An assessment of heavy metal accumulation in mangrove species of Bhitarkanika, Odisha, India. *Research in Plant Biology 3* (6): 1–5.

Parvaresh, H., Abedi, Z., Farshchi, P. et al. (2011). Bioavailability and concentration of heavy metals in the sediments and leaves of grey mangrove, Avicennia marina (Forsk.) Vierh, in Sirik Azini Creek, Iran. *Biological Trace Element Research 143* (2): 1121–1130.

Patel Pratik, A. and Dharaiya Nishith, A. (2014). Can a sub surface flow wetland be an eco tool for the treatment of detergent based effluent? *Life Sciences 4* (5): 313–314.

Pawar, P.R. (2013). Monitoring of impact of anthropogenic inputs on water quality of mangrove ecosystem of Uran, Navi Mumbai, west coast of India. *Marine Pollution Bulletin 75* (1-2): 291–300.

Paz-Alberto, A.M., Sigua, G.C., Baui, B.G., and Prudente, J.A. (2007). Phytoextraction of lead-contaminated soil using vetivergrass (*Vetiveria zizanioides* L.), cogongrass (*Imperata cylindrica* L.) and carabaograss (*Paspalum conjugatum* L.). *Environmental Science and Pollution Research International 14* (7): 498–504.

Paz-Alberto, A.M., De Dios, M.J.J., Alberto, R.T., and Sigua, G.C. (2011). Assessing phytoremediation potentials of selected tropical plants for acrylamide. *Journal of Soils and Sediments 11* (7): 1190–1198.

Paz-Alberto, A.M., Celestino, A.B., and Sigua, G.C. (2014). Phytoremediation of Pb in the sediment of a mangrove ecosystem. *Journal of Soils and Sediments 14* (1): 251–258.

Pekey, H. (2006). Heavy metal pollution assessment in sediments of the Izmit Bay, Turkey. *Environmental Monitoring and Assessment 123* (1-3): 219–231.

Perdomo, L., Ensminger, I., Fernanda Espinosa, L. et al. (1999). The mangrove ecosystem of the Ciénaga Grande de Santa Marta (Colombia): observations on regeneration and trace metals in sediment. *Marine Pollution Bulletin 37* (8-12): 393–403.

Peters, E.C., Gassman, N.J., Firman, J.C. et al. (1997). Ecotoxicology of tropical marine ecosystems. *Environmental Toxicology and Chemistry 16*: 12–40.

Prabhahar, C., Saleshrani, K., Tharmaraj, K., and Vellaiyan, M. (2012). Heavy metal pollution in mangrove region: a review. *International Journal of Pharmaceutical and Biological Archives 3* (3): 513–518.

Primavera, J.H., Sadaba, R.B., Lebata, M.J.H.L., and Altamirano, J.P. (2004). *Handbook of Mangroves in Philippines*. Iloilo, Philippines: SEAFDEC Aquaculture Department.

Pumijumnong, N. and Uppadit, B. (2012). Accumulation of heavy metals in mangrove sediments of chumphon province, Thailand. *Applied Environmental Research 34* (2): 21–38.

Qiu, Y.W., Yu, K.F., Zhang, G., and Wang, W.X. (2011). Accumulation and partitioning of seven trace metals in mangroves and sediment cores from three estuarine wetlands of Hainan Island, China. *Journal of Hazardous Materials 190* (1-3): 631–638.

Rahman, M.M., Chongling, Y., Rahman, M.M., and Islam, K.S. (2009). Accumulation, distribution and toxicological effects induced by chromium on the development of mangrove plant Kandelia candel (L.) Druce. *Ambiente & Água-An Interdisciplinary Journal of Applied Science 4* (1): 6–19.

Rai, P.K. (2008). Heavy metal pollution in aquatic ecosystems and its phytoremediation using wetland plants: an ecosustainable approach. *International Journal of Phytoremediation 10* (2): 133–160.

Raj, S.M. and Jayaprakash, M. (2008). Distribution and enrichment of trace metals in marine sediments of Bay of Bengal, off Ennore, south-east coast of India. *Environmental Geology 56* (1): 207–217.

Ramanathan, A.L., Subramanian, V., Ramesh, R. et al. (1999). Environmental geochemistry of the Pichavaram mangrove ecosystem (tropical), southeast coast of India. *Environmental Geology 37* (3): 223–233.

Ramos e Silva, C.A., da Silva, A.P., and de Oliveira, S.R. (2006). Concentration, stock and transport rate of heavy metals in a tropical red mangrove, Natal, Brazil. *Marine Chemistry 99* (1-4): 2–11.

Ranjan, R.K., Ramanathan, A.L., Singh, G., and Chidambaram, S. (2008). Assessment of metal enrichments in tsunamigenic sediments of Pichavaram mangroves, southeast coast of India. *Environmental Monitoring and Assessment 147* (1-3): 389–411.

Ravikumar, S., Williams, G.P., Shanthy, S. et al. (2007). Effect of heavy metals (Hg and Zn) on the growth and phosphate solubilising activity in halophilic phosphobacteria isolated from Manakudi mangrove. *Journal of Environmental Biology 28* (1): 109–114.

Ray, A.K., Tripathy, S.C., Patra, S., and Sarma, V.V. (2006). Assessment of Godavari estuarine mangrove ecosystem through trace metal studies. *Environment International 32* (2): 219–223.

Riffaldi, R., Levi-Minzi, R., Saviozzi, A., and Tropea, M. (1983). Sorption and release of cadmium by some sewage sludges. *Journal of Environmental Quality 12* (2): 253–256.

Saenger, P. and McConchie, D. (2004). Heavy metals in mangroves: methodology, monitoring and management. *Envis Forest Bulletin 4*: 52.

Saiyood, S., Inthorn, D., Vangnai, A.S., and Thiravetyan, P. (2013). Phytoremediation of bisphenol A and total dissolved solids by the mangrove plant, Bruguiera gymnorhiza. *International Journal of Phytoremediation 15* (5): 427–438.

Sakineh, L., Gunale, V.R., and Rajurkar, N.S. (2013). Petroleum hydrocarbons pollution in soil and its bioaccumulation in mangrove species, Avicennia marina from Alibaug mangrove ecosystem, Maharashtra, India. *International Journal of Advanced Scientific and Technical Research 2* (2): 1–7.

Sánchez-Arias, L.E., Remolina, D.A., and Alvarez-León, R. (2013). Evaluation of a recovery technique for mangrove soils affected by oil spills, using as indicator plantules of Rhizophora mangle L.(Rhizophoraceae). *Pan-American Journal of Aquatic Sciences 8* (2): 79–88.

Sarangi, R.K., Kathiresan, K., and Subramanian, A.N. (2002). Metal concentration in five mangrove species of the bhitarkanika, Orissa east coast of India. *Indian Journal of Marine Sciences 31* (3): 251–253.

Sayadi, M.H., Kargar, R., Doosti, M.R., and Salehi, H. (2012). Hybrid constructed wetlands for wastewater treatment: a worldwide review. *Proceedings of the International Academy of Ecology and Environmental Sciences* 2 (4): 204.

Schwitzguébel, J.P. (2001). Hype or hope: the potential of phytoremediation as an emerging green technology. *Remediation Journal: The Journal of Environmental Cleanup Costs, Technologies & Techniques 11* (4): 63–78.

Shanmugaarasu, V., Saravanakumar, K., and Kathiresan, K. (2013). Studies on Heavy Metal Accumulation in Mangrove Sediments of Northern Karnataka Northern Karnataka. *Journal of Biotechnological Sciences 1* (2): 146–152.

Shete, A., Gunale, V.R., and Pandit, G.G. (2007). Bioaccumulation of Zn and Pb in Avicennia marina (Forsk.) Vierh. and Sonneratia apetala Buch. Ham. from urban areas of Mumbai (Bombay), India. *Journal of Applied Sciences and Environmental Management* 11 (3): 109–112.

Shete, A., Gunale, V.R., and Pandit, G.G. (2009). Organochlorine pesticides in Avicennia marina from the Mumbai mangroves, India. *Chemosphere 76* (11): 1483–1485.

Shriadah, M.M.A. (1999). Heavy metals in mangrove sediments of the United Arab Emirates shoreline (Arabian Gulf). *Water, Air, and Soil Pollution 116* (3-4): 523–534.

Silva, L.F.F., Machado, W., Lisboa Filho, S.D., and Lacerda, L.D. (2003). Mercury accumulation in sediments of a mangrove ecosystem in SE Brazil. *Water, Air, and Soil Pollution 145* (1-4): 67–77.

Soto-Jiménez, M.F. and Páez-Osuna, F. (2001). Distribution and normalization of heavy metal concentrations in mangrove and lagoonal sediments from Mazatlan Harbor (SE Gulf of California). *Estuarine, Coastal and Shelf Science 53* (3): 259–274.

State Administration for Quality Supervision and Inspection, Quarantine (AQSIQ) (2002). *Marine Sediment Quality Guidelines (GB 18668-2002)*, 1–3. Beijing: *AQSIQ*.

Stottmeister, U., Wießner, A., Kuschk, P. et al. (2003). Effects of plants and microorganisms in constructed wetlands for wastewater treatment. *Biotechnology Advances 22* (1-2): 93–117.

Suresh, B. and Ravishankar, G.A. (2004). Phytoremediation – a novel and promising approach for environmental clean-up. *Critical Reviews in Biotechnology 24* (2-3): 97–124.

Swaminathan, M.S. (1991). Genesis of Workshop (Fore word). *Proceedings of the formulation workshop for establishing a global network of mangrove genetic resource centeres for adaptation to sea level rise.* (Sanjay, V., Deshmugh, and Rajeswari Mahalingarn., Eds.) 15–19 January, Madras, India. Proceedings No.2, CRSARD. Madras. India.

Tam, N.F. and Wong, Y.S. (1995). Mangrove soils as sinks for wastewater-borne pollutants. *Hydrobiologia 295* (1-3): 231–241.

Tam, N.F.Y. and Wong, Y.S. (1996). Retention of wastewater-borne nitrogen and phosphorus in mangrove soils. *Environmental Technology 17* (8): 851–859.

Tam, N.F. and Wong, Y.S. (1997). Accumulation and distribution of heavy metals in a simulated mangrove system treated with sewage. *Asia-Pacific Conference on Science and Management of Coastal Environment* (pp. 67–75). Springer, Dordrecht.

Tam, N.F.Y. and Wong, Y.S. (2000). Spatial variation of heavy metals in surface sediments of Hong Kong mangrove swamps. *Environmental Pollution 110* (2): 195–205.

Tam, N.F.Y. and Wong, Y.S. (2008). Effectiveness of bacterial inoculum and mangrove plants on remediation of sediment contaminated with polycyclic aromatic hydrocarbons. *Marine Pollution Bulletin 57* (6-12): 716–726.

Thomas, G. and Fernandez, T.V. (1997). Incidence of heavy metals in the mangrove flora and sediments in Kerala, India. *Asia-Pacific Conference on Science and Management of Coastal Environment* (pp. 77–87). Springer, Dordrecht.

Trefry, J.H. and Presley, B.J. (1976). Heavy metal transport from the Mississippi River to the Gulf of Mexico. *Marine Pollutant Transfer*: 39–76.

Uera, R.B., Paz-Alberto, A.M., and Sigua, G.C. (2007). Phytoremediation potentials of selected tropical plants for ethidium bromide. *Environmental Science and Pollution Research International 14* (7): 505–509.

Untawale, A.G., Wafar, S., and Bhosale, N.B. (1980). Seasonal variation in heavy metal concentration in mangrove foliage. *Oceanography 13* (3): 215–223.

United Nations Environment Programme (UNEP). (2003) IETC Freshwater Management Series 7. United Nations Environment Programme, accessed at: http://www.unep.or.jp/Ietc/Publications/Freshwater/FMS7/copyright.asp

Valipour, A., Raman, V.K., and Ghole, V.S. (2009). A new approach in wetland systems for domestic wastewater treatment using Phragmites sp. *Ecological Engineering 35* (12): 1797–1803.

Vijay, V., Birader, R.S., Inamdar, A.B. et al. (2005). Mangrove mapping and change detection around Mumbai (Bombay) using remotely sensed data. *Indian Journal of Marine Sciences 34* (3): 310–315.

Vymazal, J. (2010). Constructed Wetlands for wastewater treatment. *Water 2*: 530–549.

Wang, S.H., Yang, Z.M., and Xu, L.L. (2003a). Mechanisms of copper toxicity and resistance of plants. *Ecology and Environment 3*: 336–341.

Wang, B., Zan, Q., Zhang, W., and Wang, Y. (2003b). Accumulation and cycle of heavy metals in Sonneratia apetala and S. caseolaris mangrove community at Futian of Shenzhen, China. *Marine Science Bulletin 5* (1): 59–68.

Wang, M., Zhang, J., Tu, Z. et al. (2010). Maintenance of estuarine water quality by mangroves occurs during flood periods: a case study of a subtropical mangrove wetland. *Marine Pollution Bulletin 60* (11): 2154–2160.

Wang, H., Zhang, H., and Cai, G. (2011). An application of phytoremediation to river pollution remediation. *Procedia Environmental Sciences 10*: 1904–1907.

Wang, Y., Fang, L., Lin, L. et al. (2014). Effects of low molecular-weight organic acids and dehydrogenase activity in rhizosphere sediments of mangrove plants on phytoremediation of polycyclic aromatic hydrocarbons. *Chemosphere 99*: 152–159.

Wedepohl, K.H. (1995). The composition of the continental crust. *Geochimica et Cosmochimica Acta 59* (7): 1217–1232.

Weis, J.S. and Weis, P. (2004). Metal uptake, transport and release by wetland plants: implications for phytoremediation and restoration. *Environment International 30* (5): 685–700.

Wu, Y., Chung, A., Tam, N.F.Y. et al. (2008). Constructed mangrove wetland as secondary treatment system for municipal wastewater. *Ecological Engineering 34* (2): 137–146.

Wu, Q., Tam, N.F., Leung, J.Y. et al. (2014). Ecological risk and pollution history of heavy metals in Nansha mangrove, South China. *Ecotoxicology and Environmental Safety 104*: 143–151.

Yang, Q., Tam, N.F.Y., Wong, Y.S. et al. (2008). Potential use of mangroves as constructed wetland for municipal sewage treatment in Futian, Shenzhen, China. *Marine Pollution Bulletin 57* (6-12): 735–743.

Yunus, K., Yusuf, N.M., Shazili, N.M. et al. (2011). Heavy metal concentration in the surface sediment of Tanjung Lumpur mangrove forest, Kuantan, Malaysia. *Sains Malaysiana 40* (2): 89–92.

Zhang, Z.Y. (2010). *The Study on Nutrient Purification Effects and Mechanisms of Ecological Floating Bed for Eutrophic Lagoon*. Xiamen, China: Xiamen University.

Zhang, J.E., Liu, J.L., Ouyang, Y. et al. (2010). Removal of nutrients and heavy metals from wastewater with mangrove Sonneratia apetala Buch-Ham. *Ecological Engineering 36* (6): 807–812.

Zhang, J.E., Liu, J.L., Ouyang, Y. et al. (2011). Physiological responses of mangrove Sonneratia apetala Buch-Ham plant to wastewater nutrients and heavy metals. *International Journal of Phytoremediation 13* (5): 456–464.

Zhang, Z.W., Xu, X.R., Sun, Y.X. et al. (2014). Heavy metal and organic contaminants in mangrove ecosystems of China: a review. *Environmental Science and Pollution Research 21* (20): 11938–11950.

Zhao, B., Zhou, Y.W., and Chen, G.Z. (2012). The effect of mangrove reforestation on the accumulation of PCBs in sediment from different habitats in Guangdong, China. *Marine Pollution Bulletin 64* (8): 1614–1619.

Zheng, W.J., Zheng, F.Z., Lian, Y.W., and Lin, P. (1996a). Accumulation and dynamics of Cu. Pb, Zn and Mn elements in Kandelia candel (L.) Druce mangrove community of Jiulong river estuary of Fujian. *Acta Botanica Sinica 38*: 227–233.

Zheng, W.J., Zheng, F.Z., Lian, Y.W., and Lin, P. (1996b). Accumulation and dynamics of cadmium, chromium and nickel in Kandelia candel (L.) Drucc mangrove community of Jiulong river estuary of Fujian. *Acta Scientiae Circumstantiac 16*: 232–237.

Zheng, W.J., Chen, X.Y., and Lin, P. (1997). Accumulation and biological cycling of heavy metal elements in Rhizophora stylosa mangroves in Yingluo Bay, China. *Marine Ecology Progress Series 159*: 293–301.

12

Mangrove Forests: Distribution, Species Diversity, Roles, Threats and Conservation Strategies

Ajay Kumar[1], Thattantavide Anju[1], Vijay Archa[1], Vidya P. Warrier[1,2], Sushil Kumar[3], Ginkuntla Saikiran Goud[1], Arun Kumar Kashyap[4], Shuchita Singh[5], Komal[6], Pardeep Singh[7], Raj Kumar[8], Sanjeev Sharma[8], Arun M. Radhakrishnan[1], and Nirala Ramchiary[9]

[1] *Department of Plant Science, School of Biological Sciences, Central University of Kerala, Kasaragod, Kerala, India*
[2] *Department of Biotechnology, Indian Institute of Technology, Madras, Chennai, Tamil Nadu, India*
[3] *Department of Botany, Government Degree College, Ramnagar, Jammu and Kashmir, India*
[4] *Department of Biotechnology, Government E. Raghavendra Rao PG Science College, Bilaspur, Chhattisgarh, India*
[5] *Lachoo Memorial College of Science and Technology, Jodhpur, Rajasthan, India*
[6] *Department of Higher Education, Government Degree College, Bani, Jammu and Kashmir, India*
[7] *Department of Environmental Studies, PGDAV College, University of Delhi, New Delhi, India*
[8] *Centre for the Study of Regional Development, School of Social Sciences, Jawaharlal Nehru University, New Delhi, India*
[9] *School of Life Sciences, Laboratory of Translational and Evolutionary Genomics, Jawaharlal Nehru University, New Delhi, India*

12.1 Introduction

Mangroves are a group of trees and shrubs belonging to several families of flowering plants (Santisuk 1983, p. 63; Ricklefs and Latham 1993, p. 215). Owing to a large number of unique and special morphological, physiological, and anatomical modifications such as buttress root system, pneumatophores, root ultrafiltration systems, vivipary, salt secretion system, and cell wall composition, they can tolerate extreme saline coastal conditions and are adapted to narrow zone between the land and sea (Kathiresan and Bingham 2001; Dahdouh-Guebas et al. 2007; Liang et al. 2008). Mangroves are also known as tidal forests, marine forests, marsh forests, or ocean rain forests (Kathiresan and Bingham 2001; Naidoo 2016). The term "mangrove," in general, is used to encompass the unique plants belonging to different families having specialized characters that help them grow in a narrow intertidal zone between land and sea (Macnae 1968; Tomlinson 2016; Bibi et al. 2019). The term "mangal" is used for mangrove forest communities including other biotic components such as microbes and fungi, animals and other mangrove-associated plants (Macnae 1968; Bibi et al. 2019). Mangroves are a

Wetlands Conservation: Current Challenges and Future Strategies, First Edition.
Edited by Sanjeev Sharma and Pardeep Singh.

good source of numerous products such as food, honey, medicines (such as steroids, triterpenes, saponins, flavonoids, alkaloids, and tannin), timber and firewood, and services such as recreation, ecotourism, and aesthetics (Othman 1994, p. 277; Mazda et al. 1997; Kathiresan 2012; Guannel et al. 2016; Macreadie et al. 2017, p. 206; Owuor et al. 2019b). Timber from mangroves is used for constructing houses, furniture, house studs, rafters, joists, telegraph poles, walls, bridges, railway sleepers, fish trap poles, paddles and rafts, boats and paddleboards, and fuel (Bandaranayake 1998). Mangroves help in the protection of seagrasses, coral reefs, and shrimp by collecting riverine runoff sediments (Primavera 1998; Valiela and Cole 2002; Duke and Larkum 2008, p. 156). They are an important component of "coastal blue carbon" as they store and sequester large amounts of carbon (Donato et al. 2011; Lee et al. 2014; Alongi 2015; Estrada and Soares 2017; Kelleway et al. 2017; Rogers et al. 2019). Current studies suggest that mangroves sequester higher carbon than the mature tropical rain forests and most of it is stored below ground, suggesting their important roles in combating climate change posed by global warming (Francisco et al. 2018; Sanderman et al. 2018; National Oceanic and Atmospheric Administration 2020).

As per the IUCN red list of threatened species report, 11 out of 73 mangrove species are near threatened (IUCN 2020b) due to several reasons such as excessive exploitation, habitat destruction, mangrove forest encroachments, human settlements, and road and construction activities (Barbier and Sathirathai 2004; Polidoro et al. 2010; Barbier et al. 2011; Costanza et al. 2014). The rising global sea level due to global warming and changing salinity levels also threaten the existence of the mangroves. Despite being one of the most productive, unique, and important ecosystems, the mangroves face a serious existential crisis. Their narrow zone of existence and the shrinking of the habitat raise serious concerns (Polidoro et al. 2010; IUCN 2010). Increasing concerns about the loss of such important ecosystems of the world have shifted the focus of the global community toward taking actions in saving, conserving, and restoring mangrove ecosystems (Carugati et al. 2018). Therefore, conservation and restoration efforts must be taken to rescue the leftover habitats and the remaining species for their tremendous goods and services to the plants and people (Lefebvre and Poulin 1997; Lewis 2005; Giri et al. 2011). The protection of mangrove ecosystems requires the cooperation of local communities, NGOs, and state and central governments along with the expertise of scientists and academics (Farley et al. 2010, p. 36; Romañach et al. 2018). Concerted efforts from various corners of domains are required to protect the unique ecosystems of the world. This chapter provides an outline of the importance of the unique mangrove ecosystems, their diversity, threats, and conservation challenges.

12.2 Mangrove Species Diversity

Based on several characters, mangroves have been classified as true mangroves or obligate mangroves and mangrove associates or semi-mangroves (Wang et al. 2010, 2011; Tomlinson 2016, p. 29). True mangroves are characterized by their occurrence in mangrove forests and not in terrestrial communities, are major components of the mangrove forests possessing special characters for adaptation/survival in the saline environment and mechanisms for the salt exclusion (Tomlinson 2016, p. 29; Quadros and Zimmer 2017). True mangroves are further grouped into major and minor components of mangrove ecosystems (Spalding et al. 2010; Tomlinson 2016, p. 29; Quadros and Zimmer 2017). Spalding et al. (2010) have published an important updated book titled "Atlas of Mangroves" giving detailed maps of mangroves across the globe. They have enlisted a total of 73 true mangrove species and hybrids (Spalding et al. 2010). Mangrove species are genetically distinct but ecologically related (Bibi et al. 2019) with the common property of inhabiting the intertidal region (Duke et al. 1998). True mangrove species and hybrids including major and minor components are distributed across 28 genera belonging to 18 diverse families of flowering plants (Table 12.1). Rhizophoraceae is one of the most important and diverse families of mangroves comprising four genera, i.e. *Bruguiera* (7 species), *Ceriops* (3), *Kandelia* (2), and *Rhizophora* (10). Combretaceae comprises three genera, i.e. *Conocarpus* (1), *Laguncularia* (1), and *Lumnitzera* (3). Malvaceae, Leguminosae, Bignoniaceae, Lythraceae, and Meliaceae comprise two genera each. Malvaceae is represented by *Camptostemon* (2) and *Heritera* (3), Leguminosae by *Cynometra* (1) and *Mora* (1), Bignoniaceae by *Dolichandrone* (1), and *Tabebuia* (1) and Lythraceae by *Pemphis* (1), and *Sonneratia* (9). The family Meliaceae is represented by *Aglaia* (1) and *Xylocarpus* (2). Rest of the families are represented by one genus each, i.e. Acanthaceae by *Acanthus* (2), Pteridaceae by *Acrostichum* (3), Plumbaginaceae by *Aegialitis* (2), Primulaceae by *Aegiceras* (2), Avicenniaceae by *Avicennia* (8), Ebenaceae by *Diospyros* (1), Euphorbiaceae by *Excoecaria* (2), Arecaceae by *Nypa* (1), Myrtaceae by *Osbornia* (1), Tetrameristaceae by *Pelliciera* (1), and Rubiaceae by *Scyphiphora* (1). Diversity of true mangroves and their threat status are shown in Table 12.1. Of the total 73 species and hybrids, three are critically endangered, three are endangered, and seven are vulnerable whereas five are near threatened (Table 12.1, Figure 12.1). Although the majority of species are least concerned, some of them show a decline in their population and face several threats and need special attention for their conservation (Duke et al. 2007; IUCN 2020b; See Table 12.1).

Table 12.1 Diversity of true mangroves and current threat status as per the IUCN Red List of Threatened Species.

S. No.	Name of the species	IUCN status
Genus: *Acanthus* (2), Family: Acanthaceae		
1.	*A. ilicifolius* L.	LC
2.	*A. ebracteatus* Vahl	LC

(*Continued*)

Table 12.1 (Continued)

S. No.	Name of the species	IUCN status
Acrostichum (3), Pteridaceae		
3.	*A. aureum* Cav.	LC
4.	*A. speciosum* (Fée) C. Presl	LC
5.	*A. danaeifolium* (Fée) C. Presl	LC
Aegialitis (3), Plumbaginaceae		
6.	*A. annulata* Kurz.	LC
7.	*A. rotundifolia* Roxb.	NT
Aegiceras (2), Primulaceae		
8.	*A. corniculatum* (L.) Blanco.	LC
9.	*A. floridum* Roem. & Schult.	NT
Aglaia (1), Meliaceae		
10.	*A. cucullata* (Roxb.) Pellegr.	DD
Avicennia (8), Avicenniaceae		
11.	*A. alba* Blume.	LC
12.	*A. bicolor* Standl.	VU
13.	*A. germinans* (L.) Stearn.	LC
14.	*A. integra* N.C.Duke.	VU
15.	*A. marina* (Forssk.) Vierh.	LC
16.	*A. officinalis* L.	LC
17.	*A. rumphiana* Hallier f.	VU
18.	*A. schaueriana* Stapf & Leechm.	LC

Table 12.1 (Continued)

S. No.	Name of the species	IUCN status
Bruguiera (7), Rhizophoraceae		
19.	*B. cylindrica* (L.) Blume.	LC
20.	*B. exaristata* Ding Hou.	LC
21.	*B. gymnorhhiza* (L.) Lam.	LC
22.	*B. hainesii* C.G.Rogers	CR
23.	*B. parviflora* (Roxb.) Wight & Arn.	LC
24.	*B. rhynchopetala* (W.C.Ko) N.C.Duke & X.J.Ge.	Hybrid taxon
25.	*B. sexangula* (Lour.) Poir.	LC
Camptostemon (2), Malvaceae		
26.	*C. schultzii* Mast.	LC
27.	*C. philippinense* (Vidal) Becc.	EN
Ceriops (3), Rhizophoraceae		
28.	*C. australis* (C.T.White) Ballment.	LC
29.	*C. decandra* Ding Hou.	NT
30.	*C. tagal* (Perr.).	LC
Conocarpus (1), Combretaceae		
31.	*C. erectus* L.	LC
Cynometra (2), Leguminosae		
32.	*C. iripa* Kostel	LC
Dolichandrone (1), Bignoniaceae		
33.	*D. spathacea* (L.f.) Seem.	LC
Diospyros (1), Ebenaceae		
34.	*D. littorea* (R.Br.) Kosterm.	NA

(Continued)

Table 12.1 (Continued)

S. No.	Name of the species	IUCN status
Excoecaria (2), Euphorbiaceae		
35.	*E. agallocha* L.	LC
36.	*E. indica* (Willd.) Müll. Arg. (Now synonym of *Shirakiopsis indica (Willd) Esser)*	DD
Heritiera (3), Malvaceae		
37.	*H. fomes* Buch.-Ham.	EN
38.	*H. globosa* Kosterm	EN
39.	*H. littoralis* Aiton.	LC
Kandelia (2), Rhizophoraceae		
40.	*K. candel* (L.) Druce	LC
41.	*K. obovata* Sheue, Liu & Yong.	LC
Laguncularia (1), Combretaceae		
42.	*L. racemosa* Willd.	LC
Lumnitzera (2), Combretaceae		
43.	*L. littorea* (Jack) Voigt.	LC
44.	*L. racemosa* Willd.	LC
45.	*Lumnitzera X rosea*	NA
Mora (1), Leguminosae		
46.	*M. oleifera* (Hemsl.) Ducke	VU
Nypa (1), Arecaceae		
47.	*N. fruticans* Wurmb.	LC
Osbornia (1), Myrtaceae		
48.	*O. octodonta* F.Muell.	LC
Pelliciera (1), Tetrameristaceae		
49.	*P. rhizophorae* Planch. & Triana.	VU

Table 12.1 (Continued)

S. No.	Name of the species	IUCN status
Pemphis (1), Lythraceae		
50.	*P. acidula* J.R. Forst.	LC
Rhizophora (10), Rhizophoraceae		
51.	*R. apiculata* Blume.	LC
52.	*Rhizophora x neocaledonica*	NA
53.	*R. harrisonii* Leechm.	Species name is accepted in the genus *Rhizophora*.
54.	*R. lamarckii* Montrouz.	Species name is accepted in the genus *Rhizophora*.
55.	*R. mangle* Roxb.	LC
56.	*R. mucronata* Poir.	LC
57.	*R. racemosa* G.Mey.	LC
58.	*R. samoensis* (Hochr.) Salvoza.	NT
59.	*R. stylosa* Griff.	LC
60.	*R. selala* (Salvoza) Toml.	Hybrid taxon
Scyphiphora (1), Rubiaceae		
61.	*S. hydrophylacea* C.F. Gaertn.	LC
Sonneratia (9), Lythraceae		
62.	*S. alba* Griff.	LC
63.	*S. apetala* Buch.-Ham.	LC
64.	*S. caseolaris* Druce.	LC
65.	*S. griffithii* Kurz.	CR
66.	*S. gulngai* N.C. Duke.	Hybrid Parentage (*S. alba* × *caseolaris*).
67.	*S. hainanensis* W.C. Ko, E.Y. Chen & W.Y. Chen.	CR

(*Continued*)

Table 12.1 (Continued)

S. No.	Name of the species	IUCN status
68.	*S. lanceolata* Blume.	LC
69.	*S. urama* N.C. Duke	Hybrid taxon (NA on IUCN website)
70.	*S. ovata* Backer.	NT
Tabebuia (1), Bignoniaceae		
71.	*T. palustris* Hemsl.	VU
Xylocarpus (3), Meliaceae		
72.	*X. granatum* J. Koenig.	LC
73.	*X. moluccensis* (Lam.) M. Roem.	LC

Source: From Spalding et al. (2010) and IUCN (2020b). © John Wiley & Sons. Families have changed due to APG IV classification: Primulaceae = Myrsinaceae; Malvaceae = Bombacaceae; Leguminosae = Caesalpiniaceae; Heritiera (3), Malvaceae = Sterculiaceae; Tetrameristaceae = Pellicieraceae; Lythraceae = Sonneratiaceae. 18 families with their respective genera are: 1. Acanthaceae – Acanthus (2), 2. Pteridaceae – Acrostichum (3), 3. Plumbaginaceae – Aegialitis (2), 4. Primulaceae – Aegiceras (2), 5. Meliaceae – Aglaia (1), Xylocarpus (2), 6. Avicenniaceae – Avicennia (8), 7. Rhizophoraceae – Bruguiera (7), Ceriops (3), Kandelia (2), Rhizophora (10), 8. Malvaceae – Camptostemon (2), Heritera (3), 9. Combretaceae – Conocarpus (1), Laguncularia (1), Lumnitzera (3), 10. Leguminosae – Cynometra (1), Mora (1), 11. Bignoniaceae – Dolichandrone (1), Tabebuia (1), 12. Ebenaceae – Diospyros (1), 13. Euphorbiaceae – Excoecaria (2), 14. Arecaceae – Nypa (1), 15. Myrtaceae – Osbornia (1), 16. Tetrameristaceae – Pelliciera (1), 17. Lythraceae – Pemphis (1), Sonneratia (9), 18. Rubiaceae – Scyphiphora (1).

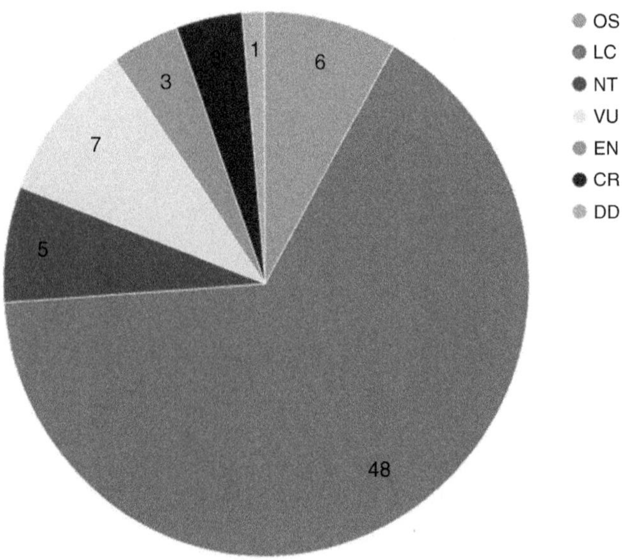

Figure 12.1 Threat status of the 73 true mangrove species and hybrids as per IUCN Red List of threatened species. *Source:* Modified from Spalding et al. (2010). IUCN (2020b). © John Wiley & Sons.

12.3 Geographical Distribution of Mangroves Across the Globe and India

Mangroves are distributed in nearly 118 countries across the globe (Spalding et al. 2010). Continuous efforts are being made to access the distribution of mangroves using various techniques. Giri et al. (2011) have reported an area of 137 760 km^2 under mangrove cover in 118 countries for the year 2000. A more recent estimate shows that mangroves occupy an area of approximately 152 000 km^2 across the world (Yeo 2014). They are present between 30° N and 30° S in the tropical and subtropical areas (Sandilyan and Kathiresan 2012), while the highest percentage of mangroves is available between 5° N and 5° S (Giri et al. 2011). Mangrove distribution is dependent on rainfall, runoff, salinity, and temperature (Jayatissa et al. 2008; Wang et al. 2011; Osland et al. 2017). The higher rainfall and runoff decrease salinity and improve nutrient availability in the estuaries which contribute to increased mangrove growth and productivity (Singh 2020). Several authors suggest that salinity is one of the most important factors determining the distribution of the mangroves (Jayatissa et al. 2008; Wang et al. 2011; Osland et al. 2017). Geographically, mangroves are located in the tropics and subtropics. Among the continents, Asia represents the highest percentage of mangrove area (42%) followed by Africa (21%), North and Central America (15%), and South America (11%). Among the countries having mangrove vegetation, Indonesia represents the highest percentage of land under mangroves (Hamilton and Casey 2016; Yong 2018; Bibi et al. 2019) whereas Southeast Asian countries contribute 33% to global mangroves (Basha 2016, p. 766). Country-wise distribution of the mangroves across the globe is given in Figure 12.2 while Table 12.2 represents the top fifteen countries having the highest mangrove cover.

India accounts for around 3% of the total mangrove cover in South Asia (Forest Survey of India 2019) harboring more than 50% of the world's mangrove species (Ragavan et al. 2016). India represents nearly 46 true mangrove species and hybrids belonging to 14 families and 22 genera, and is distributed along the coastline of nine states and three union territories: Andhra Pradesh, Gujarat, West Bengal, Odisha, Andaman and Nicobar Islands, Daman and Diu, Tamil Nadu, Kerala, Karnataka, Goa, Pondicherry, and Maharashtra (Forest Survey of India 2019; Table 12.3). Mangroves in India span an area of more than 4975 sq km (Forest Survey of India 2019; Ragavan et al. 2019, p. 257; Kumari et al. 2020, p. 1). Of the total mangrove cover in India, 29% occur along the west coast and 58% along the east coast whereas the remaining 13% occur along the Andaman and Nicobar Islands (Kumari et al. 2020, p. 1). State-wise mangrove cover of India is represented in Figure 12.3 and Table 12.3.

12.4 Important Roles of Mangroves

Mangrove forests are one of the most important ecosystems on earth (Sandilyan and Kathiresan 2012; Carugati et al. 2018). Mangroves offer a large number of goods and services to the society in addition to their important ecological roles (Barbier et al. 2011;

Figure 12.2 Extent of the global distribution of mangroves in 118 countries. *Source:* Global Mangrove Alliance Data Portal (2020) and Giri et al. (2011). © John Wiley & Sons.

Table 12.2 Top 15 countries having the maximum extent of mangrove distribution (area in km^2; Giri et al. 2011).

S. No.	Country	Area (km^2)
1.	Ashmore and Cartier Islands	9581.154059
2.	Australia	9581.154059
3.	Bangladesh	4435.329037
4.	Brazil	10521.26464
5.	Burma	5047.38696
6.	Guinea-Bissau	2733.991351
7.	India	3848.495466
8.	Indonesia	27729.48109
9.	Malaysia	5554.206441
10.	Mexico	7260.029787
11.	Mozambique	2941.053557
12.	Nigeria	6224.844022
13.	Papua New Guinea	4715.912505
14.	Philippines	2574.165676
15.	Venezuela	3334.072747

Source: Based on Giri et al. (2011). © John Wiley & Sons.

Table 12.3 State-wise distribution of mangroves in India.

S. No.	State	Area in (km^2)	No. of mangrove species
1.	Andhra Pradesh	404	22
2.	Goa	26	16
3.	Gujarat	1177	15
4.	Karnataka	10	16
5.	Kerala	9	19
6.	Maharashtra	320	22
7.	Odisha	251	34
8.	Tamil Nadu	45	17
9.	West Bengal	2,112	33
10.	A&N Islands	616	38
11.	Daman & Diu	3	4
12.	Puducherry	2	15
Total		**4975**	**46**[a]

[a] Indian mangroves represent 46 true mangrove species (42 species and four natural hybrids) belonging to 14 families and 22 genera (Ragavan et al. 2019)
Source: Forest Survey of India (2019), Ragavan et al. (2019). © John Wiley & Sons.

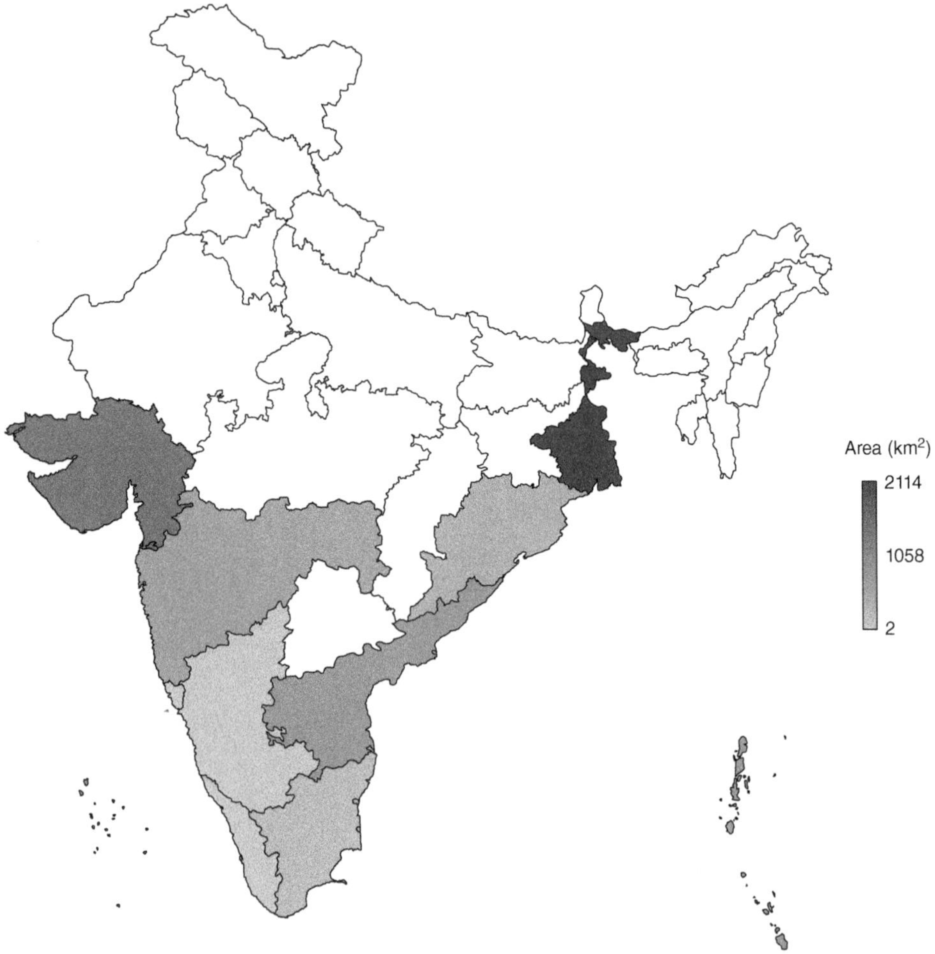

Figure 12.3 Extent of the area under mangroves in India. The mangroves are distributed in the 12 coastal states and Union Territories of India. *Source:* Forest Survey of India (2019) and Ragavan et al. (2019). © John Wiley & Sons.

Mangroves for the future 2012; IUCN 2020a). Mangroves provide goods such as timber, fuelwood, charcoal, and medicines (Uddin et al. 2013). They provide important services such as protection of coastal communities from natural disasters such as tsunami, floods, hurricanes, erosion, and landslides by serving as bioshields (Kathiresan and Rajendran 2005; Duke et al. 2007; Cheong et al. 2013; Sandilyan and Kathiresan 2015), act as habitats for several organisms including microorganisms and serve as breeding and spawning ground for many marine species such as fish, crabs, mollusks, insects, and birds (Teas 1977; Goforth and Thomas 1980; FAO 1994; Upadhyay et al. 2002; Duke et al. 2007; Giri et al. 2011; Lee et al. 2014; Buelow and Sheaves 2015; Mohd-Azlan (2014)). Unique adaptations in the mangrove communities provide many other direct and indirect benefits to the society (Mitra 2020). For example, some mangroves and even their associated flora can accumulate

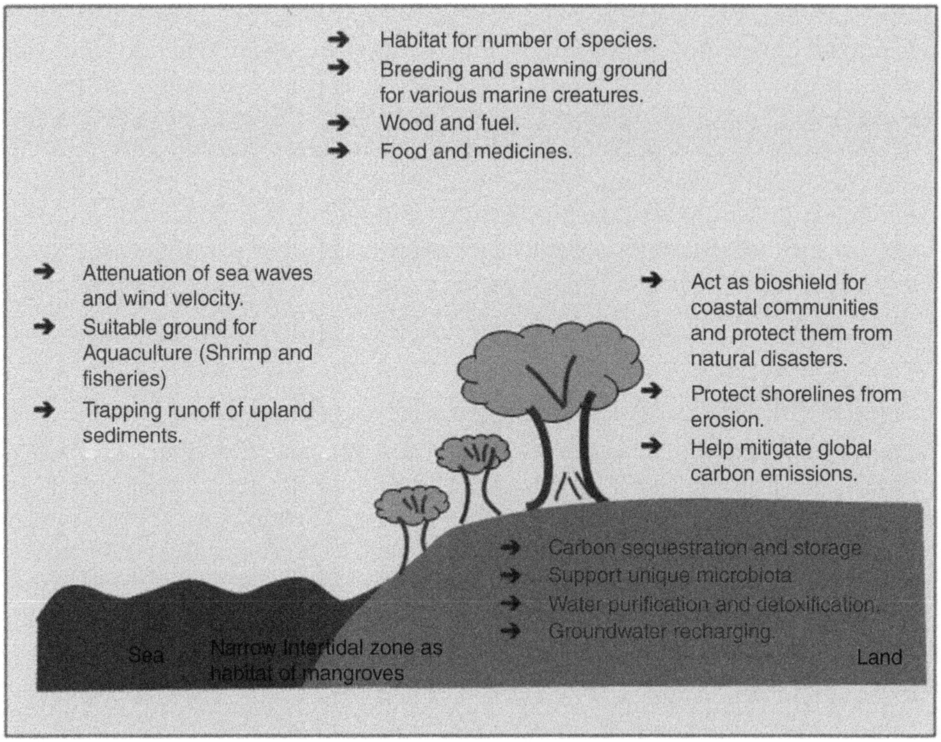

Figure 12.4 Important roles of mangroves. *Source:* Dahdouh-Guebas et al. (2005), Kathiresan and Rajendran (2005), National Parks Board Singapore (2010). Lee et al. (2014), Buelow and Sheaves (2015), Mohd-Azlan (2014). © John Wiley & Sons.

and immobilize heavy metals and other pollutants in their roots (Kumari et al. 2020). Mangrove ecosystems effectively aid in natural bioremediation and pollutant detoxification (Kumari et al. 2020). Mangroves can also absorb runoff water and waste emitted from aquaculture and therefore protect coral reefs, seagrasses, and fish habitats (Gilbert and Janssen 1998; Tam and Wong 1999; Kathiresan and Qasim 2005, p. 537). The mangroves also help in carbon sequestration and thus help combat global climate change (Alongi 2011). Following subheadings briefly explain some of the important roles that mangroves play. Figure 12.4 shows the major important roles played by mangroves and mangrove ecosystems.

12.4.1 Mangrove Forests are the Richest and Most Biodiverse Ecosystems on Earth

The mangrove ecosystem hosts diverse forms of life and therefore are reservoirs of genetic and biological diversity (Kathiresan and Bingham 2001; Sandilyan and Kathiresan 2012). They harbor huge diversity of other organisms including microorganisms, phytoplanktons, fish, crabs, and zooplanktons (Lugo and Snedaker 1974; Mumby et al. 2004; Gaos et al. 2016). A study by Upadhyay et al. (2002) stated that mangroves support a large number of organisms such as crabs, mollusks, and insects. Many termites and ants are also dependent on mangrove species (Adams and Levings 1987, p. 1069; Nielsen 2011, p. 113; Duke and Schmitt 2015, p. 1). Leaves and twigs of mangroves act as important shelters for insects (Macnae 1968). They are also breeding ground for migratory arthropods and birds (Poulin et al. 1992; Noske 1993; Noske 1995; Lefebvre and Poulin 1997; Nagelkerken et al. 2008; Florida Museum 2018; Wood 2019; Arcos et al. 2020). Studies have shown that several vertebrates are endemic to mangrove ecosystems (Rog et al. 2017, p. 221) and nearly half of endemic mangrove vertebrates are globally endangered (Luther and Greenberg 2009; Heimbuch 2011; Virata 2011; IUCN 2017). Some aquatic species thriving in mangrove environments are also facing various degrees of threats (Carugati et al. 2018).

12.4.2 Aquaculture: Shrimp and Fish Cultivation

Mangroves are important to humans as they provide a wide range of ecological services and livelihood opportunities. Of the several important services, aquaculture is an important service provided by mangrove forests. Fish and shrimp cultivation in the estuarine habitat of mangroves provides economic stability to the coastal communities and contribute to the enhancement of their lifestyles (Carrasquilla-Henao et al. 2019). Studies have reported the importance of mangroves for coastal shrimp and fish cultivation (Gunawardena and Rowan 2005; Manson et al. 2005; Primavera 2008; Lee et al. 2014). Studies indicate that increasing mangrove diversity may lead to increased fish production (Aburto-Oropez et al. 2008; Abrantes et al. 2019; Vincentius et al. 2019). However, recent studies show that excessive use of mangrove areas for aquaculture has contributed much to its destruction and is one of the major factors responsible for mangrove losses (Norris and Cargile 1998; Pattanaik and Prasad 2011; Rochmyaningsih 2017). The negative pressure posed by excessive conversion of mangrove habitats into aquaculture indicates that steps must be taken to reverse the losses incurred due to this and new methods for integrated and sustainable fish and shrimp farming must be developed keeping mangroves at the center of all other activities (Ahmed et al. 2018).

12.4.3 Protection from Natural Disasters: Mangroves Act as Natural Bioshields Against Natural Disasters

Mangroves are known as natural bioshields and protect coastal communities and other marine organisms for they act as a barrier to natural disasters such as tsunamis, hurricanes, and cyclones (Dahdouh-Guebas et al. 2005; Kathiresan and Rajendran 2005; del Valle et al. 2019). The aerial root system and the canopy architecture of the mangroves provide protective properties to them (Massel et al. 1999; Mazda et al. 2006; Barbier et al. 2008; Van Lavieren et al. 2012; Horstman et al. 2014; Yeo 2014; Hamilton and Casey 2016; Naidoo 2016). They are known to protect coastal areas from high wind velocity (Das and Crépin 2013) and storm surge (Gurib-Fakim and Brendler 2004; Krauss et al. 2009; Schneider 2011; Zhang et al. 2012; Liu et al. 2013; Gardner 2016). Several studies suggest that during Indian Ocean cyclones, human settlements lacking mangroves face more damages than those shielded by mangrove vegetation (Danielsen et al. 2005; Das and Vincent 2009).

Increased mangrove forest cover can increase the protective effect. Various characteristics of mangroves such as the dominance of the species, age, and size determine the reduction of damage by natural disasters (Patel et al. 2014, p. 29; Asbridge et al. 2018; Dasgupta et al. 2019). A large number of studies explain the important protective roles of mangrove forests. The degree of protection is dependent on the species' composition, tree size, age, and canopy architecture of the mangrove forests. Therefore, it is important to take steps and actions to protect the mangroves for stable coastal communities.

12.4.4 Medicinal Value of Mangroves

Many mangroves have been used by traditional communities based on their ethnobotanical knowledge (Ravindran et al. 2005, p. 409; Bibi et al. 2019). Therefore, like any other terrestrial medicinal plants, several species of mangroves also possess important medicinal properties because of the important metabolites they harbor in their cells (Bandaranayake 1998; Eldeen and Effendy 2013, p. 872; Saranraj and Sujitha 2015). Not only true mangroves but mangrove associates and other associated organisms, such as endophytes, also possess medicinal properties (Nurunnabi et al. 2020). Ethnobotanical studies suggest that mangroves have been used for many diseases/ailments across different countries including India (Bibi et al. 2019). Bibi et al. (2019) suggest that *Bruguiera gymnorrhiza*, *Rhizophora mucronata*, *Acanthus ilicifolius*, and *Heritiera fomes* are medicinally more important than other species. Increasing studies suggest important medicinal roles of the mangroves (Liebezeit and Rau 2006; Nurdiani et al. 2012, p. 27; Seepana et al. 2016). Bibi et al. (2019) has extensively reviewed the current knowledge on medicinal, pharmacological, and phytochemical aspects of various species of mangroves. It will not be impossible to explore new sources of drugs from mangrove ecosystems. However, future studies on the prospecting of mangroves for medicinal plants and drugs must keep in mind the principles of the sustainable development goals.

12.5 Threats to Mangroves

Mangroves, though one of the most important and productive ecosystems, face survival threats (Gilman et al. 2008; Chaudhuri et al. 2015; Makowski and Finkl 2018; Romañach et al. 2018). Large parts of mangrove forests have been converted to non-mangrove activities (Valiela et al. 2001; Thu and Populus 2007, p. 98). Many species of mangroves face varying degrees of threats and some of them are already critically endangered (IUCN 2020b). Mangroves face a large number of threats due to several factors such as human settlements, excessive use of wood, climate change, conversion of mangroves for other developmental activities, and oil spills (Duke et al. 1997; Lamparelli et al. 1997, p. 191; Walters 1997; Goldberg et al. 2020). Loss of mangroves has an indirect effect on the monumental diversity of organisms they support. Loss of mangroves directly means the loss of a large number of other organisms, also endangering the stability and survival of the coastal marine ecosystems. The following subheadings explain various threats that are responsible for the loss of mangroves and if corrective steps are not taken to reverse them, the loss of mangroves will continue putting the coastal communities and other marine biotas at risk. The various threats that mangroves currently face are summarized in Figure 12.5.

12.5.1 Human Settlements and Other Developmental Activities

Increasing population and immigration to the mangrove-rich coastal areas has resulted in the loss of mangrove cover across the globe including India (Walters 1997; Walters 2003; Nguyen et al. 2013; Munji et al. 2014; Saw and Kanzaki 2015; Bhomia et al. 2016; Sarmin et al. 2016; Chowdhury et al. 2017, p. 275; Goldberg et al. 2020). Rideout et al. (2013) have reported an increase in the population in Kenyan mangrove areas between 2000 and 2010 and found its positive correlation with mangrove loss. With immigration and increasing population density, other developmental activities such as roads, buildings, and construction activities also contribute to the loss of mangrove areas (Hirales-Cota et al. 2010, p. 147). Some mangrove regions of the world also face increased industrial development at

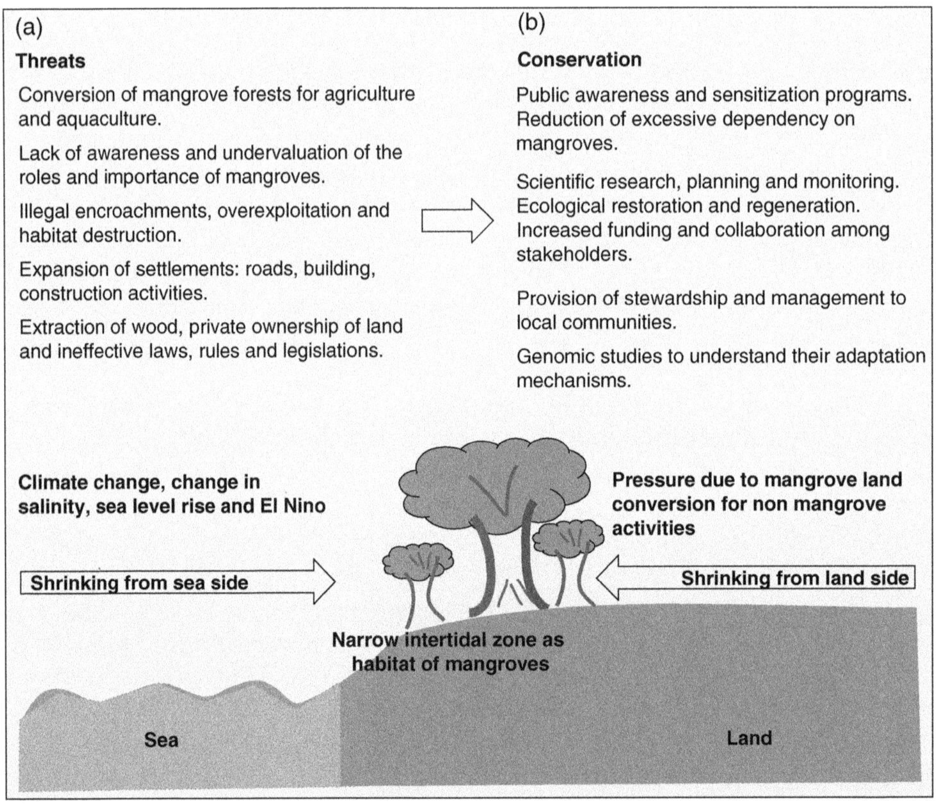

Figure 12.5 Threats to mangroves and conservation efforts. (a) Various threats mangroves facing. *Source*: Sippo et al. (2018), Munji et al. (2014), Hema and Devi (2014), Palacios and Cantera (2017), Islam and Wahab (2005), Badola et al. (2012), Krauss et al. (2014) and Lovelock et al. (2017). © John Wiley & Sons. (b) Strategies for effective mangrove conservation and restoration. *Source*: Based on Islam and Bhuiyan (2018). © John Wiley & Sons.

the cost of precious mangroves (Ferreira and Lacerda 2016). The industries and human settlements also release pollutants that endanger other marine lives in addition to harming the mangroves (Agoramoorthy et al. 2008).

12.5.2 Excessive Extraction of Wood

Timber from mangroves species is used as a construction material for various purposes (Feka and Manzano 2008; Tolangara 2014, p. 54; Palacios and Cantera 2017). Wood harvesting has helped a large number of communities but unsustainable harvesting can lead to the change of mangrove structure and composition (Semesi 1992; Rajkaran et al. 2010; Scales and Friess 2019). Recently, we have seen increased harvesting and overexploitation of the mangrove wood for various purposes (Ajonina and Usongo 2001; Rasquinha and Mishra 2020). Loss of mangroves has resulted in increased erosion of the mangrove areas, which can cause secondary issues to the coastal areas and the communities relying on them (Lara et al. 2002; Goldberg et al. 2020).

12.5.3 Conversion of Mangrove Forests for Farming and Related Activities

To feed the burgeoning population, land resources are insufficient and unsustainable practices are followed by people around the world to clear forest lands for agricultural purposes (FAO 2016; Kumari et al. 2019; Meek 2019). Even the mangrove forests, which are unique and limited, are also not spared from such activities (Alongi 2002; Richards and Friess 2016; Butler 2019). The mangroves have been converted into rice fields and oil palm plantations in several countries (Ong 1995; Ellison and Farnsworth 1996; Ong 2007; Saw and Kanzaki 2015; Kinver 2016; Richards and Friess 2016; Fauzi et al. 2019; Sustainability Times 2020). Increasing studies suggest the threats to mangroves due to excessive and unsustainable strategies adopted by farmers across the globe. There is a global consensus among the people that mangrove forest conversions must be halted by providing alternative livelihood opportunities and finding new innovative solutions to agricultural demand such as soil-less agriculture and indoor hydroponics (Saha 2010, p. 139).

12.5.4 Conversion of Mangrove Forests for Aquaculture

In addition to conversion of the mangrove forests for agriculture, larger portions of them are also converted to aquaculture activities such as shrimp and fish cultivation (De Graaf and Xuan 1998; Hein 2000, p. 48; Barbier and Cox 2004; Ashton 2008; Polidoro et al. 2010; Romañach et al. 2018). Several researchers have reported a decline in mangrove cover due to aquaculture (Dahdouh-Guebas et al. 2002; Islam and Wahab 2005; Giri and Muhlhausen 2008; Feller et al. 2017). Even excessive harvesting of the fish broods from the natural mangrove environment has resulted in disturbances to the food chains and food webs (Ellison 2008). Although aquaculture has been an important source of livelihood for the coastal communities and provides economic stability to them, the unsustainable practices of large-scale mangrove forest conversions may lead to exposure of these communities and will make them more vulnerable to natural disasters.

12.5.5 Global Warming, Climate Change, and Sea Level Rise

Global warming has contributed to global climate change (Dai 2010; McCarthy et al. 2010; Singh and Singh 2012, p. 93). Several studies suggest that global climate change will have extreme impacts on the mangroves (Field 1995, p. 75; Gilman et al. 2008; Krauss et al. 2014; Alongi 2015; Wilson 2017). Global climate change impacts global temperatures, rainfall, sea level, ocean salinity, and CO_2 levels (Najjar et al. 2000; Gilman et al. 2008; Najjar et al. 2010; Ellison and Zouh 2012; Ward et al. 2016). All climate-associated changes may affect mangrove distribution and species' diversity across the globe (Gilman et al. 2008; Record et al. 2013; Alongi 2015; Ward et al. 2016; Osland et al. 2017). Scientists have started understanding and predicting the future impacts of climate change on mangroves. Cavanaugh et al. (2019) suggested that the mangroves might move toward poles because of increasing global temperature. However, their long-term survival may not be possible despite poleward movements as they can be impacted by other climatic constraints (Osland et al. 2017). Climate change-associated changes such as global sea level rise, increased accumulation of sediments, change in salinity and availability of freshwater have started impacting mangroves (Ellison 1994, p. 11; Ellison 2012; Krauss et al. 2014; Alongi 2015; Ellison 2015; Lovelock et al. 2015; Ward et al. 2016). Few authors have studied the loss of mangroves due to sea level rise (Woodroffe and Davies 2009, p. 65; Meeder et al. 2017) and some have projected that sea level rise may result in intolerable levels of root submergence of the mangroves in future (Krauss et al. 2014; Calma 2020). Limited research is available on the impact of such changes on mangroves and future studies must focus on understanding and simulating the effects of such changes on their adaptability, physiology, and biochemistry necessary for their survival (Duke et al. 2007; Gilman et al. 2008; Yáñez-Espinosa and Flores 2011, p. 253; Godoy and de Lacerda 2015; Lovelock et al. 2016, p. 149). Ward et al. (2016) have reviewed region-specific impacts of climate change on mangroves and suggest that varying responses may be required to deal with climate change impacts on mangroves.

12.5.6 Limits to Landward Movement

The sea level rise may push the mangrove habitats landward (Semeniuk 1994, p. 1050; Di Nitto et al. 2014; Peterson and Bell 2015; Ward et al. 2016; Meeder et al. 2017; Osland et al. 2017). However, the landward movement may not be possible because human settlements and other anthropogenic activities have also extended and intruded into the mangrove habitats (Gilman et al. 2008; Di Nitto et al. 2014; IPCC 2019; See Figure 12.5 for illustration). Therefore, mangrove spaces are shrinking from the seaside due to the sea level rise and from landslides due to human activities making them vulnerable to extinction (Godoy and de Lacerda 2015; Schwartzstein 2019; Wongthong 2020). Further, the shrinking of the mangrove habitats can result in irreversible damages to the fragile yet one of the most important ecosystems. Adaptation of mangrove species to the land may not take place at the pace at which landward movement of mangroves is taking place because some of the species may adapt faster whereas others may take time (Gilman et al. 2008; Di Nitto et al. 2014). Therefore, policies should be devised based on the outcomes of the research on the mismatch in the adaptation of mangroves and the landward movement so that further mangrove losses can be prevented that could happen due to this.

12.5.7 El Niño and La Niña Events

Mangroves are also vulnerable to other weather extremes such as El Niño-Southern Oscillation (ENSO) and the La Niña events (Erftemeijer and Hamerlynck 2005, p. 228; López-Medellín et al. 2011; Heras and Soares 2017; Lovelock et al. 2017; Whitfield et al. 2019). The ENSO occurs once in two to seven years and affects global climate patterns (Scaife et al. 2019). ENSO and La Niña both have been known to impact mangrove ecosystems similar to other marine ecosystems (Drexler and Ewel 2001; Alongi 2002; Gilman et al. 2008; Heras and Soares 2017; Riascos et al. 2018). El Niño and La Niña events also result in secondary events such as flooding, excessive rains, or droughts, high temperatures, and may also adversely impact the mangroves (Rosenzweig and Hillel 2008; Cashman and Nagdee 2017, p. 155; Heras and Soares 2017). Few studies have shown several impacts of ENSO and La Niña on the mangroves; however, detailed studies are required to further understand and predict the long-term impacts of El Niño and La Niña on mangrove ecosystems.

12.6 Strategies for the Conservation of Mangroves

12.6.1 Increased and Focused Research on Understanding Mangroves

Mangroves are important components of the planet earth but research on them has picked up only recently. There are fewer studies on mangroves, especially about their distribution and species composition, primarily because many mangrove swamps are inaccessible or difficult to field survey (Giri et al. 2008; Kamal and Phinn 2011; Kuenzer et al. 2011). Recently, remote sensing technology has enabled accessing the area cover under mangroves (Giri et al. 2008, 2011) because remote sensing is an indispensable tool for assessing and monitoring mangrove forests (Abbas et al. 2013). In recent years, we have seen increasing literature for the mangrove distribution, area, species diversity, their roles, and threats (Field and Whittaker 1998; Biswas et al. 2007; Alongi 2008; Nagelkerken et al. 2008; Giri et al. 2011; Goutham-Bharathi et al. 2014; Lee et al. 2014; Giri et al. 2015). However, increased studies are needed with a special focus on region-wise threat status monitoring and taking necessary interventions, the impact of climate change, adaptations in mangroves in response to climate change (Biswas et al. 2007; Grantham et al. 2011; DasGupta and Shaw 2013; Ghosh et al. 2015; Chow 2017; Romañach et al. 2018; Singh et al. 2019), and policy-based studies to evaluate the shortcomings of the legislation and rules framed based on the mangroves (Rogers et al. 2016; Sundar 2018, p. 1). Modern genomics studies can also be carried out to evaluate their genomic basis of adaptations to prepare for the future (Dodd and Rafii 2002; Das and Strasser 2013, p. 53; Xu et al. 2017).

12.6.2 Implementation of Mangrove Conservation-Related Laws, Guidelines, and Other Initiatives

It has been found that of the 75% of mangroves located in 15 top countries, only 6.9% of them are in the protected areas (Thomas et al. 2017; Bibi et al. 2019). Several factors have contributed to the loss of mangrove species and the mangrove ecosystems and to arrest the continuous decline, laws must be reanalyzed and reframed as per the current demands of time (Duke et al. 2007; Friess et al. 2019; Bell-James et al. 2020). Various international agencies, organizations, and governments of the different countries have framed general and specific guidelines and promulgated laws for the protection of mangroves (Duangjai et al. 2013; Lundquist et al. 2017; Udoh 2016, p. 151; Suman 2019, p. 1055). Specific legislations are needed to tackle growing losses to mangroves and these legislations vary from country to country. Many countries like Tonga lack the basic guidelines (MESCAL 2011) whereas countries like Indonesia have certain guidelines for the conservation of mangroves, but are unable to reduce the deterioration of the mangroves due to the overlapping of the laws and illegal logging in these ecosystems (Sunyowati et al. 2016). To fill that gap, FAO (1994) has issued some important mangrove forest management guidelines which have been found to be successful in several countries (Field 1999; Kairo et al. 2001). Various initiatives have been taken worldwide for the conservation of mangroves. The General Conference of UNESCO held in 2015 has adopted "The International Day for the Conservation of the Mangrove Ecosystem" and it is celebrated each year on 26 July globally (UNESCO 2015). The purpose of this is to raise awareness of the importance of mangrove ecosystems as "a unique, special, and vulnerable ecosystem" and to take steps to sustainable management of mangroves (Shunula 2002). The 2030 Agenda for Sustainable Development which includes 17 goals and 169 targets has also given due importance to mangrove conservation and management (Sharrock and Jackson 2016; Neumann et al. 2017; Nagabhatla et al. 2019). The role of mangroves and the need for their conservation is also well acknowledged under the Paris Agreement of the United Nations Framework Convention on Climate Change (UNFCCC) (Burns 2015, p. 415). Save Our Mangroves Now, an initiative of the IUCN and WWF is also contributing to the conservation of the mangroves (IUCN 2018; United Nations Ocean Conference 2019; IUCN 2020a).

In India also, several measures have been taken including statutory mechanisms for the conservation of mangroves (Kumar 2000). Several legislations such as The Environment (Protection) Act of 1986, The Biological Diversity Act of 2002 are directly or indirectly linked to mangrove conservation and management (Ramachandran et al. 2018). Green India Mission which is a subcomponent of the National Action Plan on Climate Change, 2008 has prioritized conservation and restoration of mangroves (Ravindranath and Murthy 2010; DasGupta and Shaw 2013).

In spite of all these efforts, difficulties arise in the proper implementation of laws and various other guidelines and initiatives due to the undervaluation of mangroves and other factors such as fragmented geographical distribution of mangroves and acute shortage of government staff and facilities (Ashokkumar and Irfan 2018). However, to save mangroves, a proper policy and legislative framework is important but its implementation is more challenging and important. Therefore, an implementation must be given more preference and also reformation of some of the archaic legislations must be done. Table 12.4 represents a non-exhaustive and indicative list of various types of initiatives and the organizations/agencies involved in mangrove conservation and management.

Table 12.4 Some of the important global initiatives for the Conservation, Protection, and Regeneration of Mangroves.

S. No	Organization/Initiative	Effective working area	Website/URL
1.	Global Mangrove Alliance	Financing, strengthening policy, building capacity, developing a proof of concept and knowledge sharing.	http://www.mangrovealliance.org/initiatives/
2.	Food and Agriculture Organization	Various activities such mangrove forest management, afforestation and various studies on them, reports, guidelines.	http://www.fao.org/forestry/mangrove/3940/en/
3.	IUCN Mangrove Rehabilitation for Sustainably Managed Healthy Forests Project.	Building capacity and improving the sustainable management of mangroves at local and national level.	https://www.iucn.org/regions/oceania/our-work/nature-based-solutions/water-and-wetlands/pacific-mangroves-initiative/mangrove-rehabilitation-sustainably-managed-healthy-forests-marsh-project
4.	World Wildlife Fund	Protection of mangrove ecosystems using various programs and by collaborating with various organizations and governments.	https://www.wwfca.org/en/species_and_places/mangroves/
5.	The International Blue Carbon Initiative	Emphasize on the importance of the marine resources such as mangroves for their protection and use for carbon sequestration.	https://www.thebluecarboninitiative.org/
6.	International Partnership for Blue Carbon	Protection and conservation of mangroves and other ecosystems that are important for capturing carbon.	https://bluecarbonpartnership.org/
7.	Global Nature Fund	Reforestation of global degraded mangrove areas	https://www.globalnature.org/Mangroves
8.	Mangrove Action Project	US-based nonprofit which collaborates with individuals and organizations to preserve, conserve, and restore mangrove forests.	https://mangroveactionproject.org/
9.	Global Mangrove Watch	Online platform that provides the remote sensing data and tools for monitoring mangroves.	https://www.globalmangrovewatch.org/?
10.	Mangrove Science	Monitoring and modeling mangroves with remote sensing.	https://mangrovescience.org/

(*Continued*)

Table 12.4 (Continued)

S. No	Organization/Initiative	Effective working area	Website/URL
11.	The Mangrove Alliance	Community-led stewardship of mangrove forest ecosystems and restoration of mangrove ecosystems.	https://ecoviva.org/mangrove-alliance/#
12.	International Society for Mangrove Ecosystems	Knowledge, training for mangrove conservation.	http://www.mangrove.or.jp/isme/english/index.htm
13.	Turing Foundation	Mangrove rehabilitation projects.	http://www.turingfoundation.org/kw_mangrove_uk.html
14.	BirdLife International	Protection of mangroves by involving communities.	http://www.birdlife.org on 21/09/2020
14.	ZSL Institute of Zoology	Conservation and monitoring of mangrove forests.	https://www.zsl.org/science/research/mangroves
15.	International Tropical Timber Organization	It is involved in conservation and sustainable management of mangroves.	https://www.itto.int/sustainable_forest_management/mangroves/
16.	Nature Environment and Wildlife Society	Conserve ecology and environment, wildlife, natural resources.	https://naturewildlife.org/
17.	Mission Mangroves	Aims to restore Mumbai's depleted mangrove cover	https://www.unitedwaymumbai.org/mission-mangroves
18.	Australian Mangrove and Saltmarsh Network	An Australian independent informal network for protection of mangroves.	https://www.amsn.net.au/
19.	Mangroves for the Future	Promotes mangrove conservation by partnerships with organizations and communities.	https://www.mangrovesforthefuture.org/

12.6.3 Strengthening Conservation Mechanisms

Conservation and/or restoration of mangroves and degraded mangrove ecosystems must be based on the evidence-based conservation/restoration approaches (Le 2008). Ideally, conservation and restoration practices should be based on natural regeneration systems (Kairo et al. 2001; IUCN 2011; Mchenga and Ali 2014, p. 334; Romañach et al. 2018; Lee et al. 2019). For example, monocultures of one type of mangrove species to increase mangrove forest cover may not yield the desirable results but holistic approaches must be taken to recover the mangrove ecosystems along with other organisms of the ecosystem (Primavera et al. 2016; Romañach et al. 2018). Therefore, conservation activities must be revisited and holistic elements must be added to the mangrove conservation and regeneration programs (Kairo and Mwita 2020). Moreover, the focus should also be placed on high-priority species which are at high risk of extinction in the near future (Macintosh and Ashton 2002, p. 338) (See Table 12.1 for varying degrees of threats mangrove species face). Evidence-based sustainable conservation mechanisms must be strengthened to obtain desirable outcomes (Farley et al. 2010, p. 39).

12.6.4 Targeting Land Ownership-Related Issues

In several regions, mangroves exist on privately owned lands, which creates hurdles in the implementation of mangrove-related laws (Hema and Devi 2014: Suman 2019, p. 1055). In one of the Indian states, i.e. Kerala alone, about 80% of the mangrove land is under private ownership (Muraleedharan et al. 2009; Vidyasagaran and Madhusoodanan 2014, p. 38). A similar issue exists in other states of India and other countries of the world also such as Srilanka, the Philippines, and Ghana (Farley et al. 2010; Adger et al. 1997; Asante et al. 2017). Because of the ineffective imposition of the laws to private properties, mangrove forests are being converted to personal private activities such as aquaculture, agriculture and settlement purposes (Ron and Padilla, 1999, p. 297; Farley et al. 2010; Asante et al. 2017; Sathirathai and Barbier 2007; Srivastava and Mehta 2017). Several countries have allowed the conservation of plants on private lands also by sufficiently rewarding the people involved (McNeely 1993, p. 144; Spiteri and Nepalz 2006, p. 1; Pascual and Perrings 2007, p. 256; Kamal et al. 2015, p. 576). Some successful examples of mangrove conservation on private land are available. For example, in Southeast Asia, payment for ecosystem services approach through landowners ensured sustainable livelihoods and mangrove conservation (Friess et al. 2016). Therefore, a similar practice of promotion of conservation on private land can be taken up wherever this issue exists.

12.6.5 Involvement of Local Communities

Involvement of local communities is important for the effective implementation of the projects aimed at the conservation of the mangroves because in several parts of the country and globe, mangroves are located in the private lands of the people. Engagement of local communities can improve the conservation success rates (Mangora 2011). Badola et al. (2012) studied the attitudes of local communities toward the conservation of mangroves in the East Coast of India and found that many participants provided positive feedback and agreed to participate in mangrove conservation programs. A similar study by Roy (2016) from Bangladesh suggests that more than 50% of the people agreed to participate in mangrove conservation programs. Studies also suggest that the unavailability of alternative employment opportunities also push people toward illegal harvesting of mangrove resources (Badola et al. 2012; Roy 2016). Local coastal communities are connected to the mangrove forests and somehow know the values of mangrove–human connections and their ecological and economic importance (Hussain and Badola 2010; Kusmana and Sukristijiono 2016). Spreading more awareness and sensitizing them about the roles and importance of mangroves can yield positive results in the participatory management of the mangroves (Shunula 2002). Since mangroves provide a large number of economic benefits to the communities associated with them, excessive usage and overexploitation of mangrove resources can result in loss of jobs and lead to poverty among the coastal communities relying on them (Hoanh et al. 2006; Aye et al. 2019). It has been suggested that educating coastal communities about sustainable consumption and production and harvesting practices can help in effective management and sustainable exploitation of mangrove and associated resources (Glaser 2003; Ramirez and Townsend 2006). The bottom-up approaches to conservation and management of mangroves involving the local communities have been successful so far (Owuor et al. 2019a). Therefore, policies should be designed keeping both mangroves as well as local communities relying on them at the center in which locals also derive economic and livelihood benefits and mangroves are also sustainably managed (Badola et al. 2012). Alternatively, policies can be designed to provide alternative sources of income to the locals by reducing their overdependency on mangroves (Mitra et al. 2006; Sathirathai and Barbier 2007; Leal Filho et al. 2019). However, excessive restrictions on the locals and prohibiting them from using mangroves do not yield positive results without providing alternative income and livelihood resources (Mitra et al. 2006). A very successful model of sustainable conservation involving local communities is often cited from the two coastal sites in the Philippines, i.e. Bais Bay and Banacon Island (Walters 2003, 2004). The local communities in these two islands started participating in the conservation and restoration of mangroves using local knowledge even before the participation of governments and non-governmental organizations (Walters 2003, 2004). Several other studies have also indicated the important roles of local community knowledge in the sustainable management of mangroves (Kumar 2000, p. 41; Zorini et al. 2004; Iftekhar 2008; Badola et al. 2012; Datta et al. 2012; Roy and Alam 2012; Febryano et al. 2014; Lee et al. 2019).

12.7 Conclusion

This chapter explored various important aspects of mangroves such as species diversity, global distribution, threats, challenges, and the recent conservation initiatives. The chapter also discussed various goods and services provided by the mangroves that contribute toward the overall prosperity of mankind. It is well established that they provide numerous ecological services that could be harnessed to mitigate the effects of global climate change. For example, they are one of the most important storehouses of carbon sequestration and storage. Some species of mangroves face serious survival threats.

Various threats and the conservation challenges that impede conservation and restoration efforts have been discussed in detail. The chapter will be an important and useful material for the students and researchers who wish to start their research in the diverse areas of mangroves. It will also act as an awareness material for people from nonscientific and nontechnical backgrounds.

Acknowledgments

Ajay Kumar acknowledges the funding support provided by Science and Engineering Research Board (SERB), Department of Science and Technology (DST) vide Sanction order File No. EEQ/2019/000441 dated 2 December 2019 and Central University of Kerala for extending other facilities.

References

Abbas, S., Qamer, F.M., Ali, G. et al. (2013). An assessment of status and distribution of mangrove forest cover in Pakistan. *Journal of Biodiversity and Environmental Sciences* 3 (6): 64–78.

Abrantes, K.G., Sheaves, M., and Fries, J. (2019). Estimating the value of tropical coastal wetland habitats to fisheries: Caveats and assumptions. *PLoS One* 14: 1–23. https://doi.org/10.1371/journal.pone.0215350.

Aburto-Oropez, O., Ezcurra, E., Danemann, G. et al. (2008). Mangroves in the Gulf of California increase fishery yields. *Proceedings of the National Academy of Sciences of the United States of America* 105: 10456–10459. https://doi.org/10.1073/pnas.0804601105.

Adams, E.S. and Levings, S.C. (1987). Territory size and population limits in mangrove termites. *Journal of Animal Ecology* 56 (3): 1069–1081.

Adger, W.N., Kelly, M., Ninh, N.H. et al. (1997). Property rights and the social incidence of mangrove conversion in Vietnam. CSERGE Working Paper GEC 97-21.

Agoramoorthy, G., Chen, F.A., and Hsu, M.J. (2008). Threat of heavy metal pollution in halophytic and mangrove plants of Tamil Nadu, India. *Environmental Pollution* 155: 320–326. https://doi.org/10.1016/j.envpol.2007.11.011.

Ahmed, N., Thompson, S., and Glaser, M. (2018). Integrated mangrove-shrimp cultivation: potential for blue carbon sequestration. *Ambio: A Journal of the Human Environment* 47: 441–452. https://doi.org/10.1007/s13280-017-0946-2.

Ajonina, G. and Usongo, L. (2001). Preliminary quantitative impact assessment of wood extraction on the mangroves of Douala-Edea Forest Reserve, Cameroon. *Tropical Biodiversity* 7 (2-3): 137–149.

Alongi, D.M. (2002). Present state and future of the world's mangrove forests. *Environmental Conservation* 29: 331–349. https://doi.org/10.1017/s0376892902000231.

Alongi, D.M. (2008). Mangrove forests: Resilience, protection from tsunamis, and responses to global climate change. *Estuarine, Coastal and Shelf Science* 76: 1–13. https://doi.org/10.1016/j.ecss.2007.08.024.

Alongi, D.M. (2011). Carbon payments for mangrove conservation: ecosystem constraints and uncertainties of sequestration potential. *Environmental Science & Policy* 14: 462–470. https://doi.org/10.1016/j.envsci.2011.02.004.

Alongi, D.M. (2015). The impact of climate change on mangrove forests. *Current Climate Change Reports* 1: 30–39. https://doi.org/10.1007/s40641-015-0002-x.

Arcos, C.D., Badillo-Alemán, M., Arceo-Carranza, D. et al. (2020). Feeding ecology of the waterbirds in a tropical mangrove in the southeast Gulf of Mexico. *Studies on Neotropical Fauna and Environment* 55: 1–9. https://doi.org/10.1080/01650521.2019.1682232.

Asante, W.A., Acheampong, E., Boateng, K. et al. (2017). The implications of land tenure and ownership regimes on sustainable mangrove management and conservation in two Ramsar sites in Ghana. *Forest Policy and Economics* 85: 65–75. https://doi.org/10.1016/j.forpol.2017.08.018.

Asbridge, E., Lucas, R., Rogers, K. et al. (2018). The extent of mangrove change and potential for recovery following severe Tropical Cyclone Yasi, Hinchinbrook Island, Queensland, Australia. *Ecology and Evolution* 8: 10416–10434. https://doi.org/10.1002/ece3.4485.

Ashokkumar, S. and Irfan, Z.B. (2018). Current status of mangroves in India: benefits, rising threats policy and suggestions for the way forward. Madras School of Economics Working Paper- 174/2018. http://www.mse.ac.in/wp-content/uploads/2018/08/Working-Paper-174.pdf.

Ashton, E.C. (2008). The impact of shrimp farming on mangrove ecosystems. *CAB Reviews: Perspectives in Agriculture, Veterinary Science, Nutrition and Natural Resources* 3: 1–12. https://doi.org/10.1079/PAVSNNR20083003.

Aye, W.N., Yali, W., Marin, K. et al. (2019). Contribution of mangrove forest to the livelihood of local communities in Ayeyarwaddy Region, Myanmar. *Forests* 10: 1–12. https://doi.org/10.3390/f10050414.

Badola, R., Barthwal, S., and Hussain, S.A. (2012). Attitudes of local communities towards conservation of mangrove forests: a case study from the east coast of India. *Estuarine, Coastal and Shelf Science* 96: 188–196. https://doi.org/10.1016/j.ecss.2011.11.016.

Bandaranayake, W.M. (1998). Traditional and medicinal uses of mangroves. *Mangroves and Salt Marshes* 2: 133–148. https://doi.org/10.1023/a:1009988607044.

Barbier, E. and Sathirathai, S. (2004). *Shrimp Farming and Mangrove Loss in Thailand.* Cheltenham: Edward Elgar Publishing.

Barbier, E.B. and Cox, M. (2004). An economic analysis of shrimp farm expansion and mangrove conversion in Thailand. *Land Economics* 80: 389–407. https://doi.org/10.2307/3654728.

Barbier, E.B., Hacker, S.D., Kennedy, C. et al. (2011). The value of estuarine and coastal ecosystem services. *Ecological Monographs* 81: 169–193. https://doi.org/10.1890/10-1510.1.

Barbier, E.B., Koch, E.W., Silliman, B.R. et al. (2008). Coastal ecosystem-based management with nonlinear ecological functions and values. *Science* 319: 321–323. https://doi.org/10.1126/science.1150349.

Basha, C.S.K. (2016). An overview on global mangrove distribution. *Indian Journal of Geo-marine Sciences* 47 (4): 766–772.

Bell-James, J., Boardman, T., and Foster, R. (2020). Can't see the (mangrove) forest for the trees: Trends in the legal and policy recognition of mangrove and coastal wetland ecosystem services in Australia. *Ecosystem Services* 45: 1–12. https://doi.org/10.1016/j.ecoser.2020.101148.

Bhomia, R.K., MacKenzie, R.A., Murdiyarso, D. et al. (2016). Impacts of land use on Indian mangrove forest carbon stocks: Implications for conservation and management. *Ecological Applications* 26: 1396–1408. https://doi.org/10.1890/15-2143.

Bibi, N.S., Fawzi, M.M., Gokhan, Z. et al. (2019). Ethnopharmacology, phytochemistry, and global distribution of mangroves- a comprehensive review. *Marine Drugs* 17: 1–82. https://doi.org/10.3390/md17040231.

Biswas, S.R., Choudhury, J.K., Nishat, A. et al. (2007). Do invasive plants threaten the Sundarbans mangrove forest of Bangladesh? *Forest Ecology and Management* 245: 1–9. https://doi.org/10.1016/j.foreco.2007.02.011.

Buelow, C. and Sheaves, M. (2015). A birds-eye view of biological connectivity in mangrove systems. *Estuarine, Coastal and Shelf Science* 152: 33–43. https://doi.org/10.1016/j.ecss.2014.10.014.

Burns, W. (2015). Loss and damage and the 21st Conference of the Parties to the United Nations Framework Convention on Climate Change. *International Law Students Association Journal of International and Comparative Law* 22: 415–433.

Butler (2019). Mongabay. https://rainforests.mongabay.com/08-deforestation.html (accessed 29 October 2020).

Calma, J. (2020). Rising sea levels could wipe out mangroves by 2050. https://www.theverge.com/2020/6/4/21280580/sea-level-rise-mangroves-climate-change-2050 (accessed 29 October 2020)

Carrasquilla-Henao, M., Ban, N., Rueda, M. et al. (2019). The mangrove-fishery relationship: a local ecological knowledge perspective. *Marine Policy* 108: 1–11. https://doi.org/10.1016/j.marpol.2019.103656.

Carugati, L., Gatto, B., Rastelli, E. et al. (2018). Impact of mangrove forests degradation on biodiversity and ecosystem functioning. *Scientific Reports* 8: 1–11. https://doi.org/10.1038/s41598-018-31683-0.

Cashman, A. and Nagdee, M.R. (2017). Impacts of climate change on settlements and infrastructure in the coastal and marine environments of Caribbean small island developing states (SIDS). *Science Review*: 155–173.

Cavanaugh, K.C., Dangremond, E.M., Doughty, C.L. et al. (2019). Climate-driven regime shifts in a mangrove–salt marsh ecotone over the past 250 years. *Proceedings of the National Academy of Sciences* 116: 21602–21608. https://doi.org/10.1073/pnas.1902181116.

Cheong, S.M., Silliman, B., Wong, P.P. et al. (2013). Coastal adaptation with ecological engineering. *Nature Climate Change* 3: 787–791. https://doi.org/10.1038/nclimate1854.

Chow, J. (2017). Mangrove management for climate change adaptation and sustainable development in coastal zones. *Journal of Sustainable Forestry* 37: 139–156. https://doi.org/10.1080/10549811.2017.1339615.

Chaudhuri, P., Ghosh, S., Bakshi, M. et al. (2015). A review of threats and vulnerabilities to mangrove habitats: with special emphasis on east coast of India. *Journal of Earth Science and Climatic Change* 6: 1–9. https://doi.org/10.4172/2157-7617.1000270.

Chowdhury, R.R., Uchida, E., Chen, L. et al. (2017). Anthropogenic drivers of mangrove loss: geographic patterns and implications for livelihoods. In: *Mangrove Ecosystems: A Global Biogeographic Perspective* (eds. V.H. Rivera-Monroy et al.), 275–300. New York: Springer.

Costanza, R., de Groot, R., Sutton, P. et al. (2014). Changes in the global value of ecosystem services. *Global Environmental Change* 26: 152–158. https://doi.org/10.1016/j.gloenvcha.2014.04.002.

Dahdouh-Guebas, F., Jayatissa, L.P., Di Nitto, D. et al. (2005). How effective were mangroves as a defence against the recent tsunami? *Current Biology* 15: 443–447. https://doi.org/10.1016/j.cub.2005.06.008.

Dahdouh-Guebas, F., Kairo, J.G., De Bondt, R. et al. (2007). Pneumatophore height and density in relation to micro-topography in the grey mangrove *Avicennia marina*. *Belgian Journal of Botany* 140: 213–221. https://doi.org/10.2307/20794640.

Dahdouh-Guebas, F., Zetterström, T., Rönnbäck, P. et al. (2002). Recent changes in land-use in the Pambala–Chilaw lagoon complex (Sri Lanka) investigated using remote sensing and GIS: conservation of mangroves vs. development of shrimp farming. *Environment, Development and Sustainability* 4: 185–200. https://doi.org/10.1023/A:1020854413866.

Dai, A. (2010). Drought under global warming: a review. *Wiley Interdisciplinary Reviews: Climate Change* 2: 45–65. https://doi.org/10.1002/wcc.81.

Danielsen, F., Sørensen, M.K., Olwig, M.F. et al. (2005). The Asian tsunami: a protective role for coastal vegetation. *Science* 310: 643–643. https://doi.org/10.1126/science.1118387.

Das, A.B. and Strasser, R.J. (2013). Salinity-induced genes and molecular basis of salt-tolerant strategies in mangroves. In: *Molecular Stress Physiology of Plants* (eds. R. Rout and A.B. Das), 53–86. New Delhi, India: Springer.

Das, S. and Crépin, A.-S. (2013). Mangroves can provide protection against wind damage during storms. *Estuarine, Coastal and Shelf Science* 134: 98–107. https://doi.org/10.1016/j.ecss.2013.09.021.

Das, S. and Vincent, J.R. (2009). Mangroves protected villages and reduced death toll during Indian super cyclones. *Proceedings of the National Academy of Sciences* 106: 7357–7360. https://doi.org/10.1073/pnas.0810440106.

DasGupta, R. and Shaw, R. (2013). Changing perspectives of mangrove management in India–an analytical overview. *Ocean and Coastal Management* 80: 107–118. https://doi.org/10.1016/j.ocecoaman.2013.04.010.

DasGupta, S., Islam, M.S., Huq, M. et al. (2019). Quantifying the protective capacity of mangroves from storm surges in coastal Bangladesh. *PLoS One* 14: 1–14. https://doi.org/10.1371/journal.pone.0214079.

Datta, D., Chattopadhyay, R.N., and Guha, P. (2012). Community based mangrove management: a review on status and sustainability. *Journal of Environmental Management* 107: 84–95. https://doi.org/10.1016/j.jenvman.2012.04.013.

De Graaf, G.J. and Xuan, T.T. (1998). Extensive shrimp farming, mangrove clearance and marine fisheries in the southern provinces of Vietnam. *Mangroves and Salt Marshes* 2: 159–166. https://doi.org/10.1023/A:1009975210487.

del Valle, A., Eriksson, M., Ishizawa, O.A. et al. (2019). Mangroves protect coastal economic activity from hurricanes. *Proceedings of the National Academy of Sciences* 117: 265–270. https://doi.org/10.1073/pnas.1911617116.

Di Nitto, D., Neukermans, G., Koedam, N. et al. (2014). Mangroves facing climate change: landward migration potential in response to projected scenarios of sea level rise. *Biogeosciences* 11: 857–871. https://doi.org/10.5194/bg-11-857-2014.

Dodd, R.S. and Rafii, Z.A. (2002). Evolutionary genetics of mangroves: continental drift to recent climate change. *Trees* 16: 80–86. https://doi.org/10.1007/s00468-001-0142-6.

Donato, D.C., Kauffman, J.B., Murdiyarso, D. et al. (2011). Mangroves among the most carbon-rich forests in the tropics. *Nature Geoscience* 4: 293–297. https://doi.org/10.1038/ngeo1123.

Drexler, J.Z. and Ewel, K.C. (2001). Effect of the 1997–1998 ENSO-related drought on hydrology and salinity in a Micronesian wetland complex. *Estuaries* 24: 347–356. https://doi.org/10.2307/1353237.

Duangjai, W., Ngamniyom, A. et al. (2013). The guideline development for sustainable livelihood indicators of village marginal mangrove forest in the Satun Province, Thailand. *Asian Social Science* 9 (9): 123–130.

Duke, N.C., Pinzon, Z.S., and Prada, M.C.T. (1997). Large-scale damage to Mangrove forests following two large oil spills in Panama. *Biotropica* 29: 2–14. https://doi.org/10.1111/j.1744-7429.1997.tb00001.x.

Duke, N.C., Meynecke, J.O., Dittmann, S. et al. (2007). A world without mangroves? *Science* 317: 41–43. https://doi.org/10.1126/science.317.5834.41b.

Duke, N., Ball, M., and Ellison, J. (1998). Factors influencing biodiversity and distributional gradients in mangroves. *Global Ecology and Biogeography Letters* 7: 27–47. https://doi.org/10.1111/j.1466-8238.1998.00269.x.

Duke, N.C. and Larkum, A.W.D. (2008). Mangroves and seagrasses. In: *The Great Barrier Reef: Biology, Environment and Management* (eds. P.A. Hutchings, M. Kingsford and O. Hoegh-Guldberg), 156–170. Clayton: CSIRO Publishing.

Duke, N.C. and Schmitt, K. (2015). Mangroves: unusual forests at the seas edge. In: *Tropical Forestry Handbook* (eds. L. Pancel and M. Kohl), 1–24. New York: Springer.

Eldeen, I.M. and Effendy, M.A. (2013). Antimicrobial agents from mangrove plants and their endophytes. In: *Microbial Pathogens and Strategies for Combating Them: Science, Technology and Education*, 9e (ed. A. Méndez-Vilas), 872–882. Spain: Formatex Research Center.

Ellison, A.M. (2008). Managing mangroves with benthic biodiversity in mind: moving beyond roving banditry. *Journal of Sea Research* 59: 2–15. https://doi.org/10.1016/j.seares.2007.05.003.

Ellison, A.M. and Farnsworth, E.J. (1996). Anthropogenic disturbance of Caribbean Mangrove ecosystems: past impacts, present trends, and future predictions. *Biotropica* 28: 549–565. https://doi.org/10.2307/2389096.

Ellison, J.C. (1994). Climate change and sea level rise impacts on mangrove ecosystems. *Impacts of climate change on ecosystems and species*, A marine conservation and development report, 11-30. Gland: IUCN.

Ellison, J.C. (2012). Climate change vulnerability assessment and adaptation planning for mangrove systems, 5–36. Washington, DC: World Wildlife Fund (WWF). ISBN 978-92-990069-0-0

Ellison, J.C. (2015). Vulnerability assessment of mangroves to climate change and sea-level rise impacts. *Wetlands Ecology and Management* 23: 115–137. https://doi.org/10.1007/s11273-014-9397-8.

Ellison, J.C. and Zouh, I. (2012). Vulnerability to climate change of mangroves: assessment from Cameroon, Central Africa. *Biology* 1: 617–638. https://doi.org/10.3390/biology1030617.

Erftemeijer, P.L.A. and Hamerlynck, O. (2005). Die-Back of the mangrove *Heritiera littoralis* Dryand, in the Rufiji Delta (Tanzania) following El Niño floods. *Journal of Coastal Research* 42: 228–235.

Estrada, G.C.D. and Soares, M.L.G. (2017). Global patterns of aboveground carbon stock and sequestration in mangroves. *Anais da Academia Brasileira de Ciências* 89: 973–989. https://doi.org/10.1590/0001-3765201720160357.

FAO (1994). Mangrove forests management guidelines, 169-191. http://www.fao.org/3/ap428e/ap428e00.pdf (accessed 21 August 2019).

FAO (2016). State of the World's Forests. http://www.fao.org/3/a-i5588e.pdf (accessed 29 October 2020)

Farley, J., Batker, D., de la Torre, I. et al. (2010). Conserving Mangrove ecosystems in the Philippines: transcending disciplinary and institutional borders. *Environmental Management* 45 (1): 39–51.

Fauzi, S., Sakti, A., Yayusman, L. et al. (2019). Contextualizing mangrove forest deforestation in Southeast Asia using environmental and socio-economic data products. *Forests* 10: 1–18. https://doi.org/10.3390/f10110952.

Febryano, I.G., Suharjito, D., Darusman, D. et al. (2014). The roles and sustainability of local institutions of mangrove management in Pahawang island. *Jurnal Manajemen Hutan Tropika*: 69–76. https://doi.org/10.7226/jtfm.20.2.69.

Feka, N.Z. and Manzano, M.G. (2008). The implications of wood exploitation for fish smoking on mangrove ecosystem conservation in the South West Province, Cameroon. *Tropical Conservation Science* 1: 222–241. https://doi.org/10.1177/194008290800100305.

Feller, I.C., Friess, D.A., Krauss, K.W. et al. (2017). The state of the world's mangroves in the 21st century under climate change. *Hydrobiologia* 803: 1–12. https://doi.org/10.1007/s10750-017-3331-z.

Ferreira, A.C. and Lacerda, L.D. (2016). Degradation and conservation of Brazilian mangroves, status and perspectives. *Ocean and Coastal Management* 125: 38–46. https://doi.org/10.1016/j.ocecoaman.2016.03.011.

Field, C.B. and Whittaker, R.J. (1998). Biodiversity and function of mangrove ecosystems. *Global Ecology and Biogeography Letters (United Kingdom)* 7: 3–14. https://doi.org/10.1111/j.1466-8238.1998.00278.x.

Field, C.D. (1999). Rehabilitation of mangrove ecosystems: an overview. *Marine Pollution Bulletin* 37: 383–392. https://doi.org/10.1016/s0025-326x(99)00106-x.

Field, C.D. (1995). Impact of expected climate change on mangroves. In: *Asia-Pacific Symposium on Mangrove Ecosystems* (eds. Y.S. Wong and N.F.Y. Tam), 75–81. New York: Springer.

Florida Museum (2018). https://www.floridamuseum.ufl.edu/southflorida/habitats/mangroves/mangrove-life/

Forest Survey of India (2019). India State Of Forest Report. *Ministry of Environment, Environment, Forests & Climate Change Government of India*, India

Francisco, P.M., Mori, G.M., Alves, F.M. et al. (2018). Population genetic structure, introgression, and hybridization in the genus Rhizophora along the Brazilian coast. *Ecology and Evolution* 8: 3491–3504. https://doi.org/10.1002/ece3.3900.

Friess, D.A., Rogers, K., Lovelock, C.E. et al. (2019). The state of the world's mangrove forests: past, present, and future. *Annual Review of Environment and Resources* 44: 89–115. https://doi.org/10.1146/annurev-environ-101718-033302.

Friess, D.A., Thompson, B.S., Brown, B. et al. (2016). Policy challenges and approaches for the conservation of mangrove forests in Southeast Asia. *Conservation Biology* 30 (5): 933–949.

Gaos, A.R., Lewison, R.L., Liles, M.J. et al. (2016). Hawksbill turtle terra incognita: conservation genetics of eastern Pacific rookeries. *Ecology and Evolution* 6: 1251–1264. https://doi.org/10.1002/ece3.1897.

Gardner, C.J. (2016). Use of mangroves by lemurs. *International Journal of Primatology* 37: 317–332. https://doi.org/10.1007/s10764-016-9905-1.

Ghosh, S., Bakshi, M., Bhattacharyya, S. et al. (2015). A review of threats and vulnerabilities to mangrove habitats: with special emphasis on East Coast of India. *Journal of Earth Science & Climatic Change* 6: 1–9. https://doi.org/10.4172/2157-7617.1000270.

Gilbert, A.J. and Janssen, R. (1998). Use of environmental functions to communicate the values of a mangrove ecosystem under different management regimes. *Ecological Economics* 25: 323–346. https://doi.org/10.1016/s0921-8009(97)00064-5.

Gilman, E.L., Ellison, J., Duke, N.C. et al. (2008). Threats to mangroves from climate change and adaptation options: a review. *Aquatic Botany* 89: 237–250. https://doi.org/10.1016/j.aquabot.2007.12.009.

Giri, C. and Muhlhausen, J. (2008). Mangrove forest distributions and dynamics in Madagascar (1975–2005). *Sensors* 8: 2104–2117. https://doi.org/10.3390/s8042104.

Giri, C., Zhu, Z., Tieszen, L.L. et al. (2008). Mangrove forest distributions and dynamics (1975–2005) of the tsunami-affected region of Asia. *Journal of Biogeography* 35: 519–528. https://doi.org/10.1111/j.1365-2699.2007.01806.x.

Giri, C., Ochieng, E., Tieszen, L.L. et al. (2011). Status and distribution of mangrove forests of the world using earth observation satellite data. *Global Ecology and Biogeography* 20: 154–159. https://doi.org/10.1111/j.1466-8238.2010.00584.x.

Giri, C., Long, J., Abbas, S. et al. (2015). Distribution and dynamics of mangrove forests of South Asia. *Journal of Environmental Management* 148: 101–111. https://doi.org/10.1016/j.jenvman.2014.01.020.

Glaser, M. (2003). Interrelations between mangrove ecosystem, local economy and social sustainability in Caeté Estuary, North Brazil. *Wetlands Ecology and Management* 11: 265–272. https://doi.org/10.1023/A:1025015600125.

Global Mangrove Alliance Data Portal (2020). https://gma-panda.opendata.arcgis.com/ (accessed on 30 October 2020)

Godoy, M.D.P. and de Lacerda, L.D. (2015). Mangroves response to climate change: a review of recent findings on mangrove extension and distribution. *Anais da Academia Brasileira de Ciências* 87: 651–667. https://doi.org/10.1590/0001-3765201520150055.

Goforth Jr, H.W. and Thomas, J.R. (1980). Plantings of red Mangroves (*Rhizophora mangle* L.) for stabilization of Marl shorelines in the Florida keys (No. NOSC/TR-506). NAVAL OCEAN SYSTEMS CENTER SAN DIEGO CA.

Goldberg, L., Lagomasino, D., Thomas, T. et al. (2020). Global declines in human-driven mangrove loss. *Global Change Biology* 26: 5844–5855. https://doi.org/10.1111/gcb.15275.

Goutham-Bharathi, M.P., Roy, S.D., Krishnan, P. et al. (2014). Species diversity and distribution of mangroves in Andaman and Nicobar Islands, India. *Botanica Marina* 57: 421–432. https://doi.org/10.1515/bot-2014-0033.

Grantham, H.S., McLeod, E., Brooks, A. et al. (2011). Ecosystem-based adaptation in marine ecosystems of tropical Oceania in response to climate change. *Pacific Conservation Biology* 17: 241–258. https://doi.org/10.1071/pc110241.

Guannel, G., Arkema, K., Ruggiero, P. et al. (2016). The power of three: Coral reefs, seagrasses and mangroves protect coastal regions and increase their resilience. *PLoS One* 11: 1–22. https://doi.org/10.1371/journal.pone.0158094.

Gunawardena, M. and Rowan, J.S. (2005). Economic valuation of a mangrove ecosystem threatened by shrimp aquaculture in Sri Lanka. *Environmental Management* 36: 535–550. https://doi.org/10.1007/s00267-003-0286-9.

Gurib-Fakim, A. and Brendler, T. (2004). *Medicinal and Aromatic Plants of Indian Ocean Islands: Madagascar, Comoros, Seychelles and Mascarenes*. Stuttgart, Germany: Medpharm GmbH Scientific Publishers.

Hamilton, S.E. and Casey, D. (2016). Creation of a high spatio-temporal resolution global database of continuous mangrove forest cover for the 21st century (CGMFC-21). *Global Ecology and Biogeography* 25: 729–738. https://doi.org/10.1111/geb.12449.

Heimbuch, J. (2011). Hawksbill sea turtles discovered living in mangroves. Climate action. http://www.climateaction.org/news/endangered_hawksbill_turtles_living_in_mangroves.

Hein, L. (2000). Impact of shrimp farming on mangroves along India's East Coast. *Unasylva* 51: 48–55.

Hema, M. and Devi, I.P. (2014). Factors of mangrove destruction and management of mangrove ecosystem of Kerala, India. *Journal of Aquatic Biology and Fisheries* 2: 184–196.

Heras, S.R. and Soares, M.O. (2017). Effects of El niño on the coastal ecosystems and their related services. *Mercator* 16: 1–16. https://doi.org/10.4215/rm2017.e16030.

Hirales-Cota, M., Espinoza-Avalos, J., Schmook, B. et al. (2010). Drivers of mangrove deforestation in Mahahual-Xcalak, Quintana Roo, southeast Mexico. *Ciencias Marinas* 36 (2): 147–159.

Hoanh, C.T., Tuong, T.P., Gowing, J.W. et al. (2006). *Environment and Livelihoods in Tropical Coastal Zones*. UK: CAB International.

Horstman, E.M., Dohmen-Janssen, C.M., Narra, P.M.F. et al. (2014). Wave attenuation in mangroves: a quantitative approach to field observations. *Coastal Engineering* 94: 47–62. https://doi.org/10.1016/j.coastaleng.2014.08.005.

Hussain, S.A. and Badola, R. (2010). Valuing mangrove benefits: contribution of mangrove forests to local livelihoods in Bhitarkanika Conservation Area, East Coast of India. *Wetlands Ecology and Management* 18 (3): 321–331. https://doi.org/10.1007/s11273-009-9173-3.

Iftekhar, M.S. (2008). An overview of mangrove management strategies in three South Asian countries: Bangladesh, India and Sri Lanka. *International Forestry Review* 10: 38–51. https://doi.org/10.1505/ifor.10.1.38.

IPCC (2019). Sea level rise and implications for low lying islands, coasts and communities. https://www.ipcc.ch/srocc/chapter/chapter-4-sea-level-rise-and-implications-for-low-lying-islands-coasts-and-communities/ (accessed 29 October 2020).

Islam, M.S. and Wahab, M.A. (2005). A review on the present status and management of mangrove wetland habitat resources in Bangladesh with emphasis on mangrove fisheries and aquaculture. *Hydrobiologia* 542: 165–190. https://doi.org/10.1007/s10750-004-0756-y.

Islam, S.M.D.U. and Bhuiyan, M.A.H. (2018). Sundarbans mangrove forest of Bangladesh: causes of degradation and sustainable management options. *Environmental Sustainability* 1: 113–131. https://doi.org/10.1007/s42398-018-0018-y.

IUCN (2010). Mangrove forests in worldwide decline. https://www.iucn.org/content/mangrove-forests-worldwide-decline (accessed 29 October 2020)

IUCN (2011). Mangroves to receive huge boost from new carbon credit rules. https://www.iucn.org/content/mangroves-receive-huge-boost-new-carbon-credit-rules-0.

IUCN (2017). Mangroves: nurseries for the world's seafood supply. www.iucn.org/news/forests/201708/mangroves-nurseries-world%E2%80%99s-seafood-supply (accessed 27 October 2020).

IUCN (2018). Save Our Mangroves Now! https://www.iucn.org/news/environmental-law/201811/iucn-and-wwf-unite-enhance-implementation-legal-tools-protecting-mangroves (accessed 31 October 2020).

IUCN (2020a). Mangroves and coastal ecosystems. https://www.iucn.org/theme/marine-and-polar/our-work/climate-change-and-ocean/mangroves-and-coastal-ecosystems (accessed 29 October 2020)

IUCN (2020b). IUCN Red List of Threatened Species. www.iucnredlist.org/en (accessed 24 October 2020).

Jayatissa, L.P., Wickramasinghe, W.A.A.D.L., Dahdouh-Guebas, F. et al. (2008). Interspecific variations in responses of mangrove seedlings to two contrasting salinities. *International Review of Hydrobiology* 93: 700–710. https://doi.org/10.1002/iroh.200711017.

Kairo, J.G. and Mwita, M.M. (2020). Western Indian Ocean Mangrove Network, and Western Indian Ocean Marine Science Association. *Guidelines on Mangrove Ecosystem Restoration for the Western Indian Ocean Region-Western Indian Ocean Ecosystem Guidelines and Toolkits.* https://www.nairobiconvention.org/CHM%20Documents/WIOSAP/guidelines/GuidelinesonMangroveRestorationForTheWIO.pdf. (accessed 29 October 2020).

Kairo, J.G., Dahdouh-Guebas, F., Bosire, J. et al. (2001). Restoration and management of mangrove systems – a lesson for and from the East African region. *South African Journal of Botany* 67: 383–389. https://doi.org/10.1016/s0254-6299(15)31153-4.

Kamal, M. and Phinn, S. (2011). Hyperspectral data for mangrove species mapping: a comparison of pixel- based and object-based approach. *Remote Sensing* 3: 2222–2242. https://doi.org/10.3390/rs3102222.

Kamal, S., Grodzińska-Jurczak, M., and Brown, G. (2015). Conservation on private land: a review of global strategies with a proposed classification system. *Journal of Environmental Planning and Management* 58 (4): 576–597.

Kathiresan, K. (2012). Importance of mangrove ecosystem. *International Journal of Marine Science* 2: 70–89. https://doi.org/10.5376/ijms.2012.02.0010.

Kathiresan, K. and Bingham, B.L. (2001). Biology of mangroves and mangrove ecosystems. *Advances in Marine Biology* 40: 81–251. https://doi.org/10.1016/S0065-2881(01)40003-4.

Kathiresan, K. and Qasim, S.Z. (2005). *Biodiversity of Mangrove Ecosystems*, 537–539. New Delhi: Hindustan Publication Corporation.

Kathiresan, K. and Rajendran, N. (2005). Coastal mangrove forests mitigates tsunami. *Estuarine, Coastal and Shelf Science* 65: 601–606. https://doi.org/10.1016/j.ecss.2005.06.022.

Kelleway, J.J., Cavanaugh, K., Rogers, K. et al. (2017). Review of the ecosystem service implications of mangrove encroachment into salt marshes. *Global Change Biology* 23 (10): 3967–3983. https://doi.org/10.1111/gcb.13727.

Kinver, M. (2016). Rice and palm oil risk to mangroves. *BBC News* (4 January).

Krauss, K.W., Doyle, T.W., Doyle, T.J. et al. (2009). Water level observations in mangrove swamps during two hurricanes in Florida. *Wetlands* 29: 142–149. https://doi.org/10.1672/07-232.1.

Krauss, K.W., McKee, K.L., Lovelock, C.E. et al. (2014). How mangrove forests adjust to rising sea level. *New Phytologist* 202: 19–34. https://doi.org/10.1111/nph.12605.

Kuenzer, C., Bluemel, A., Gebhardt, S. et al. (2011). Remote sensing of mangrove ecosystems: a review. *Remote Sensing* 3: 878–928. https://doi.org/10.3390/rs3050878.

Kumar, R. (2000). Conservation and management of mangroves in India, with special reference to the State of Goa and the Middle Andaman Islands. *Unasylva* 51 (203): 41–46.

Kumari, P., Singh, J.K., and Pathak, B. (2020). Potential contribution of multifunctional mangrove resources and its conservation. In: *Biotechnological Utilization of Mangrove Resources* (ed. J.K. Patra), 1–26. London: Academic Press.

Kumari, R., Banerjee, A., Kumar, R. et al. (2019). Deforestation in India: Consequences and sustainable solutions. In: *Deforestation Around the World* (ed. P. Moutinho), 1–18. United Kingdom: IntechOpen.

Kusmana, C. and Sukristijiono, S. (2016). Mangrove resource uses by local community in Indonesia. *Jurnal Pengelolaan Sumberdaya Alam dan Lingkungan* 6 (2) https://doi.org/10.19081/jpsl.2016.6.2.217.

Lamparelli, C.C., Rodrigues, F.O., and Moura, D.O. (1997). Long-term assessment of an oil-spill in a mangrove forest in Sao Paulo, Brazil. In: *Mangrove Ecosystem Studies in Latin America and Africa* (eds. B. Kjerfve et al.), 191–203. Paris: UNESCO.

Lara, R., Szlafsztein, C., Cohen, M. et al. (2002). Implications of mangrove dynamics for private land use in Bragança, North Brazil: a case study. *Journal of Coastal Conservation* 8: 97–102. https://doi.org/10.1007/BF02806589.

Leal Filho, W., Barbir, J., and Preziosi, R. (2019). *Handbook of Climate Change and Biodiversity*. New York: Springer.

Lee, S.Y., Hamilton, S., Barbier, E.B. et al. (2019). Better restoration policies are needed to conserve mangrove ecosystems. *Nature Ecology & Evolution* 3: 870–872. https://doi.org/10.1038/s41559-019-0861-y.

Lee, S.Y., Primavera, J.H., Dahdouh-Guebas, F. et al. (2014). Ecological role and services of tropical mangrove ecosystems: a reassessment. *Global Ecology and Biogeography* 23: 726–743. https://doi.org/10.1111/geb.12155.

Lefebvre, G. and Poulin, B. (1997). Bird communities in Panamanian black mangroves: potential effects of physical and biotic factors. *Journal of Tropical Ecology* 13: 97–113. https://doi.org/10.1017/S0266467400010282.

Le, H. (2008). Economic reforms and mangrove forests in central Vietnam. *Society and Natural Resources* 21: 106–119. https://doi.org/10.1080/08941920701617775.

Lewis, R.R. III (2005). Ecological engineering for successful management and restoration of mangrove forests. *Ecological Engineering* 24: 403–418. https://doi.org/10.1016/j.ecoleng.2004.10.003.

Liang, S., Zhou, R., Dong, S. et al. (2008). Adaptation to salinity in mangroves: Implication on the evolution of salt-tolerance. *Chinese Science Bulletin* 53: 1708–1715. https://doi.org/10.1007/s11434-008-0221-9.

Liebezeit, G. and Rau, M.T. (2006). New Guinean mangroves-Traditional usage and chemistry of natural products. *Senckenbergiana Maritima* 36: 1–10. https://doi.org/10.1007/bf03043698.

Liu, H., Zhang, K., Li, Y. et al. (2013). Numerical study of the sensitivity of mangroves in reducing storm surge and flooding to hurricane characteristics in southern Florida. *Continental Shelf Research* 64: 51–65. https://doi.org/10.1016/j.csr.2013.05.015.

Lovelock, C., Cahoon, D., Friess, D. et al. (2015). The vulnerability of Indo-Pacific mangrove forests to sea-level rise. *Nature* 526: 559–563. https://doi.org/10.1038/nature15538.

Lovelock, C.E., Feller, I.C., Reef, R. et al. (2017). Mangrove dieback during fluctuating sea levels. *Scientific Reports* 7: 1–8. https://doi.org/10.1038/s41598-017-01927-6.

Lovelock, C.E., Krauss, K.W., Osland, M.J. et al. (2016). The physiology of mangrove trees with changing climate. In: *Tropical Tree Physiology* (eds. G. Goldstein and L.S. Santiago), 149–179. New York: Springer.

Lugo, A.E. and Snedaker, S.C. (1974). The ecology of mangroves. *Annual Review of Ecology and Systematics* 5: 39–64. https://doi.org/10.1146/annurev.es.05.110174.000351.

Lundquist, C., Carter, K., Hailes, S.F. et al. (2017). *Guidelines for managing mangrove (Mānawa) expansion in New Zealand*. National Institute of Water & Atmospheric Research Ltd. http://www.niwa.co.nz/managingmangroveguide (accessed 31 October 2020).

Luther, D.A. and Greenberg, R. (2009). Mangroves: a global perspective on the evolution and conservation of their terrestrial vertebrates. *BioScience* 59: 602–612. https://doi.org/10.1525/bio.2009.59.7.11.

López-Medellín, X., Ezcurra, E., González-Abraham, C. et al. (2011). Oceanographic anomalies and sea-level rise drive mangroves inland in the Pacific coast of Mexico. *Journal of Vegetarian Science* 22: 143–151. https://doi.org/10.1111/j.1654-1103.2010.01232.x.

Macintosh, D.J. and Ashton, E.C. (2002). A review of mangrove biodiversity conservation and management. *Centre for Tropical Ecosystems Research* 2 (11): 338–341.

Macnae, W. (1968). A general account of the fauna and flora of mangrove swamps and forests in the Indo-West-Pacific region. *Advances in Marine Biology* 6: 73–270. https://doi.org/10.1016/S0065-2881(08)60438-1.

Macreadie, P.I., Nielsen, D.A., Kelleway, J.J. et al. (2017). Can we manage coastal ecosystems to sequester more blue carbon? *Frontiers in Ecology and the Environment* 15 (4): 206–213.

Makowski, C. and Finkl, C.W. (2018). *Threats to Mangrove Forests: Hazards, Vulnerability, and Management*. New York: Springer.

Mangora, M.M. (2011). Poverty and institutional management stand-off: a restoration and conservation dilemma for mangrove forests of Tanzania. *Wetlands Ecology and Management* 19: 533–543. https://doi.org/10.1007/s11273-011-9234-2.

Mangroves for the future (2012). India. http://www.mangrovesforthefuture.org/countries/members/india/ (accessed 29 October 2020)

Manson, F.J., Loneragan, N.R., Harch, B.D. et al. (2005). A broad-scale analysis of links between coastal fisheries production and mangrove extent: a case-study for northeastern Australia. *Fisheries Research* 74: 69–85. https://doi.org/10.1016/j.fishres.2005.04.001.

Massel, S.R., Furukawa, K., and Brinkman, R.M. (1999). Surface wave propagation in mangrove forests. *Fluid Dynamics Research* 24: 219–249. https://doi.org/10.1016/s0169-5983(98)00024-0.

Mazda, Y., Magi, M., Ikeda, Y. et al. (2006). Wave reduction in a mangrove forest dominated by *Sonneratia* sp. *Wetlands Ecology and Management* 14: 365–378. https://doi.org/10.1007/s11273-005-5388-0.

Mazda, Y., Magi, M., Kogo, M. et al. (1997). Mangroves as a coastal protection from waves in the Tong King delta, Vietnam. *Mangroves and Salt Marshes* 1: 127–135. https://doi.org/10.1023/A:1009928003700.

McCarthy, M.P., Best, M.J., and Betts, R.A. (2010). Climate change in cities due to global warming and urban effects. *Geophysical Research Letters* 37: 1–5. https://doi.org/10.1029/2010gl042845.

Mchenga, I.S.S. and Ali, A.I. (2014). Natural regeneration of mangroves in a degraded and non-degraded tropical forest of Zanzibar island. *Journal of Global Biosciences* 3 (1): 334–344.
McNeely, J.A. (1993). Economic incentives for conserving biodiversity: lessons for Africa. *Ambio* 22: 144–150.
Meeder, J.F., Parkinson, R.W., Ruiz, P.L. et al. (2017). Saltwater encroachment and prediction of future ecosystem response to the Anthropocene Marine Transgression, Southeast Saline Everglades, Florida. *Hydrobiologia* 803: 29–48. https://doi.org/10.1007/s10750-017-3359-0.
Meek (2019). Effects of Deforestation: How Does Agriculture Cause Deforestation? https://sentientmedia.org/how-does-agriculture-cause-deforestation/ (accessed 29 October 2020).
MESCAL (2011). Review of Policy and Legislation Relating to Mangroves, the Use and Management of Mangroves in Tonga. https://www.iucn.org/sites/dev/files/content/documents/tonga_policy_and_legislative_review_report.pdf (accessed 31 October 2020).
Mitra, A. (2020). *Mangrove Forests in India*. New York: Springer.
Mitra, R., Bhattacharya, R.N., Hazra, S. et al. (2006). Mangrove conservation efforts and the local economy: a case study. *Economic and Political Weekly* 41: 3612–3616. https://doi.org/10.2307/4418592.
Mohd-Azlan, J., Noske, R.A. and Lawes, M.J. (2014). Resource Partitioning by Mangrove Bird Communities in North Australia. Biotropica, 46: 331–340. https://doi.org/10.1111/btp.12108
Mumby, P.J., Edwards, A.J., Arias-González, J.E. et al. (2004). Mangroves enhance the biomass of coral reef fish communities in the Caribbean. *Nature* 427: 533–536. https://doi.org/10.1038/nature02286.
Munji, C.A., Bele, M.Y., Idinoba, M.E. et al. (2014). Floods and mangrove forests, friends or foes? Perceptions of relationships and risks in Cameroon coastal mangroves. *Estuarine, Coastal and Shelf Science* 140: 67–75. https://doi.org/10.1016/j.ecss.2013.11.017.
Muraleedharan, P.K., Swarupanandan, K., Anitha, V. et al. (2009). The conservation of mangroves in Kerala: economic and ecological linkages. Division of Forestry and Human Dimension, Kerala Forest Research Institute. Peechi: 24.
Nagelkerken, I., Blaber, S.J.M., Bouillon, S. et al. (2008). The habitat function of mangroves for terrestrial and marine fauna: a review. *Aquatic Botany* 89: 155–185. https://doi.org/10.1016/j.aquabot.2007.12.007.
Naidoo, G. (2016). The mangroves of South Africa: an ecophysiological review. *South African Journal of Botany* 107: 101–113. https://doi.org/10.1016/j.sajb.2016.04.014.
Najjar, R., Walker, H., Anderson, P. et al. (2000). The potential impacts of climate change on the mid-Atlantic coastal region. *Climate Research* 14: 219–233. https://doi.org/10.3354/cr014219.
Najjar, R.G., Pyke, C.R., Adams, M.B. et al. (2010). Potential climate-change impacts on the Chesapeake Bay. *Estuarine, Coastal and Shelf Science* 86: 1–20. https://doi.org/10.1016/j.ecss.2009.09.026.
National Oceanic and Atmospheric Administration (2020). Coastal Blue Carbon. https://oceanservice.noaa.gov/ecosystems/coastal-blue-carbon/ (accessed 06 October 2020).
National Parks Board Singapore (2010). 4th National Report to the Convention on Biological Diversity. https://www.cbd.int/doc/world/sg/sg-nr-04-en.pdf (accessed 18 October 2020).
Neumann, B., Ott, K. et al. (2017). Strong sustainability in coastal areas: a conceptual interpretation of SDG 14. *Sustainability Science* 12 (6): 1019–1035.

Nagabhatla, N., Hung, N.T., Tuyen, L.T. et al. (2019). Ecosystem-based approach for planning research and capacity development for integrated coastal zone management in Southeast Asia. *Archives of Psychiatric Nursing* 9 (1): 3–9.

Nguyen, H.H., McAlpine, C., Pullar, D. et al. (2013). The relationship of spatial–temporal changes in fringe mangrove extent and adjacent land-use: case study of Kien Giang coast, Vietnam. *Ocean and Coastal Management* 76: 12–22. https://doi.org/10.1016/j.ocecoaman.2013.01.003.

Nielsen, M.G. (2011). Ants (Hymenoptera: Formicidae) of mangrove and other regularly inundated habitats: life in physiological extreme. *Myrmecological News* 14: 113–121.

Norris, R. and Cargile, M. (1998). Aquaculture and mangrove destruction. http://darwin.bio.uci.edu/~sustain/issueguides/mangrove/. (accessed 29 October 2020).

Noske, R.A. (1993). *Bruguiera hainesii*: another bird-pollinated mangrove? *Biotropica 25*: 481–483. https://doi.org/10.2307/2388873.

Noske, R.A. (1995). The ecology of mangrove forest birds in Peninsular Malaysia. *International Journal of Avian Science* 137: 250–263. https://doi.org/10.1111/j.1474-919X.1995.tb03247.x.

Nurdiani, R., Firdaus, M., and Prihanto, A.A. (2012). Phytochemical screening and antibacterial activity of methanol extract of mangrove plant (*Rhizophora mucronata*) from Porong river estuary. *Journal of Basic Science and Technology* 1 (2): 27–29.

Nurunnabi, T.R., Sabrin, F., Sharif, D.I. et al. (2020). Antimicrobial activity of endophytic fungi isolated from the mangrove plant *Sonneratia apetala* (Buch.-Ham) from the Sundarbans mangrove forest. *Advances in Traditional Medicine* 20: 419–425. https://doi.org/10.1007/s13596-019-00422-9.

Ong, J.E. (1995). The ecology of Mangrove conservation and management. *Hydrobiologia* 295: 343–351.

Ong, J.E. (2007). Food and Agriculture Organization of the United Nations regional office for Asia and the Pacific. http://www.fao.org/tempref/docrep/fao/010/ag124e/ag124e_full.pdf#page=142.

Osland, M.J., Feher, L.C., Griffith, K.T. et al. (2017). Climatic controls on the global distribution, abundance, and species richness of mangrove forests. *Ecological Monographs* 87: 341–359. https://doi.org/10.1002/ecm.1248.

Othman, M.A. (1994). Value of mangroves in coastal protection. *Hydrobiologia* 285: 277–282.

Owuor, M.A., Icely, J., and Newton, A. (2019a). Community perceptions of the status and threats facing mangroves of Mida Creek, Kenya: implications for community based management. *Ocean and Coastal Management* 175: 172–179. https://doi.org/10.1016/j.ocecoaman.2019.03.027.

Owuor, M.A., Mulwa, R., Otieno, P. et al. (2019b). Valuing Mangrove biodiversity and ecosystem services: a deliberative choice experiment in Mida Creek, Kenya. *Ecosystem Services* 40: 1–12. https://doi.org/10.1016/j.ecoser.2019.101040.

Palacios, M.L. and Cantera, J.R. (2017). Mangrove timber use as an ecosystem service in the Colombian Pacific. *Hydrobiologia* 803: 345–358. https://doi.org/10.1007/s10750-017-3309-x.

Pascual, U. and Perrings, C. (2007). Developing incentives and economic mechanisms for in situ biodiversity conservation in agricultural landscapes. *Agriculture, Ecosystems & Environment* 121 (3): 256–268.

Patel, D.M., Patel, V.M., Katriya, B. et al. (2014). Performance of mangrove in tsunami resistance. *International Journal of Emerging Technology and Research* 1 (3): 29–32.

Pattanaik, C. and Prasad, N.S. (2011). Assessment of aquaculture impact on mangroves of Mahanadi delta (Orissa), East coast of India using remote sensing and GIS. *Ocean and Coastal Management* 54: 789–795. https://doi.org/10.1016/j.ocecoaman.2011.07.013.

Peterson, J.M. and Bell, S.S. (2015). Saltmarsh boundary modulates dispersal of mangrove propagules: implications for mangrove migration with sea-level rise. *PLoS One* 10: 1–15. https://doi.org/10.1371/journal.pone.0119128.

Polidoro, B.A., Carpenter, K.E., Collins, L. et al. (2010). The loss of species: mangrove extinction risk and geographic areas of global concern. *PLoS One* 5: 1–10. https://doi.org/10.1371/journal.pone.0010095.

Poulin, B., Lefebvre, G., and McNeil, R. (1992). Tropical avian phenology in relation to abundance and exploitation of food resources. *Ecology* 73: 2295–2309. https://doi.org/10.2307/1941476.

Primavera, J.H. (1998). Mangroves as nurseries: shrimp populations in Mangrove and non-mangrove habitats. *Estuarine, Coastal and Shelf Science* 46: 457–464. https://doi.org/10.1006/ecss.1997.0275.

Primavera, J.H. (2008). Socio-economic impacts of shrimp culture. *Aquaculture Research* 28: 815–827. https://doi.org/10.1046/j.1365-2109.1997.00946.x.

Primavera, J.H., dela Cruz, M., Montilijao, C. et al. (2016). Preliminary assessment of post-Haiyan mangrove damage and short-term recovery in Eastern Samar, central Philippines. *Marine Pollution Bulletin* 109: 744–750. https://doi.org/10.1016/j.marpolbul.2016.05.050.

Quadros, A.F. and Zimmer, M. (2017). Dataset of "true mangroves" plant species traits. *Biodiversity Data Journal* 5: 1–21. https://doi.org/10.3897/BDJ.5.e22089.

Ragavan, P., Dubey, S.K., Dagar, J.C. et al. (2019). Current understanding of the mangrove forests of India. In: *Research Developments in Saline Agriculture* (ed. J.C. Dagar), 257–304. New York: Springer nature.

Ragavan, P., Saxena, A., Jayaraj, R.S.C. et al. (2016). A review of the mangrove floristics of India. *Taiwania* 61: 224–242. https://doi.org/10.6165/tai.2016.61.224.

Rajkaran, A., Adams, J.B., and du Preez, D.R. (2010). A method for monitoring mangrove harvesting at the Mngazana estuary, South Africa. *African Journal of Aquatic Science* 29: 57–65. https://doi.org/10.2989/16085910409503792.

Ramachandran, R., Banerjee, K., Paneer Selvam, A. et al. (2018). Legislation and policy options for conservation and management of seagrass ecosystems in India. *Ocean and Coastal Management* 159: 46–50. https://doi.org/10.1016/j.ocecoaman.2017.12.025.

Ramirez, D.R. and Townsend, W. (2006). Seaweed and mangroves: improving environmental practices in coastal communities to secure sustainable livelihoods. In: *Coastal Resource Management in the Wider Caribbean: Resilience, Adaptation, and Community Diversity* (eds. Y. Breton, D. Brown, B. Davy, et al.), 172–190. Jamaica: Ian Randle Publishers.

Rasquinha, D.N. and Mishra, D.R. (2020). Impact of wood harvesting on mangrove forest structure, composition and biomass dynamics in India. *Estuarine, Coastal and Shelf Science* 106974 https://doi.org/10.1016/j.ecss.2020.106974.

Ravindran, K.C., Venkatesan, K., Balakrishnan, V. et al. (2005). Ethnomedicinal studies of Pichavaram mangroves of East coast, Tamil nadu. *Indian Journal of Traditional Knowledge* 4 (4): 409–411.

Ravindranath, N.H. and Murthy, I.K. (2010). Greening India Mission. *Current Science* 99 (4): 444–449.

Record, S., Charney, N.D., Zakaria, R.M. et al. (2013). Projecting global mangrove species and community distributions under climate change. *Ecosphere* 4 (3): 1–23.

Riascos, J.M., Cantera, J.R., and Blanco-Libreros, J.F. (2018). Growth and mortality of mangrove seedlings in the wettest neotropical mangrove forests during ENSO: implications for vulnerability to climate change. *Aquatic Botany* 147: 34–42. https://doi.org/10.1016/j.aquabot.2018.03.002.

Richards, D.R. and Friess, D.A. (2016). Rates and drivers of mangrove deforestation in Southeast Asia, 2000–2012. *Proceedings of the National Academy of Sciences* 113: 344–349. https://doi.org/10.1073/pnas.1510272113.

Ricklefs, R.E. and Latham, R.E. (1993). Global patterns of diversity in mangrove floras. In: *Species Diversity in Ecological Communities: Historical and Geographical Perspectives* (eds. R.E. Ricklefs and D. Schluter), 215–229. Chicago: University of Chicago Press.

Rideout, A.J.R., Joshi, N.P., Viergever, K.M. et al. (2013). Making predictions of mangrove deforestation: a comparison of two methods in Kenya. *Global Change Biology* 19: 3493–3501. https://doi.org/10.1111/gcb.12176.

Rochmyaningsih, D. (2017). Aquaculture is main driver of mangrove losses. https://www.eco-business.com/news/aquaculture-is-main-driver-of-mangrove-losses/ (accessed 29 October 2020).

Rog, S.M., Clarke, R.H., and Cook, C.N. (2017). More than marine: revealing the critical importance of mangrove ecosystems for terrestrial vertebrates. *Diversity and Distributions* 23 (2): 221–230.

Rogers, K., Boon, P.I., Branigan, S. et al. (2016). The state of legislation and policy protecting Australia's mangrove and salt marsh and their ecosystem services. *Marine Policy* 72: 139–155. https://doi.org/10.1016/j.marpol.2016.06.025.

Rogers, K., Saintilan, N., Mazumder, D. et al. (2019). Mangrove dynamics and blue carbon sequestration. *Biology Letters* 15 (3): 1–5. https://doi.org/10.1098/rsbl.2018.0471.

Romañach, S.S., DeAngelis, D.L., Koh, H.L. et al. (2018). Conservation and restoration of mangroves: global status, perspectives, and prognosis. *Ocean and Coastal Management* 154: 72–82. https://doi.org/10.1016/j.ocecoaman.2018.01.009.

Ron, J. and Padilla, J.E. (1999). Preservation or conversion? Valuation and evaluation of a mangrove forest in the Philippines. *Environmental and Resource Economics* 14 (3): 297–331.

Rosenzweig, C. and Hillel, D. (2008). *Climate Variability and the Global Harvest: Impacts of El Niño and other Oscillations on Agro-ecosystems*. United Kingdom: Oxford University Press.

Roy, A.K.D. (2016). Local community attitudes towards mangrove forest conservation: lessons from Bangladesh. *Marine Policy* 74: 186–194. https://doi.org/10.1016/j.marpol.2016.09.021.

Roy, A.K.D. and Alam, K. (2012). Participatory forest management for the sustainable management of the Sundarbans mangrove forest. *American Journal of Environmental Sciences* 8: 549–555. https://doi.org/10.3844/ajessp.2012.549.555.

Saha, S.K. (2010). Soilless cultivation for landless people: an alternative livelihood practice through indigenous hydroponic agriculture in flood-prone Bangladesh. *Ritsumeikan Asia Pacific University* 27: 139–151.

Sanderman, J., Hengl, T., Fiske, G. et al. (2018). A global map of mangrove forest soil carbon at 30 m spatial resolution. *Environmental Research Letters* 13: 1–12. https://doi.org/10.1088/1748-9326/aabe1c.

Sandilyan, S. and Kathiresan, K. (2012). Mangrove conservation: a global perspective. *Biodiversity and Conservation* 21: 3523–3542. https://doi.org/10.1007/s10531-012-0388-x.

Sandilyan, S. and Kathiresan, K. (2015). Mangroves as bioshield: an indisputable fact. *Ocean and Coastal Management* 103: 94–96. https://doi.org/10.1016/j.ocecoaman.2014.11.011.

Santisuk, T. (1983). Taxonomy and distribution of terrestrial trees and shrubs in the mangrove formations in Thailand. *Natural History Bulletin of the Siam Society* 31 (1): 63–91.

Saranraj, P. and Sujitha, D. (2015). Mangrove medicinal plants: a review. *American-Eurasian Journal of Toxicological Sciences* 7: 146–156. https://doi.org/10.5829/idosi.aejts.2015.7.3.94150.

Sarmin, N.S., Hasmadi, I.M., Pakhriazad, H.Z. et al. (2016). The DPSIR framework for causes analysis of mangrove deforestation in Johor, Malaysia. *Environmental Nanotechnology Monitoring & Management* 6: 214–218. https://doi.org/10.1016/j.enmm.2016.11.002.

Sathirathai, S. and Barbier, E.B. (2007). Valuing mangrove conservation in southern Thailand. *Contemporary Economic Policy* 19: 109–122. https://doi.org/10.1111/j.1465-7287.2001.tb00054.x.

Saw, A.A. and Kanzaki, M. (2015). Local livelihoods and encroachment into a mangrove forest reserve: a case study of the Wunbaik reserved mangrove forest, Myanmar. *Procedia Environmental Sciences* 28: 483–492. https://doi.org/10.1016/j.proenv.2015.07.058.

Scaife, A., Guilyardi, E., Cain, M. et al. (2019). What is the El Niño–Southern oscillation? *Weather* 74: 250–251. https://doi.org/10.1002/wea.3404.

Scales, I.R. and Friess, D.A. (2019). Patterns of mangrove forest disturbance and biomass removal due to small-scale harvesting in southwestern Madagascar. *Wetlands Ecology and Management* 27: 609–625. https://doi.org/10.1007/s11273-019-09680-5.

Schneider, P. (2011). The discovery of tropical mangroves in graeco-roman antiquity: science and wonder. *The Journal of the Hakluyt Society*: 1–16. https://www.hakluyt.com/downloadable_files/Journal/mangroves1.pdf.

Schwartzstein, P. (2019). This vanishing forest protects the coasts – and lives – of two countries. https://www.nationalgeographic.com/magazine/2019/07/sundarbans-mangrove-forest-in-bangladesh-india-threatened-by-rising-waters-illegal-logging/ (accessed 30 October 2020).

Seepana, R., Perumal, K., Kada, N.M. et al. (2016). Evaluation of antimicrobial properties from the mangrove *Rhizophora apiculata* and *Bruguiera gymnorrhiza* of Burmanallah coast, South Andaman, India. *Journal of coastal life medicine* 4: 475–478. https://doi.org/10.12980/jclm.4.2016J6-52.

Semeniuk, V. (1994). Predicting the effect of sea-level rise on mangroves in northwestern Australia. *Journal of Coastal Research* 10 (4): 1050–1076.

Semesi, A.K. (1992). Developing management plans for the mangrove forest reserves of mainland Tanzania. *Hydrobiologia* 247: 1–10. https://doi.org/10.1007/BF00008199.

Sharrock, S. and Jackson, P.W. (2016). Plant Conservation and the Sustainable Development Goals. 1-19. *Conference of the Global Partnership for Plant Conservation on the theme 'Plant Conservation and the Sustainable Development Goals (SDGs)'*, Missouri Botanical Garden, St Louis, U.S.A. (28–29 June 2016).

Shunula, J. (2002). Public awareness, key to mangrove management and conservation: the case of Zanzibar. *Trees* 16: 209–212. https://doi.org/10.1007/s00468-001-0147-1.

Singh, B.R. and Singh, O. (2012). Study of impacts of global warming on climate change: eise in sea level and disaster frequency. In: *Global Warming – Impacts and Future Perspectives* (ed. B.R. Singh), 93–118. United Kingdom: IntechOpen.

Singh, J.K. (2020). Structural characteristics of mangrove forest in different coastal habitats of Gulf of Khambhat arid region of Gujarat, west coast of India. *Heliyon* 6: 1–7. https://doi.org/10.1016/j.heliyon.2020.e04685.

Singh, P.K., Papageorgiou, K., Chudasama, H. et al. (2019). Evaluating the effectiveness of climate change adaptations in the world's largest Mangrove ecosystem. *Sustainability* 11: 1–17. https://doi.org/10.3390/su11236655.

Sippo, J.Z., Lovelock, C.E., Santos, I.R. et al. (2018). Mangrove mortality in a changing climate: an overview. *Estuarine, Coastal and Shelf Science* 215: 241–249. https://doi.org/10.1016/j.ecss.2018.10.011.

Spalding, M., Kainuma, M., and Collins, L. (2010). *World Atlas of Mangroves*. London: Earthscan.

Spiteri, A. and Nepalz, S.K. (2006). Incentive-based conservation programs in developing countries: a review of some key issues and suggestions for improvements. *Environmental Management* 37 (1): 1–14.

Srivastava, S. and Mehta, L. (2017). The social life of Mangroves: Resource complexes and contestations on the industrial coastline of Kutch, India. STEPS Working Paper 99, Brighton: STEPS Centre.

Suman, D.O. (2019). Mangrove management: challenges and guidelines. In: *Coastal Wetlands* (eds. G. Perillo et al.), 1055–1079. Netherlands: Elsevier.

Sundar, I. (2018). Environmental conservation policies, programmes and legislations in India. *International Journal of Management, IT & Engineering* 8 (9): 1–11.

Sunyowati, D., Hastuti, L., and Butar, F.B. (2016). The regulation of sustainable mangroves and coastal zones management in Indonesia. *Journal of Civil & Legal Sciences* 6 (1) https://doi.org/10.4172/2169-0170.1000220.

Sustainability Times (2020). Myanmar's mangrove forests are in grave peril. https://www.sustainability-times.com/environmental-protection/myanmars-mangrove-forests-are-in-grave-peril/ (accessed 29 October 2020).

Tam, N.F.Y. and Wong, Y.S. (1999). Mangroves soils in removing pollutants from municipal waste water of different salinities. *Journal of Environmental Quality* 28 (2): 556–564.

Teas, H.J. (1977). Ecology and restoration of mangrove shorelines in Florida. *Environmental Conservation* 4: 51–58. https://doi.org/10.1017/S0376892900025042.

Thomas, N., Lucas, R., Bunting, P. et al. (2017). Distribution and drivers of global mangrove forest change, 1996–2010. *PLoS One* 12: 1–14. https://doi.org/10.1371/journal.pone.0179302.

Thu, P.M. and Populus, J. (2007). Status and changes of mangrove forest in Mekong Delta: case study in Tra Vinh, Vietnam. *Estuarine, Coastal and Shelf Science* 71 (1-2): 98–109.

Tolangara, A.R. (2014). Forest destruction, wood utilization and mangrove area in district Jailolo, West Halmahera Regency, province of North Mollucas and the conservation education. *International Journal of Engineering Research and Development* 10 (1): 54–60.

Tomlinson, P.B. (2016). Floristics. In: *The Botany of Mangroves*, 29–42. Cambridge, UK: Cambridge University Press.
Udoh, J.P. (2016). Sustainable nondestructive mangrove-friendly aquaculture in Nigeria II: models, best practices and policy frame work. *Aquaculture, Aquarium, Conservation & Legislation* 9 (1): 151–173.
Uddin, M.S., van Steveninck, E.D.R., Stuip, M. et al. (2013). Economic valuation of provisioning and cultural services of a protected mangrove ecosystem: a case study on Sundarbans Reserve Forest, Bangladesh. *Ecosystem Services* 5: 88–93. https://doi.org/10.1016/j.ecoser.2013.07.002.
UNESCO (2015). Proclamation of the International day for the conservation of the mangrove ecosystem. 38th general Conference. https://unesdoc.unesco.org/ark:/48223/pf0000235350. (accessed on 31 October 2020).
United Nations Ocean Conference (2019). Save our mangroves now! by Germany, Federal Ministry for Economic Cooperation and Development (BMZ) (Government). https://oceanconference.un.org/commitments/?id=27534.
Upadhyay, V.P., Ranjan, R., and Singh, J.S. (2002). Human-mangrove conflicts: the way out. *Current Science* 83: 1328–1336.
Valiela, I. and Cole, M.L. (2002). Comparative evidence that salt marshes and mangroves may protect seagrass meadows from land-derived nitrogen loads. *Ecosystems* 5: 92–102. https://doi.org/10.1007/s10021-001-0058-4.
Valiela, I., Bowen, J.L., and York, J.K. (2001). Mangrove Forests: one of the world's threatened major tropical environments. *BioScience* 51: 807–815.
Van Lavieren, H., Spalding, M., Alongi, D.M. et al. (2012). *Securing the Future of Mangroves*. Hamilton, ON, Canada: United Nations University, Institute for Water, Environment and Health.
Vidyasagaran, K. and Madhusoodanan, V.K. (2014). Distribution and plant diversity of mangroves in the west coast of Kerala, India. *Journal of Biodiversity and Environmental Sciences* 4 (5): 38–45.
Vincentius, A., Nessa, M.N., Jompa, J. et al. (2019). Complex relationship between mangrove ecosystem variables and fish assemblages at Maumere Bay, Indonesia. *IOP Conference Series: Earth and Environmental Science* 253: 1–10. https://doi.org/10.1088/1755-1315/253/1/012035.
Virata, J.B. (2011). Hawksbill sea turtles found foraging and living in mangrove estuaries. https://www.reptilesmagazine.com/hawksbill-sea-turtles-found-foraging-and-living-in-mangrove-estuaries/#:~:text=Gaos%20says%20it%20is%20still,rock%20outcrops%20within%20the%20estuaries.
Walters, B.B. (1997). Human ecological questions for tropical restoration: experiences from planting native upland trees and mangroves in the Philippines. *Forest Ecology and Management* 99: 275–290. https://doi.org/10.1016/S0378-1127(97)00211-9.
Walters, B.B. (2003). People and mangroves in the Philippines: fifty years of coastal environmental change. *Environmental Conservation* 30: 293–303. https://doi.org/10.1017/S0376892903000298.
Walters, B.B. (2004). Local management of mangrove forests in the Philippines: successful conservation or efficient resource exploitation? *Human Ecology* 32: 177–195. https://doi.org/10.1023/b:huec.0000019762.36361.48.

Wang, L., Mu, M., Li, X. et al. (2010). Differentiation between true mangroves and mangrove associates based on leaf traits and salt contents. *Journal of Plant Ecology* 4: 292–301. https://doi.org/10.1093/jpe/rtq008.

Wang, W., Yan, Z., You, S. et al. (2011). Mangroves: obligate or facultative halophytes? A review. *Trees* 25: 953–963. https://doi.org/10.1007/s00468-011-0570-x.

Ward, R.D., Friess, D.A., Day, R.H. et al. (2016). Impacts of climate change on mangrove ecosystems: a region by region overview. *Ecosystem Health and Sustainability* 2: 1–25. https://doi.org/10.1002/ehs2.1211.

Whitefield, P., Beauchamp, E., Boyd, D.S. et al. (2019). Exploring temporality in socio-ecological resilience through experiences of the 2015–16 El Niño across the Tropics. *Global Environmental Change* 55: 1–14. https://doi.org/10.1016/j.gloenvcha.2019.01.004.

Wilson, R. (2017). Impacts of climate change on mangrove ecosystems in the coastal and marine environments of Caribbean small island developing states (SIDS). *Caribbean Climate Change Report Card: Science Review*, (pp. 60–82).

Wongthong, P. (2020). Potential impacts of sea level rise on mangroves. http://www.coastalwiki.org/wiki/Potential_Impacts_of_Sea_Level_Rise_on_Mangroves (accessed 30 October 2020).

Wood, J. (2019). World Economic Forum. https://www.weforum.org/agenda/2019/02/5-reasons-to-protect-mangrove-forests-for-the-future/ (accessed 29 October 2020).

Woodroffe, C.D. and Davies, G. (2009). The morphology and development of Coastal wetlands in the Tropics. In: *Coastal Wetlands: An Integrated Ecosystem Approach* (eds. G.M.E. Perello et al.), 65–88. Netherlands: Elsevier.

Xu, S., He, Z., Zhang, Z. et al. (2017). The origin, diversification and adaptation of a major mangrove clade (Rhizophoreae) revealed by whole genome sequencing. *National Science Review* https://doi.org/10.1093/nsr/nwx065.

Yeo, S (2014). Save mangroves for people, planet and the economy, says UN. https://www.climatechangenews.com/2014/09/30/save-mangroves-for-people-planet-and-the-economy-says-un/ (accessed 02 August 2018).

Yong, J. (2018). Origin of mangroves & mangrove diversity. https://mangroveactionproject.org/origin-of-mangrove-mangrove-diversity/ (accessed 02 May 2018)

Yáñez-Espinosa, L. and Flores, J. (2011). A review of sea-level rise effect on mangrove forest species: anatomical and morphological modifications. In: *Global Warming Impacts: Case Study on the Economy, Human Health and on Urban and Natural Environments* (ed. S. Casalegno), 253–276. United Kingdom: IntechOpen.

Zhang, K., Liu, H., Li, Y. et al. (2012). The role of mangroves in attenuating storm surges. *Estuarine, Coastal and Shelf Science* 102: 11–23. https://doi.org/10.1016/j.ecss.2012.02.021.

Zorini, L.O., Contini, C., Jiddawi, N. et al. (2004). Participatory appraisal for potential community-based mangrove management in East Africa. *Wetlands Ecology and Management* 12 (2): 87–102. https://doi.org/10.1023/b:wetl.0000021672.15252.54.

13

Wetland Conservation and Restoration

Kajal Patel[1], Meenakshi Chaurasia[1], InduTripathi[2], and Shilpi Nagar[2,3]

[1] *Department of Botany, University of Delhi, New Delhi, India*
[2] *Department of Environmental Studies, University of Delhi, New Delhi, India*
[3] *Centre for Fire, Explosive and Environment Safety (CFEES), Defence Research Development Organization (DRDO), New Delhi, India*

13.1 Introduction

Wetlands are natural ecosystems which have innumerable benefits and global purport which also makes them stand out as the most productive ecosystems. Their wide range of plant and animal diversity that includes bird diversity as well is one of the major reasons behind their conservation studies. Wetland ecosystems offer many services to mankind including improving water quality, flood mitigation, nutrient cycling, groundwater recharge, etc. Wetlands act as transition zones between land and water, providing natural protection from floods and storms. Wetlands have gained more attention since the past few decades due to their distinct ecology. As per the classical classification of these ecosystems, wetlands are categorized into different categories, viz. marine, estuarine, lacustrine (lakes), and palustrine (swamps, marshes, and bogs) which are characterized on the grounds of their hydrology, ecology, and geology.

According to Ramsar Convention on Wetlands (1971), which is a joint large-scale collaboration initiative by various nations and international organizations of the world for the conservation and sustainable use of wetlands and their resources, wetlands are defined (Article 1.1) as "areas of marsh, fen, peatland, or water, whether natural or artificial, permanent or temporary, with water that is static or flowing, fresh, brackish, or salt, including areas of marine water the depth of which at low tide does not exceed six metres" (Figure 13.1a). The sites which are included as Ramsar sites and are of great importance are enlisted as: 1052 sites in Europe; 289 sites in Asia; 359 sites in Africa; 175 sites in South America; 211 sites in North America; and 79 sites in the Oceania region (Figure 13.1b) (Ramsar Secretariat 2013). Wetland could be any water body including natural (such as lakes, coastal lagoons, river basins, mangroves, peatland, and coral reefs) and manmade wetlands (such as ponds, farm ponds, irrigated fields, sacred groves, salt pans, reservoirs, gravel pits, sewage farms, and canals) constitute these vital ecosystems according to the

Wetlands Conservation: Current Challenges and Future Strategies, First Edition.
Edited by Sanjeev Sharma and Pardeep Singh.

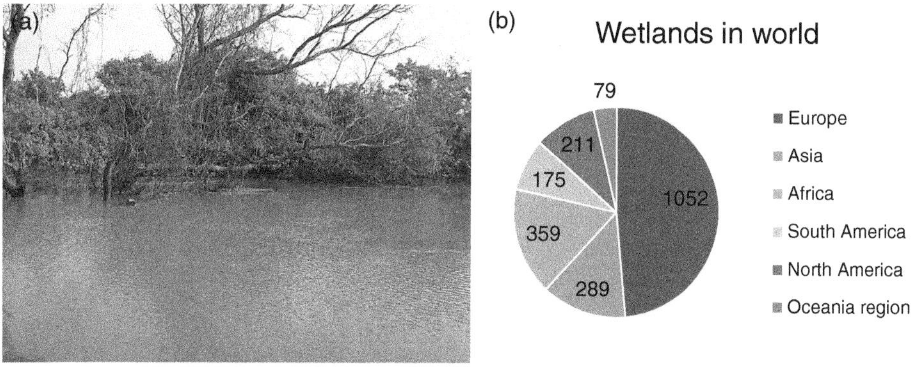

Figure 13.1 (a) Wetland in Bhitarakanika National Park, Rajnagar, Kendrapada, Odisha. *Source:* Wikimedia Commons; https://commons.wikimedia.org/wiki/File:Bhitarkanika_National_Park.jpg#filehistory. (b) Number of wetlands in different regions of the world comprise Europe (49%), Asia (13%), Africa (16%), South America (8%), North America (10%), and the Oceania region (4%).

definition of Ramsar Convention. Due to several limitations, among the vast range of wetlands, a mere 26 are enlisted as Ramsar sites (Ramsar 2013), numbers of freshwater wetland ecosystems are threatened and many of them have already degraded and lost. The reasons for the degradation include urbanization, population growth, and increased economic activities (Central Pollution Control Board 2008).

Some of the distinctive attributes such as topographical, climatic, hydrological, and vast biodiversity of the Indian subcontinent has ensured a significant diversity of its wetlands. These include both temporary seasonally flooded as well as permanent marshes and swamps in shallow lakes, large river basins, and littoral zones of water bodies and reservoirs. There are several water lagoons, estuarine backwaters, and extensive mangroves in the deltas and estuaries of numerous rivers. Along the edges of coastal areas, there are also marine wetlands that constitute coastal beds of marine algae and coral reefs. Multidimensional studies related to the ecology of both the freshwater wetlands and mangroves of the Indian subcontinent have been done in order to deeply understand and protect these indispensable ecosystems (Denny 1991; Gopal 1982; Gopal and Krishnamurthy 1992). Keoladeo Ghana at Bharatpur, Kaziranga in Assam, and Sunderbans in Ganga-Brahmaputra delta are some of the most important sites were designated as sanctuaries for the protection of their biological diversity long before the need for wetland conservation was accounted by international organizations. Whereas the distribution, habitat characteristics, and biota of the mangroves have been relatively corroborated (Gopal and Krishnamurthy 1992) but the inland wetlands (freshwater and saline) are poorly documented. Various inventories in context to resources and classification of wetland have been prepared recently which was an unprecedented gesture. These inventories vastly include every aspect related to wetlands such as the current status of resources, biodiversity, their functions, importance, and future qualitative and quantitative changes which may arise due to the impact of natural and human activities. The inventories can play a significant role in their classification based on comparison of functions and values of different kinds of wetlands, selection of representative sites of different wetland types for conservation and to formulate scientifically sound management strategies.

Since the time of initiation of industrialization, there is a huge increment in anthropogenic activities, which leads to environmental degradation especially of the aquatic ecosystems. Wetlands act as the most important part of the aquatic ecosystem as they play an important role in water storage, groundwater replenishment, shoreline stabilization, water purification, reservoirs of biodiversity, pollination, climate change mitigation, tourism, and cultural and societal importance. Hence, there is a need for immediate actions and strategies for the conservation of wetlands to prevent environmental degradation. In order to prepare a strategy for a long term, it is very important to understand the wetland degradation process, their significance, the consequences of their degradation in terms of ecological imbalance, the current status of management strategies globally and thus, finally, plan for their restoration and sustainability. Hence, in the coming sections, all the above-mentioned points will be discussed in an integrative and holistic way.

13.2 Wetlands: Role and Importance

The unique and most significant transition zones between terrestrial and aquatic ecosystems can be simply described as wetlands. Some of their characteristic features make these systems play a key role in many processes, which ultimately helps maintain the integrity of the ecosystem and, thus, the ecological balance. Therefore, these ecosystems are generally referred to as the "kidneys" of the earth. Some of the important processes include the regulation of nutrients, water cycle, groundwater recharge and purification, food and water supply which fulfil all the vital needs of the wildlife, i.e. for both flora and fauna. Thus, they served as one of the most essential aspects of tourism, and cultural and societal values, which promote sustainable development for all. Further, the importance of wetlands is mentioned below:

i) Wetlands can be referred to as multiple-value systems as they do not just play a single role whereas they execute several functions simultaneously and, therefore, can provide a suite of values to human welfare. Optimizing for one is usually at the expense of another.
ii) Mostly, the valuable "products" of wetlands are public amenities that provide economic "payoff" to individuals and also most of the services of wetlands accrue to the public on a large scale. Thus, wetland protection is the domain of a representative government working in partnership with private landowners. Hence, nowadays, the compensation for these values has become a controversial topic.
iii) It's really complicated and important to study the relationship between wetland area and their marginal value. Traditional economics concept elucidated that "the less there is of some commodity, the more valuable it is," also known as the general economic law of scarcity. This concept can be applied to wetlands in some cases. Like, when there are extensive wetlands in a given area (e.g. the Louisiana delta, the Everglades area in Florida), they purposefully got converted to other land uses to fulfill social necessity because there are many more available wetlands elsewhere. In areas where almost all natural wetlands have been drained (e.g. Europe, Midwestern USA, etc.), the desire to conserve what few wetlands are left with is often very strong. However, the idea of

wetland conservation does not always work even in areas where almost all natural wetlands have been drained, and where critical thresholds exist that define the minimum ecologically acceptable number of wetlands in a given region. If wetlands are too small, functions such as support of certain mammals or storage of floodwater (for mitigation of floods) no longer exist.

iv) Commercial values are limited, whereas wetlands provide values in perpetuity of wetland functions and thus values have the potential to last for a very long time. Modern agriculture or industrial/commercial activity are generally unsustainable and resource-depleting (soil loss; use of fossil fuels). So, the lifetime of these human-based alternatives is short-lived. Even large public works projects have time spans of 50–100 years. A corollary of this point is that once wetlands are lost through development, the loss of their functions and values is often irreversible.

v) Comparison of short-term economic gain from a non-wetland commercial use of a piece of landscape, with long-term wetland values, is often not appropriate. The social trap of assigning value to wetlands, in addition to the danger of comparing ecosystems, is that it will always be possible to find a more profitable use for the land if short-term economic analyses are made. For example, wetland value estimates cannot compete very well with the economic return of corn and soybeans in the Midwestern USA. So, drainage will always be economically favored over wetland conservation if the traditional cost-benefit analysis is carried out.

vi) The social as well as economical values are influenced by cultural bias and economic system. There are good reasons why we wish to protect nature; developed countries, having taken care of the basic needs of their citizens, are particularly involved in protecting ecosystems, including wetlands, for their aesthetic as well as more functional attributes, not all of which translate into direct economic benefit. Other cultures, where the basic needs of food and shelter cannot be taken for granted, have a different view of the economics of wetlands. Many cultures do live in and among wetlands and use them for daily subsistence – the production of food and fiber. Yet they generally leave the normal wetland functions intact. The values that we ascribe to wetlands are not separate from the institutions and culture from which we come (Mitsch and Gosselink 2000). Ascribing economic value to wetlands has been an exercise frequently practiced in the past 25 years.

13.3 Wetland Loss Leading to Ecological Imbalance

Wetlands are one of the most vital ecosystems on the earth which can be defined as the distinct areas of marsh, fen, peatland, or water whether natural or constructed, permanent or temporarily flooded with water. These areas are specialized with static or flowing, fresh, brackish water with a depth of not more than six meters. They play a critical role in the regulation of oxygen, thus functioning as the "kidneys" of the earth and hence are the key players in maintaining ecological balance and also as the bank for the world's biological and genetic material pool (Bai et al. 2013; Dong et al. 2013). A study by Costanza et al. (1997) estimated that the value of wetland ecosystem services occupies the topmost rank among different kinds of the ecosystem and they account for 47% of the value of the global ecosystem. Thus, it can be concluded that these ecosystems are the most productive. There are a total of 2388 wetland sites across the world with 37 wetlands (also called Ramsar sites) in India (Ramsar Secretariat 2020).

According to the Global Wetland Outlook (2018), there are almost 20 000 wetland-dependent species assessed globally and 25%, i.e. almost 4000 of them are threatened with extinction due to several reasons. Despite their great socioeconomic values, wetlands are still facing the menaces of loss or degradation. The mammoth wetland areas of the global wetland areas are in Asia (32%), North America (27%), and Latin America and the Caribbean (16%) while, in Europe (13%), Africa (10%), and Oceania (3%) are relatively smaller (Davidson et al. 2018). Studies with data available between 1970 and 2015 evaluated a 35% reduction of global wetlands (Gardner and Finlayson 2018). Ramsar Convention Bureau report (2015) also insisted that the world had lost more than 50% of the total wetlands and the loss mainly took place in the twentieth century, which is considered as the century of urbanization and modernization at a large scale. Around 84% of Ramsar's listed wetlands experienced threatening due to ecological changes (Moser et al. 1996).

The major threats to wetlands include several anthropogenic activities, viz. hunting, pollution, agricultural drainage, settlements, and urbanization, which affect almost 85% of wetlands in the Asian region (Scott and Poole 1989; Moser et al. 1996). Besides, anthropogenic activities, some natural threats are also responsible for wetland losses like erosion, subsidence, droughts, hurricanes, and other storms (Fretwell et al. 1996). The major human activities which pose threat to the wetlands can be categorized in hydrologic alterations, pollution inputs, and vegetation damage. The hydrologic alterations represent the yearly variation in saturation of groundwater which evolves wetland's characteristics. The common hydrologic alterations include filling, drainage, diking, and damming to form ponds and diversion of flow, which are practiced at a large scale for development purposes. Pollution inputs include the additions of hazardous material such as sediment, fertilizer, human sewage, animal waste, road salts, pesticides, and heavy metals. The sources of water pollution input include runoff from urban, rural, agricultural, and mining areas, and inputs of air pollution are cars, factories, power plants, and others. Lastly, the vegetation damage which is the most critical one includes grazing by domestic animals, introduction of non-native plants (which compete with other plants) and removal of vegetation for construction and mining (USEPA 2001). Based on the reports of USEPA (2001), the major causes can be categorized into three classes on the basis of process and products:

1) Biological alterations: Include removal of wetland and vegetations.
2) Chemical alterations: Include the release of pollutants and toxic chemicals.
3) Physical alterations: Include construction and mining activities like filling, draining, dredging, stream channelization, peat mining, recreational uses, and changes in sediments.

The main functions of wetlands include improvement of water quality, harvesting of floodwater, habitat availability for fishes and wildlife, aesthetics, and biological productivity also (USEPA 2002). Besides it, services provided by wetlands also include erosion control, carbon sequestration, and biodiversity conservation. Consequently, it is evident that the species that directly or indirectly depend on wetland have become either extinct or globally threatened. In terms of quantitative values, there are bird species which account for 21%, mammal species which account for 37%, and freshwater fish species which account

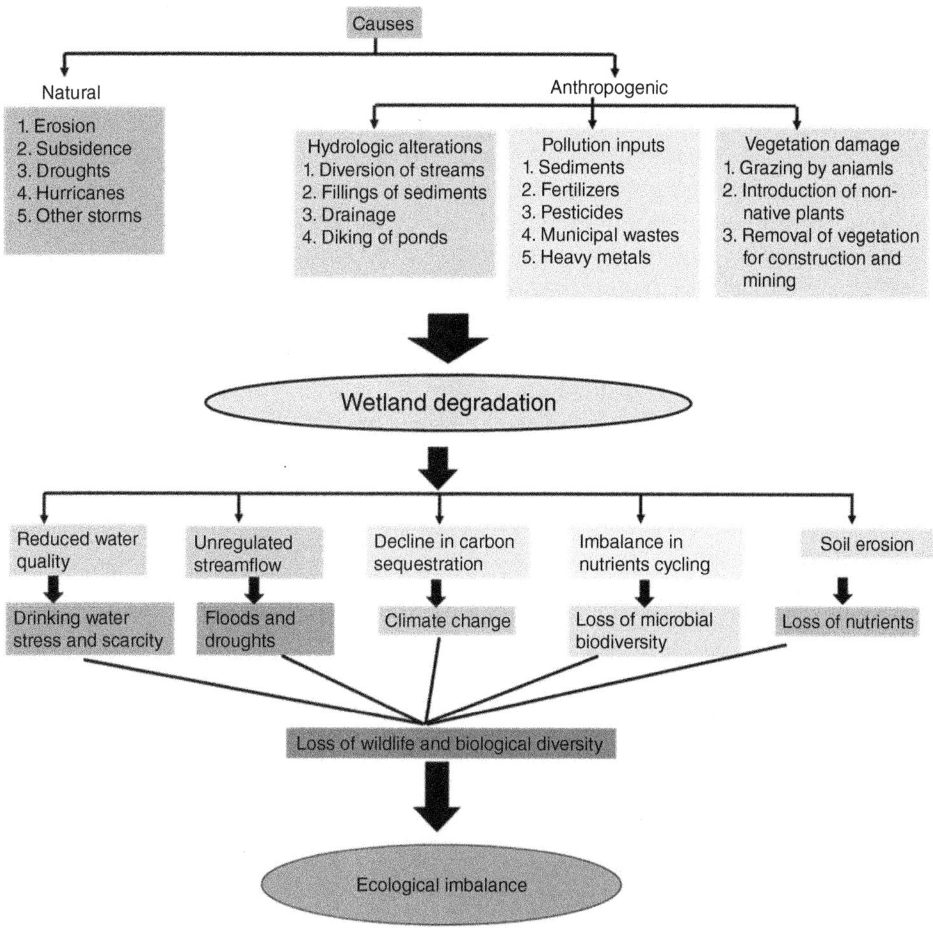

Figure 13.2 Summary of how wetland degradation leads to ecological imbalance.

for 27%. These functions and services are directly linked to climate changes, the entire environment, and the global ecology which includes both biotic, i.e. flora and fauna and the abiotic factors, i.e. soil, water (Chandra and Kumar 2020; Lolu et al. 2020). Additionally, these functions and services indirectly help in flood attenuation, regulation of streamflow, sediment trapping, nutrients cycling and balancing. All these functions and services are interrelated to each other and help maintain an ecological balance and functioning so as to conserve nature and life on this earth (Figure 13.2) (Gupta et al. 2020).

13.4 Wetland Management Strategies: Current Status

Wetland systems are most productive, yet greatly threatened due to their multifunctional nature. Due to exponential growth in overexploitation of resources and pollution from anthropogenic sources, wetland ecosystems around the world are facing the threat of loss or degradation. With the realization of the importance of these systems, wetland management is receiving far increasing attention worldwide. The concept of wetland management is not new. It has differed with period across the different regions (Xu et al. 2019). In due course of time, wetland management evolved from wetland drainage to flood control, coastal protection, and water quality improvement. Today, wetland management means a systematic, continuous process of setting management objectives on the basis of the description of ecological, socioeconomic and cultural features and current environmental regulations, accordingly formulation and implementation of the action plan, followed by continuous monitoring and evaluation to assess its effectiveness and further revisions in the management process.

The process may sound simple, but there are many downsides to consider such as lack of comprehensive knowledge of these systems and coordination impact of actions on local communities, sustainability of the plan, making the plan inflexible and not allocating resources to ensure the plan can be implemented. Thus, a centralized management strategy with a multidisciplinary approach is needed for the restoration of these systems and to maintain their productivity and biodiversity. The establishment of Ramsar Convention has benefited these ecologically sensitive systems. To help site managers, a framework for planning wetland management has been issued by Ramsar Convention secretariat (2010) (Figure 13.3). The framework includes the following guidelines:

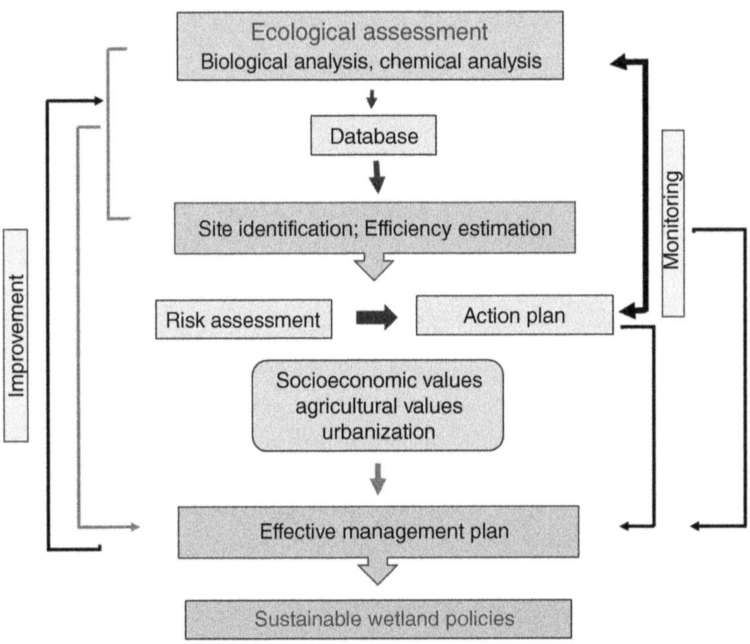

Figure 13.3 Summary of the process for sustainable wetland restoration and conservation. Key strategies involved are ecological assessment of the system and goal definition based upon the obtained database. The next step is risk assessment followed by the formation of an action plan for the effective management program. All these processes are continuously monitored for improvement in the management plan. Based upon the proper assessment and monitoring, wetland policies are prepared. *Source:* Modified from Gokce (2018).

- Identifying the existing and possible impact factor affecting the site.
- Identifying objectives for site management.
- Identifying the required actions to achieve management objectives.
- Continuous monitoring of management actions.
- Obtaining funds from stakeholders.
- Communications and negotiations with stakeholders and the local community.
- Periodic review for assessment of the effectiveness of management policies.
- Ensuring compliance with policies at local, national, and international level.

In addition to Ramsar Convention, other forms of international policies like World Heritage Convention and Bonn Convention have been made for wetland recognition and protection. More than 200 central and state policies exist for the management of wetlands around the world, but still currently more than half of wetland site action plans need to be completed in North and South America, around 49% sites in Africa and 47% sites in Asia lack management plans. Though Europe has well-established management plans, 41% of these sites are threatened by several factors due (Xu et al. 2019). The main reasons behind the inefficiency of these plans are the lack of proper integration of policies, coordination between different stakeholders, and conflict regarding land-use planning (Smardon and Faust 2006). In developing countries like India, the government disclosed Wetlands (Conservation and Management) Rules in 2010. A Central Wetlands Regulatory Authority (CWRA) has also been constituted under the chairmanship of the Secretary of the Environment and Forests to regulate activities such as hydraulic alterations, unsustainable grazing, harvesting of resources, releasing treated effluents, aquaculture, and agriculture (Acharya and Adak 2009). An effective and efficient management plan is imperative for the maintenance of ecological characters and wise use of resources. Following are the key strategies that need to be followed to make a sustainable management plan (Gardner and Finlayson 2018):

1) **Management policy:** They help ensure the trouble-free implementation of wetland management that includes scientific laws, regulations, and policies on wetland protection which should be issued in accordance with the local development.
2) **Monitoring:** A dynamic or real-time monitoring system for assessment of status of a global wetland is a prerequisite for wetland management and conservation. Monitoring can be done using remote sensing technique and conducting field observations.
3) **Restoration:** Restoration of degrading or vanishing wetlands using scientific ecological technology and engineering. This involves the protection of biodiversity, controlling pollution sources, increasing ecological water quantity, and prohibiting the overuse of harvesting of biological resources.
4) **Knowledge:** Awareness among people, research-oriented studies in context to protection, and restoration of wetlands should be promoted to improve wetland management to a higher level.
5) **Funding:** Safeguarding the funding for wetland conservation and restoration from different sources is a precondition to successfully achieve the above task.

13.5 Wetland Restoration and Sustainability

The wetland system regulates and links water in our landscape. It acts as a source, sink of water, supports a wide range of biodiversity, greatest natural carbon stores and is vital to mitigate and adapt to climate change. In short, wetlands are crucial to inclusive economic growth, climate mitigation, and amelioration and thus important to human well-being and adaptation. Conserving, restoring, and sustainable utilization of wetlands can help countries in achieving their sustainable development goals (SDGs) like (i) zero hunger; (ii) clean water and sanitation; (iii) sustainable cities and communities; (iv) responsible consumption and production; (v) climate action; (vi) life below water; and (vii) life on land. The fourth Strategic Plan (2016–2024) by Ramsar Convention identified four principal goals and 19 specific targets that can directly contribute to achieving SDGs as well as the Aichi Targets set by the Convention on Biological Diversity.

13.5.1 4th Ramsar Strategic Plan 2016–2024 (Source: Ramsar Secretariat 2016)

1) Address the drivers of wetland loss and degradation by engaging with relevant stakeholders to reduce threats to wetlands, influence trends, restore wetlands, and communicate good practices;
2) Effective management and conservation of Ramsar site network;
3) Wise use of all wetland resources, beyond Ramsar sites, including integrated management and restoration of a river basin; and
4) Active implementation including measures such as scientific guidance, communication, international cooperation, and capacity building.

However, the actual implementation of these strategic plans requires an integral approach and their practical implementation on the ground level. Recognition of wetland, their conservation, and wise use of wetland resources by the parties committed to the Ramsar Convention are crucial in achieving the goals of sustainable development by 2030 (https://www.ramsar.org/sites/default/files/documents/library/wetlands_sdgs_e.pdf). For this, required key actions are:

- Integrated approaches: Ensuring the efforts to implement plans of action to achieve Ramsar Strategic Plan and its result are directly linked to their national agenda of sustainable development goals.
- Effective and scaled up advanced policies and practices: Adopt scaled up wetland interventions for a broader and better result on ecological processes and human well-being and including these conservation and restoration plans in the policies of other sectors, i.e. agriculture, forestry, and urban planning to meet multiple objectives.
- Cooperation and collaborations: Promote partnership among different sectors of society for planning and implementing plans related to conservation and sustainable development of Ramsar Sites and other protected areas such as World Heritage Sites. This will also provide funding for completing these comprehensive actions.

13.6 Conclusion

In spite of interface and overlapping with aquatic and terrestrial systems, wetlands are distinct from other ecosystems in their structural and functional features. And these unique features make them highly productive but most pregnable to changing environmental and anthropogenic factors. In the past few decades, the wetland system has deteriorated exponentially due to climate change, pollution emission, loss of biodiversity, and change in land use and land cover. Wetlands provide a number of ecosystem services. Any disruption to their ecological structure and functioning ultimately leads to an ecological imbalance that may further aggravate the situation. There is an urgent need for the implementation of a comprehensive and sustainable management plan and restoration programs at the local, regional, and global level to safeguard these imperative systems. Adopting the conservation measures and restoring these systems can help in achieving the goal of sustainable development across the world.

Although, in the past few decades, the importance and value of wetlands have been recognized by the national and international community and, accordingly, several measures have been taken to protect, conserve, and ensure their sustainable utilization. However, wetland management and their wise use continue to be issues both for developing and developed countries. For this, we need further scientific knowledge and the development of efficient policies. This chapter provides an overview of the importance of wetland system and the need for their restoration and conservation for sustainable development. This chapter further suggests the following key points that need to be considered for future advancement in our understanding of these complex systems and better approaches for wetland protection in the changing world.

- Far-reaching assessment of socioeconomic values of wetland ecosystem and measurement of the values of benefits derived from these systems.
- Comprehensive assessment and inventory of biodiversity associated with these systems for their protection.
- Conservation of habitats for inhabiting plant and animal community and other species.
- Continuous evolving management programmes and policies to identify the long-term impact of these actions on the ecosystem.

Acknowledgments

The authors are grateful to Professor K.S. Rao, Natural Resources Management Lab, Department of Botany, University of Delhi, New Delhi, India for his kind support. Financial support by the UGC, New Delhi is gratefully acknowledged.

References

Acharya, S.O.M.E.N. and Adak, T. (2009). Wetland management for sustainable development. *Journal of Soil and Water Conservation* 8 (4): 25–30.

Bai, J., Cui, B., Cao, H. et al. (2013). Wetland degradation and ecological restoration. *The Scientific World Journal* 2013: 523632.

Central Pollution Control Board, CPCB. (2008). Status of water supply, wastewater generation and treatment in Class-I Cities and Class-II Towns on India, *Control of Urban Pollution Series: CUPS/70/2009-10*, New Delhi.

Chandra, P. and Kumar, M. (2020). Contribution of microbes in the renovation of Wetlands. In: *Restoration of Wetland Ecosystem: A Trajectory Towards a Sustainable Environment* (eds. A. Upadhyay, R. Singh and D. Singh), 101–124. Singapore: Springer.

Costanza, R., d'Arge, R., de Groot, R. et al. (1997). The value of the world's ecosystem services and natural capital. *Nature 387*: 253–260.

Davidson, N.C., Fluet-Chouinard, E., and Finlayson, C.M. (2018). Global extent and distribution of wetlands: trends and issues. *Marine and Freshwater Research 69* (4): 620–627.

Denny, P. (1991. B. Gopal (ed.). 1990.). *Ecology and management of aquatic vegetation in the Indian Sub-continent*. Dordrecht: Kluwer Academic Publishers Group. *Journal of Tropical Ecology*, Cambridge University Press, 7(4), pp.540–540.

Dong, X., Hengjie, Z., Haiting, J., and Shuqing, A. (2013). Ecological restoration of degraded wetlands in China. *Journal of Resources and Ecology 4* (1): 63–69.

Fretwell, J.D., Williams, J.S., and Redman, P.J. (1996). *National Water Summary of Wetland Resources*. USGPO.

Gardner, R.C. and Finlayson, M. (2018). *Global Wetland Outlook: State of the World's Wetlands and Their Services to People 2018*. Gland, Switzerland: Secretariat of the Ramsar Convention.

Gokce, D. (2018). Introductory chapter: Wetland importance and management. In: *Wetlands Management-Assessing Risk and Sustainable Solutions* (ed. D. Gokce), 3–10. Intech Open.

Gopal, B. and Krishnamurthy, K. (1992). Wetlands of south Asia. In: *Wetlands of the World. I. Inventory, Ecology and Management, Handbook of Vegetation Science*, vol. 15, no. 2 (eds. D.F. Wigham, D. Dykyjova and S. Hejny), 345–416. Dordrecht, The Netherlands: Kluwer Academic Publishers.

Gopal, B. (ed.) (2012). *Ecology and Management of Aquatic Vegetation in the Indian Subcontinent*, vol. 16. Springer Science & Business Media.

Gopal, B. (1982). Ecology and management of fresh-water wetlands in India. *Proceedings of the Fresh Water Ecosystem Dynamics in Wetlands and Shallow Waters*, Scope.

Gupta, G., Khan, J., Upadhyay, A.K., and Singh, N.K. (2020). Wetland as a sustainable reservoir of ecosystem services: prospects of threat and conservation. In: *Restoration of Wetland Ecosystem: A Trajectory Towards a Sustainable Environment* (eds. A. Upadhyay, R. Singh and D. Singh), 31–43. Singapore: Springer.

Lolu, A.J., Ahluwalia, A.S., Sidhu, M.C. et al. (2020). Carbon sequestration and storage by wetlands: implications in the climate change scenario. In: *Restoration of Wetland Ecosystem: A Trajectory Towards a Sustainable Environment* (eds. A.K. Upadhyay, R. Singh and D.P. Singh), 45–58. Singapore: Springer.

Mitsch, W.J. and Gosselink, J.G. (2000). The value of wetlands: importance of scale and landscape setting. *Ecological Economics* 35 (1): 25–33.

Moser, M., Prentice, C., and Frazier, S. (1996). A global overview of wetland loss and degradation. *Wetland International, Proceedings of the 6th Meeting of the Conference of Contracting Parties*; Brisbane, Australia.

Ramsar Convention Bureau (2006). *The Ramsar Convention Manual: A Guide to the Convention on Wetlands, Ramsar, Iran, 1971*. Ramsar Convention Bureau.

Ramsar Convention Bureau (2015). *Wetlands for Our Future: Act Now to Prevent, Stop, and Reserve Wetland Loss*. Gland, Switzerland: *Ramsar Convention Bureau- Press Release.*

Ramsar Convention Secretariat (2010). Managing wetlands: frameworks for managing Wetlands of international importance and other wetland sites. In: *Ramsar Handbooks for the Wise Use of Wetlands*, 4e, vol. 18. Gland, Switzerland: Ramsar Convention Secretariat.

Ramsar Convention Secretariat (2013). *The Ramsar Convention Manual: A Guide to the Convention on Wetlands (Ramsar, Iran, 1971)*, 6e. Gland, Switzerland: *Ramsar Convention Secretariat.*

Ramsar Secretariat (2016). *An Introduction to the Ramsar Convention on Wetlands*, 7e. The Ramsar Convention Manual. Gland: Ramsar Convention Secretariat.

Ramsar Convention on Wetlands (2018). *Global Wetland Outlook: State of the World's Wetlands and their Services to People*. Gland, Switzerland: Ramsar Convention Secretariat.

Ramsar Secretariat. (2020). The list of Wetlands of international importance. The Secretariat of the Convention on Wetlands, Gland, Switzerland.

Ramsar. (2013). Scaling up wetland conservation, wise use and restoration to achieve the sustainable Development Goals. Wetlands and the SDGs. July 2018. *Ramsar Convention On Wetlands*. at:https://www.ramsar.org/sites/default/files/documents/library/wetlands_sdgs_e.pdf (accessed 23 April 2020).

Scott, D.A. and Poole, C.M. (1989). A status overview of Asian wetlands, based on "A Directory of Asian Wetlands". Asian Wetland Bureau, Malaysia. *UNESCAP. 1996. Coastal environmental management plan for Pakistan*. ST/ESCAP/1360, Bangkok.

Smardon, R.C. and Faust, B.B. (2006). Introduction: International policy in the biosphere reserves of Mexico's Yucatan peninsula. *Landscape and Urban Planning* 74: 160–192.

USEPA. (2001). Threats to wetlands (EPA 843-F-01-002d). United States Environmental Protection Agency. http://nepis.epa.gov (accessed 23 April 2020).

USEPA. (2002). Functions and values of wetlands (EPA 843-F-01-002c). United States Environmental Protection Agency. http://nepis.epa.gov (accessed 23 April 2020).

Xu, T., Weng, B., Yan, D. et al. (2019). Wetlands of international importance: status, threats, and future protection. *International Journal of Environmental Research and Public Health* 16 (10): 1818.

Zhang, J.F. and Sun, Q.X. (2005). Causes of wetland degradation and ecological restoration in the Yellow River Delta Region. *Forestry Studies in China* 7 (2): 15–18.

Zhao, Q., Bai, J., Huang, L. et al. (2016). A review of methodologies and success indicators for coastal wetland restoration. *Ecological Indicators 60*: 442–452.

Index

a

Acanthus ilicifolius 243
Aegicers corniculatum 213
Africa 9, 17, 26, 237, 276, 279
Agenda 2030 103, 108, 248
agriculture 8, 9, 169, 202
  floating 190
  sustainable 105–106
Aichi Targets 280
algae 45, 68, 74, 112
algal blooms 10
Almarina 213
angiosperms 45
animals 45, 71, 98–99
  see also biodiversity; wetland fauna
anthropogenic activities 41, 75–77, 123, 133–134, 152–153, 166–169, 276
anthropogenic change 5, 8, 9, 11
aquaculture 9, 123, 172
  and mangroves 242, 245–246
aquatic plants 45–46
  see also biodiversity; wetland flora
artificial wetlands 44–45
Asia 9, 26, 237, 276, 279
  see also South Asia
Assam Remote Sensing Application Centre (ARSAC) 97
Australia 9
Australian Convention 2
Avicennia marina 210, 214
Avicennia schaueriana 213
Avicennia species 214
Azolla spp. 45, 67, 191
Azraq Oasis wetland 168

b

Bais Bay 252
Banacon Island 252
Bangladesh 153, 172, 252
Baraila Jheel 60
Barataria Bay 165
barrier islands 165
Bhagar Beel 64
Bhutan 104
Bihar Wasteland (Reclamation, Cultivation and Improvement) Act 1946, 64
bioaccumulation 210
biodiversity 8–10, 45–46, 78, 122, 170, 187
  mangrove forests 242
  North Bihar wetlands 65–71
Biodiversity Liaison Group (BLG) 34
Biological Diversity Act of 2002 248
bioremediation 68, 210, 241
BirdLife International 33
birds 70, 76–77, 98, 114–115, 119, 121
bisphenol A (BPA) 213
Bonaire, Island of 37
Bonn Convention 279
Brahmaputra 58, 97, 143–144, 154, 273
Brazil 141, 145
Bruguiera gymnorhiza 213, 243

c

cadmium (Cd) 75, 77, 210, 214
Camargue wetland 47

Wetlands Conservation: Current Challenges and Future Strategies, First Edition.
Edited by Sanjeev Sharma and Pardeep Singh.

Cambodia 105
Cameroon 104
Canadian wetland classification 2
canals 5
capacity-building strategies 11, 108
carbon dioxide 5, 72, 121, 153, 184
carbon emission 149
carbon sequestration 5, 72, 130, 171, 241
carbon sinks 5, 145
Caribbean, the 276
CBD *see* Convention on Biological Diversity
Central America 237
Central Wetlands Regulatory Authority (CWRA) 279
CEPA *see* Communication, Education, Participation, and Awareness
Chandubi Festival 99
Chandubi Wetland 97–102, 109
Chilika Lagoon 112–117, 124
Chilwa Wetland 104
China 9, 18, 149
cities 106–107
climate change 5, 6, 8, 9, 11, 42, 48–49, 184
  action to combat 107
  and degradation of wetlands 166
  and loss of wetlands 171
  and mangroves 246
  and North Bihar wetlands 71–77
  and reservoir-based wetlands (RBWs) 153–154
coastal marshes 43
Coastal Regulation Zone (CRZ) 123
coastal wetlands 8, 20, 163
  see also lagoon wetlands
cobalt (Co) 45, 75
Communication, Education, Participation, and Awareness (CEPA) 35
complete economic benefit (TEV) 89
Conference of Parties (COP) 22, 24, 25, 29, 32
constructed wetlands (CWs) technology 214–215
contingent choice method 96–97
contingent valuation method 89, 95–96
Convention on Biological Diversity (CBD) 34
Convention on Conservation of Migratory Species of Wild Animals (CMS) 34
Convention on the International Trade in Endangered Species (CITE) 34
Convention on Wetlands of International Significance (Ramsar Convention) 10, 11, 41, 135, 272–273
  budget 32–33
  classification of wetlands 3–4, 20–21
  cooperation with other organizations 33–35
  definition of wetlands 2–3, 19–20
  education and outreach 35–36
  effectiveness of 37–38
  Fourth Strategic Plan 31, 103–108, 280
  framework for planning wetland management 278–279
  legal status 36–37
  membership 25–26
  mission of 22
  operational framework 25–33
  Paris Protocol 19
  Regina Amendments 19
  regions under 26–30
  structural framework 22–24
  three pillars of 31–32
copper (Cu) 75, 77, 210, 214
coral reef 48–49, 104, 107
crabs 65, 69, 114, 120
crayfish 10, 169
Crown Court Estate 37
cultural bias 275
cultural practices 46
cultural services 46–47, 92, 97, 113, 116, 184
cyclones 48, 90, 106, 242, 243

d

damage cost method 94–95
dams 140–141, 145, 147–151, 154–156
  removal of 142
  run-of-the-river (RoR) 144, 147, 150–151
desertification 9
devegetation 170

diffuse recharging 41
disaster risk reduction (DRR) 90, 94, 98, 106, 108
  ecosystem-based (EcoDRR) 98, 103, 106–107, 109
ditches 5, 167, 170
dolphins 70, 71, 77, 79, 98
draining 5, 169
dredging 58, 124, 169, 202
droughts 5, 6
dumping 169

e

Eastern Himalayan river basin 140–156
ecological character 31, 58, 88, 134
ecological imbalance 275–277
economic growth 106
Economics of Ecosystems and Biodiversity (TEEB) 99
economic system 275
eco-remediation technologies 217
ecosystems 129
  see also ecosystem services (ES); ecosystem valuation
ecosystem services (ES) 97, 100, 105
  Chilika Lagoon 112–117
  Pulicat Lagoon 119–123
  reservoir-based wetlands (RBWs) 140–142 145–148
ecosystem valuation 87–109
  methodology 90–94
ecotourism 47–48, 98, 99
Eichornia crassipes 217
El Niño-Southern Oscillation (ENSO) 247
employment 106
ENSO *see* El Niño-Southern Oscillation
environmental impact assessment (EIA) 155, 156
Environmental Protection Act of 1986 123, 248
erosion 43, 172
estuaries 8
Europe 9, 17, 26, 276, 279
European Environmental Agency 35
eutrophication 10, 63, 73, 75, 168, 178
evidence-based approaches 251
Excoecaria agallocha 214
extreme climate events 48

f

farming 245
  see also agriculture; aquaculture
fertilizers 44, 72, 73, 123, 168, 176, 189, 190
fish 69, 98, 152, 242
fisheries 114, 119–120
fishing 47, 64, 69
fish ladders 155
fish passes 155
floating bed technique 216–217
floating gardens 182, 190
flood control 40–41, 185
flooding 5, 6, 8, 9, 40, 76, 98, 165, 172, 183
floodplain wetlands 9, 72, 167
flood regulation 113, 115, 120
food,
  production 46, 105, 141
  resources 185, 187
  security 46, 105, 112, 154
Food and Agriculture Organization (FAO) 33
forests 107
  rain 45, 46, 230
  sub-arctic 8
  tidal 204, 229
  tropical 8
  see also mangroves
France 47
freshwater marshes 3, 4, 164, 168
freshwater wetlands 8, 42, 133, 166, 167, 273

g

Ganga 58, 63, 64, 67, 68, 70–72, 74, 76–77, 89, 143
gastropods 69, 70
Gazaldoba 147–148, 151, 153
gender equality 105
genomics studies 247
Ghana 251
glaciers 8

global climate 184
global warming 9, 246
see also climate change
Gogabeel Pakshi Vihar 64
Grand Reservoir and Dam Database (GRaND) 141
grazing 169
green economy 106
greenhouse gases 5, 7, 107, 149, 153, 171–172
Green India Mission 248
green technologies 212
groundwater 41–42 184
ground-water pumping 168
Guxin River 216

h

habitats, loss of 8, 49, 63, 74, 168, 170, 171
head race tunnels (HRTs) 150
heavy metals 44–45, 148
and mangrove ecosystem 201–218
in mangrove plants 210, 214
in the sediment 204–209, 213
hedonic pricing method 89, 92–93
herbivory 166
Heritiera fomes 243
high-altitude wetlands (HAWs) 178, 182, 194
high wind velocity 243
human-made wetlands 5, 20, 149, 164
human settlements 244–245
hunting 191
Hurricane Sandy 107, 165
Hydrilla verticellata 190
hydrologic alterations 169, 276
hydrophytes 45, 140
hydropower 141–143, 150–152
see also dams

i

Indian Ocean 243
Indonesia 172, 237, 248
industrial development 178, 245
industrialization 106, 201, 274
industrial waste 10, 119, 185
infrastructure development 8, 9, 106, 167
inland wetlands 20, 163–164
insurance value 88–89
Intergovernmental Panel on Climate Change (IPCC) 74
international cooperation 32–35
International Organization Partners (IOPs) 33
International Union for Conservation of Nature (IUCN) 33, 34
see also IUCN Red List
International Water Management Institute (IWMI) 33
intertidal marshes 163
invasive species 10, 73, 168–168
Irrawaddy 114, 116, 143, 167
irrigated lands 164
IUCN *see* International Union for Conservation of Nature
IUCN Red List 9, 230

j

Jagatpur Wetland 62–63
Joint Liaison Group (JLG) of the Rio Conventions 34
Jordan 168
Junam wetland 47

k

Kabartal 60, 63
Kandelia candel 213
Kashmir Himalaya 175–195
current status 176–178
distribution and extent of wetlands 182–184
wetland classification 178–182
wetland conservation 191–195
Kaziranga 273
Kenya 47, 245
Keoladeo National Park 25, 89, 273
Kusheshwar Asthan 62, 67–70, 76

l

lagoon wetlands 111–124
Lake Chad 74
lakes 20, 41

land,
  conversion 167, 170
  ownership 252
  reclamation 191
  system changes 188–189
La Niña 247
Latin America 276
lead (Pb) 77, 210, 214
legal framework 78,135–136, 192
  mangrove conservation 248
local communities 251–252
Lolium perenne 217
Louisiana marsh 166

m

macroalgae 111, 115, 120
macrobenthos 69
macroinvertebrate 168
macrophytes 65–67, 69, 72, 76, 114
malaria 10
Malawi 104
manganese 210
mangroves 3, 104, 107, 229–230
  and biodiversity 242
  and climate change 246
  geographical distribution 237–240
  heavy metal pollution 191–195
  medicinal value 243
  role of 240–243
  species diversity 230–237
  strategies for conservation 247–252
  threats to 243–247
  true 230–231
  wood harvesting 245
Manguo wetland 47
marine resources 107
marine wetlands 20, 163, 273
market price method 91
market prices 90–94
MAR project 18
marshes,
  coastal 43
  freshwater 3, 4, 164, 168
  intertidal 163
  salt 43, 107, 113, 168
  subtypes 3
  tidal 3, 104
medicinal plants 46, 186, 243
metallothioneins 210
methane 5, 7, 130, 153
mid-altitude wetlands (MAWs) 178, 182
Millennium Ecosystem Assessment 46, 47, 113, 166
mining 169, 202, 204, 276
Mississippi River 165 166, 172
monsoons 55, 58–59, 63–64, 69, 73–74, 112, 121, 142–143, 145, 148, 151
Montreux Record 31
Moti Jheel 63
mountain snowfields 8
Myriophyllum spicatum 190

n

Nabugabo wetland 47
Nariva Swamp 47
National Action Plan on Climate Change 248
national action plans 136
National Ramsar Committees 30–31
National Wetland Conservation and Management Programme (NWCMP) 60
natural capital 106
natural disasters 72, 241, 242–243
  see also disaster risk reduction (DRR)
navigation 115
Nelumbo nucifera 186
Netherlands 37
New Zealand 2
nickel (Ni) 45, 75
nitrogen 44, 72, 75, 76, 116, 121, 205, 216
nitrous oxide 153
non-governmental organizations (NGOs) 33, 155
non-native species 147
North America 9, 17, 18, 26, 45, 237, 276, 279
North Bihar wetlands 55–80
  and biodiversity 65–71
  and climate change 71–77

North Bihar wetlands (*Cont'd*)
  distribution of wetlands in 58–63
  legal framework and policies 77–79
  pollution 71–77
  sustainable livelihood and services 64–65
  urbanization 71–77
nutrient cycling 187
nutrient retention 43–44, 168
Nymphea alba 186
Nymphoides peltatum 190

o
Oceania 276
oceans 107
oil pollution 202
Ondiri wetland 47
Organization of American States (OAS) 33
overexploitation 168, 191, 245
overharvesting 167, 168
oyster shells 65

p
Pallikarnai Marshlands 89
Paris Agreement 108, 248
payment for ecosystem services (PES) 105
pearl button industry 65
peatlands 3, 107
peat soil 3
periphyton 168
pesticides 10, 44, 63, 72, 77, 123, 176, 185
Philippines 172, 251, 252
phosphorus 44, 75, 145, 168, 216
Phragmites communis 186
phytochelatins 210
phytoextraction 212
phytoplankton 65, 68, 74
phytoremediation 210–217
pilgrimages 47
plantations 190
plastic particles 10
poaching 191
pollution 6, 9–10, 130, 168, 202
  coastal 214–217
  Kashmir Himalaya wetlands 189
  inputs 276
  North Bihar wetlands 71–77
  see also heavy metals; pollution load index (PLI)
pollution load index (PLI) 75
Pomacea insularum 169
ponds 5, 41
Potamogeton lucens 190
prairie wetlands 42
precipitation 48
production value 88
productivity method 91–92
property markets 93
Pulicat Lagoon 118–124

q
Queensland wetlands 1

r
railways 167, 190
Ramsar Advisory Mission 31
Ramsar Convention *see* Convention on Wetlands of International Significance
Ramsar Information Sheets (RIS) 26
Ramsar Sites 108, 112, 130, 132, 192, 272–273, 276
raw materials 186
recreation 47–48
recreational sites 93–94
regional governments 155
replacement cost method 94–95
reservoir-based wetlands (RBWs) 140–156
  adverse impact of 149–150
  management and conservation 154–155
  significance of 141–142
  threats to 152–153
  in the Tista Basin 144–145, 147–148, 150–152
  water-level fluctuations 153
Rhizophora apiculate 214
Rhizophora mucronata 243
Rhizophora stylosa 210
river deltas 8, 166, 172
rivers 20, 73, 149
  and reservoir-based wetlands (RBWs) 147, 149
  see also river deltas
roads 167, 171, 190

s

salinity 8, 10, 165
salt exploitation sites 5
salt marshes 43, 107, 113, 168
Salvinia spp. 168, 191
sanitation 105–106
Schistura inglisi 152
Schoenoplectus 166
Scientific and Technical Review Panel (STRP) 24, 25
scientific investigations 186–187
SDGs *see* sustainable development goals
sea-grass beds 107
sea level rise 42, 48, 246
seas 107
sediment 204–209, 213
  see also sedimentation
sedimentation 43–44, 116, 122, 149, 151, 153, 185
semi-mangroves 230
Sesuvium portulacastrum 216
shellfish 45, 111, 113, 114, 116
shoreline stabilization 42–43
shrimp 114, 230, 242, 245
silicon 145
siltation 10, 75, 123, 133, 190
Sonneratia apetala 213
South Africa 106
South America 9, 237, 279
South Asia 237, 251
South Korea 47, 48
Spartina 166
Sri Lanka 252
Srinagar 183
storm protection 42–43
storm surge 165, 243
substitute cost method 94–95
subtidal wetlands 20
Sunderbans 273
supporting services 116–117, 121
surveys 95–97
sustainable agriculture 105–106
sustainable development 134–135
  see also sustainable development goals (SDGs)
sustainable development goals (SDGs) 103–108, 280
sustainable livelihood 64–65
sustainable practices 129–137
swamps 3

t

technology intervention 136
terrestrial ecosystems 107–108
Thailand 172
tidal forests 204, 229
  see also mangroves
tidal marshes 3, 104
Tista Barrage Project (TBP) 147
Tista Basin 144–145, 147–148, 150–152, 155, 156
Tonga 248
Tonle Sap Lake 105
total petroleum hydrocarbons (TPHs) 213
tourism 47–48, 106, 186
  and reservoir-based wetlands (RBWs) 152–153
trace metals 75–77, 210
Trapa natans 186, 190
travel cost method 89, 93–94
Trinidad 47
Trista Low Dam Project (TLDP) 151
Tsomoriri wetland 47
Typha angustata 186

u

Uganda 47
UNECE "Water Convention" 35
UNESCO Man and Biosphere Programme (UNESCO-MAB) 34–35
UNESCO World Heritage Convention 34–35, 279
United Nations (UN) 32–33
United Nations Convention to Combat Desertification (UNCCD) 34
United Nations Educational, Scientific and Cultural Organization (UNESCO) 25, 248
United Nations Environment Programme (UNEP) 35, 214
United Nations Framework Convention on Climate Change (UNFCCC) 34, 248

United States 107
  definition of wetlands 2, 162
  wetland tourism 47
Upo wetland 47
urbanization
  North Bihar wetlands 71–77
  Tista basin 153
U. S. Fish and Wildlife Service (USFWS) 2, 162

v

vegetation damage 276
Vietnam 106
Vikramshila Biodiversity Research and Education Centre (VBREC) 77

w

wastewater treatment 5, 20, 44–45, 164, 213, 215
water,
  availability and management 105–106
  consumption 170
  diversion of flow 170
  level 171
  purification 44–45, 185
  quality 8, 10, 72
  storage areas 21
  withdrawal 168
  see also groundwater; wastewater treatment; water-related disasters
water-related disasters 106–107
wave attenuation 43
Waza Foodplain 104
weed infestation 191
wetland conversion 17, 18
wetland-dependent species 276
wetland fauna 8, 10, 45, 48, 65, 114, 119
  see also biodiversity
wetland flora 45–46, 48, 65, 114
  see also biodiversity
wetland management 10, 134–136, 194–195, 277–279
wetland products 46, 187, 274
Wetlands (Conservation and Management) Rules 279
Wetlands International 33, 41
Wetlands of International Importance 10
wetlands,
  as 'biological supermarkets' 17
  categorization and classification 1, 3–5, 20–21, 129, 162–164
  definition 1–2, 19–20, 40, 129
  degradation of 133, 162–172, 175–176, 187–191
  ecological importance 40–49, 274–275
  functions and value 184–187, 274–275
  identification 32
  as the "kidneys of the Earth" 6, 184, 274, 275
  loss of 8, 9, 12, 17–18, 90, 133–135, 166–171, 183–184, 275–277
  as multiple-value systems 274
  restoration 280
  role of 5–7, 130–133, 274–275
  sociocultural values 46–47
  spatial distribution 7–8
  status and impact of 8–10
  sustainability 280
  "wise use" approach 10, 11, 19, 22, 31–32, 35, 36, 79, 88, 103, 108, 137
women 105
World Health Organization (WHO) 33
World Heritage Sites 108
World Tourism Organization (WTO) 3
World Wetlands Day 36
WWF International 33

y

Yangtze River 172

z

zinc (Zn) 77, 210, 214
zooplankton 68